HIV and Tuberculosis

Irini Sereti • Gregory P. Bisson
Graeme Meintjes
Editors

HIV and Tuberculosis

A Formidable Alliance

Springer

Editors
Irini Sereti
National Institute of Health
National Institute of Allergy
and Infectious Diseases
Bethesda, MD, USA

Gregory P. Bisson
Perelman School of Medicine
at the University of Pennsylvania
Philadelphia, PA, USA

Graeme Meintjes
Wellcome Centre for Infectious Diseases
Research in Africa, Institute of Infectious
Disease and Molecular Medicine
Department of Medicine
University of Cape Town
Cape Town, South Africa

Division of Infectious Diseases and HIV
Medicine, Department of Medicine
University of Cape Town
Cape Town, South Africa

ISBN 978-3-030-29110-5 ISBN 978-3-030-29108-2 (eBook)
https://doi.org/10.1007/978-3-030-29108-2

This Springer imprint is published by the registered company Springer Nature Switzerland AG
The registered company address is: Gewerbestrasse 11, 6330 Cham, Switzerland

Preface

Prior to the emergence of the HIV epidemic, the global TB burden was steadily declining over time, largely due to strengthening TB treatment programs delivering short-course TB chemotherapy, which could cure most patients in 6 months. With the emergence of HIV in the 1990s, trends in progress in controlling TB, including reductions in TB incidence and decreasing TB mortality rates, began to be tragically reversed, particularly in sub-Saharan Africa but in highly resourced countries such as the United States as well [1, 2]. Evaluating changes in TB incidence over time by global region throughout the 1990s and early 2000s revealed what has now been repeatedly demonstrated in large-scale epidemiologic analyses: HIV infection profoundly increases the risk of active TB disease, and regions with the highest HIV burden experience dramatic increases in TB case notification rates at the population level [3]. Initial metrics describing the interactions between HIV and TB were nothing short of astounding. Studies from certain populations with high rates of HIV infection and intensive exposure to TB, such as South African miners, for example, began reporting some of the highest TB case notification rates on record, approaching 7000 cases per 100,000 individuals [4]. Early analytic studies provided further grim details revealing that, in areas where TB is common, not only does the risk of TB essentially double in the initial months after HIV acquisition [5] but also TB risk continues to increase as cellular immune competence, measured by CD4+ T cell counts, declines [6]. Not surprisingly, substantial rates of TB among HIV-infected individuals have pushed TB to the top of the list of causes of death among HIV-infected individuals, despite being grossly underdiagnosed [7].

Furthermore, while hope in the fight against TB has been provided by the dual triumphs of the discovery and delivery of highly active antiretroviral therapy (ART), subsequent studies revealed that ART does not completely reverse the heightened TB risk among HIV-infected individuals [8]. Further compounding the problem, the use of ART in patients with TB is complex, often triggering toxicity, with untoward drug-drug interactions or with pathologic inflammation via the immune reconstitution inflammatory syndrome [9]. Indeed, in TB meningitis, rapid provision of ART, versus a careful delay, actually appears harmful [10].

Confronting this "formidable alliance" between HIV and TB has resulted in nearly three decades of dedicated research and public health efforts that have revealed insights into how HIV-1 and *Mycobacterium tuberculosis* interact in cells, host tissues, and communities. While each area of HIV/TB research and care is associated with substantial challenges, these challenges are being met creatively in ways that are continually advancing our understanding of not only the interaction between the two diseases but also of each of the infections. For example, the study of progression from latent to active TB disease is facilitated by the higher rate of incident TB events among latently infected, HIV-positive individuals over time [11]. This has underscored the concept, recently illuminated by positron emission tomography (PET)-computed tomography (CT) studies, that the clinical space between latent infection and TB disease is most likely a spectrum, not a dichotomy, defined by increasing bacterial replication and inflammation [12, 13]. In addition, the role of monocytes/macrophages and inflammasomes in TB-associated inflammation has been facilitated by evaluating patients longitudinally early during immune reconstitution on ART [14, 15]. More practically, the low bacillary burden of mycobacteria in the sputum of HIV-infected individuals with pulmonary TB has driven the intensive search for a more sensitive TB diagnostic for use in resource-limited settings, which culminated in the approval and scale-up of a novel desktop PCR platform for detecting both *Mtb* and rifampin resistance in patient samples in as little as 2 hours [16]. This and other diagnostic innovations are fueling, in turn, large-scale efforts aimed at detecting and treating TB in heavily affected communities.

This volume will introduce the reader to the main clinical, pathophysiologic, and public health topics within the scope of HIV/TB. Global epidemiology contributions provide an orientation to the determinants and distribution of HIV and TB disease internationally, highlighting the characteristics of regions of intense concentrations of coinfected individuals in areas such as sub-Saharan Africa. The chapter on modeling builds on the epidemiology sections and provides details on the effects of HIV, as well as public health interventions, on TB transmission and TB burden at the population level. Chapters on immunology of HIV and TB and on the TB immune reconstitution inflammatory syndrome summarize the current understanding of how HIV affects the immune system to influence host susceptibility to and manifestations of TB and how immune restoration on ART can lead to immune pathology. Aspects related to the increased risk of progression to active TB in latently infected individuals with HIV, and new and conventional treatments for latent TB, are covered in a chapter that leads into sections covering the presentation, diagnosis, and management of both drug-sensitive and drug-resistant active TB disease, including the important issue of pharmacology and drug-drug interactions as well as the diagnosis and treatment of TB meningitis, the most life-threatening form of TB disease.

We would like to take the opportunity to thank all of the chapter authors of this book for their excellent contributions. We hope that this volume stimulates further interest in the interaction of these two globally important diseases and inspires future investigations to overcome their impact on human health.

Bethesda, MD, USA — Irini Sereti
Philadelphia, PA, USA — Gregory P. Bisson
Cape Town, South Africa — Graeme Meintjes

References

1. Floyd K, Wilkinson D (1997) Tuberculosis in the HIV/AIDS era: interactions, impacts and solutions. AIDS Anal Afr 7(5):5–7
2. Barnes RF et al. (2011) Trends in mortality of tuberculosis patients in the United States: the long-term perspective. Ann Epidemiol 21(10):791–795
3. Corbett EL et al. (2006) Tuberculosis in sub-Saharan Africa: opportunities, challenges, and change in the era of antiretroviral treatment. Lancet 367(9514):926–937.
4. Anonymous (2007) Tuberculosis Strategic Plan for South Africa: 2007–2011. Department of Health, Republic of South Africa.
5. Sonnenberg P et al. (2005) How soon after infection with HIV does the risk of tuberculosis start to increase? A retrospective cohort study in South African gold miners. J Infect Dis 191(2):150–158
6. Lawn SD et al. (2009) Short-term and long-term risk of tuberculosis associated with CD4 cell recovery during antiretroviral therapy in South Africa. AIDS 23(13):1717–1725
7. Anonymous (2018) Global TB Report 2018. World Health Organization: Geneva
8. Badri M, Wilson D, Wood R (2002) Effect of highly active antiretroviral therapy on incidence of tuberculosis in South Africa: a cohort study. Lancet 359(9323):2059–2064
9. Schutz C et al. (2010) Clinical management of tuberculosis and HIV-1 co-infection. Eur Respir J 36(6):1460-1481
10. Torok, ME et al. (2011) Timing of initiation of antiretroviral therapy in human immunodeficiency virus (HIV)--associated tuberculous meningitis. Clin Infect Dis 52(11):1374–1383
11. Petruccioli E et al. (2016) Correlates of tuberculosis risk: predictive biomarkers for progression to active tuberculosis. Eur Respir J 48(6):1751–1763
12. Esmail, H et al. (2016) Characterization of progressive HIV-associated tuberculosis using 2-deoxy-2-[(18)F]fluoro-D-glucose positron emission and computed tomography. Nat Med 22(10):1090–1093
13. Barry CE 3rd et al. (2009) The spectrum of latent tuberculosis: rethinking the biology and intervention strategies. Nat Rev Microbiol **7**(12):845–855.
14. Andrade BB et al. (2014) Mycobacterial antigen driven activation of CD14++CD16- monocytes is a predictor of tuberculosis-associated immune reconstitution inflammatory syndrome. PLoS Pathog 10(10):e1004433
15. Lai RPJ et al. (2015) HIV-tuberculosis-associated immune reconstitution inflammatory syndrome is characterized by Toll-like receptor and inflammasome signalling. Nat Commun 6:8451
16. Boehme CC et al. (2010) Rapid molecular detection of tuberculosis and rifampin resistance. N Engl J Med 363(11):1005–1015

Contents

The original version of this book was revised. The correction is available at https://doi.org/10.1007/978-3-030-29108-2_14

Overview of the HIV-Associated Tuberculosis Epidemic

Constance A. Benson

Abstract Globally, tuberculosis is the leading infectious cause of death and the most common opportunistic infection in people living with HIV (PLWH) (World Health Organization. Global Tuberculosis Report 2018). TB incidence has actually declined in the past 5 years both overall and for PLWH (World Health Organization. Global Tuberculosis Report 2018). However, efforts to achieve the target goals of the "End TB Strategy" both for people with and without HIV infection, will require more aggressive interventions aimed at each of the three pillars of TB control, including increased screening and diagnosis of TB infection and disease, rapid initiation of effective TB treatment, and more effective prevention of TB disease. The last decade has seen an explosion of new diagnostic technologies, development of new or novel antimycobacterial drugs, and the evolution of shorter course treatment for latent TB infection and drug resistant TB disease. While the next 5 years is likely to see a sea-change in our approaches to more effective treatment of TB, there are numerous barriers to the scale-up of new diagnostic tests and treatment regimens for PLWH that must be overcome to reach the rates of reduction in TB incidence that will be required to achieve the 2035 TB elimination goals.

Keywords HIV · TB · Opportunistic infection · TB diagnosis · TB treatment · TB prevention · TB elimination

Introduction

Tuberculosis is now the number one leading infectious cause of death worldwide and the most common opportunistic infection and cause of death globally in people living with HIV (PLWH) [1]. There were an estimated ten million new incident cases of TB and 1.6 million TB deaths in 2017, the most recent year for which the

C. A. Benson (✉)
Division of Infectious Diseases and Global Public Health,
University of California San Diego, San Diego, CA, USA
e-mail: cbenson@ucsd.edu

I. Sereti et al. (eds.), *HIV and Tuberculosis*,
https://doi.org/10.1007/978-3-030-29108-2_1

World Health Organization (WHO) has calculated TB case notification rates [1]. An estimated 464,633 incident TB cases occurred in PLWH and there were 300,000 TB deaths in PLWH, representing 18.8% of TB deaths in 2017. Based on a recent update of global models by Houben and Dodd an estimated one quarter of the world's population, 1.7 billion people, are latently infected with *Mycobacterium tuberculosis* (MTB), although the regional distribution varies, with Southeast Asia, the western Pacific region, and the African region having the highest rates of latent TB infection [2].

Although the notification rates for new and relapsed cases of TB continue to rise slowly, TB incidence has actually declined in the past 5 years both overall and for PLWH [1]. The number of TB deaths has declined at a faster rate, particularly among PLWH, mostly attributed to the increased access to and implementation of earlier antiretroviral therapy in areas with the highest burden of TB and HIV [1]. For the first time in modern history, a United Nations General Assembly High Level Meeting took place in 2018 aimed at engaging the world's politicians and public health leaders in the efforts to end TB [3]. New global TB elimination targets were adopted and included, among others, the goals of a 95% reduction in the number of TB deaths and a 90% reduction in TB incidence rate by 2035, accompanied by "zero TB-affected households that experience catastrophic costs resulting from TB" [1, 3]. However, with current reduction rates ranging from only 1.5–2.0% in TB incidence and TB deaths, respectively, meeting these targets would require substantially more aggressive reduction rates of 10% by 2020 and 17% by 2025 to reach these new targets [1, 4].

Efforts to achieve the target goals of the "End TB Strategy" both for people with and without HIV coinfection, will require more aggressive interventions aimed at each of the three pillars of TB control, namely (1) increased systematic screening of TB contacts and high-risk groups, using improved diagnostic tests for detecting active and latent TB infection; (2) rapid evaluation and more effective treatment for active TB, including drug resistant TB, making use of universal drug susceptibility testing and more active, better tolerated, and shorter duration antimycobacterial therapies, and; (3) more effective prevention of active TB by identifying better predictors of risk of progression, implementing shorter course, more effective regimens to prevent TB, including among those exposed to drug resistant TB, and continued development of vaccines capable of preventing active TB, whether in the context of preventing infection or preventing disease. For PLWH, additional steps needed to more effectively address the epidemic include integration of HIV and TB screening and testing of contacts and high risk groups; assuring that new drugs and regimens being tested in clinical trials or used in programmatic settings have appropriate assessment of drug-drug interactions with antiretroviral drugs as well as development of formulations that can be used in children and pregnant women with HIV, and; assuring that clinical trials evaluating new drugs and regimens and new diagnostic modalities for active and latent TB infection are appropriately tested in persons with HIV coinfection, i.e., that PLWH are enrolled in all such clinical trials.

The last decade has seen an explosion of new diagnostic technologies aimed at detecting active TB at or near the point of care. The most widely accessible of these is the GeneXpert MTB/RIF, a rapid (less than 2-h turnaround time) polymerase chain reaction (PCR)-based assay that can detect the presence of MTB and the *rpoB* gene that confers resistance to rifamycins in sputum and other body fluids with a sensitivity and specificity in persons with acid fast smear (AFB) positive and culture positive TB of 85% and 96%, respectively [5, 6]. Recent modifications of this technology resulted in the development of the MTB/RIF Ultra assay, which has improved sensitivity, particularly in those with sputum smear negative disease, comparable to that of sputum culture [6]. The Xpert MTB/RIF has revolutionized the rapid diagnosis of active TB in many settings, and in some settings, has replaced smear microscopy. Another rapid diagnostic test used as an adjunct screening test is the urine lipoarabinomannan antigen detection test; this test is more sensitive in seriously ill PLWH with advanced immunosuppression, and when used to trigger earlier initiation of ART can improve mortality in this population [7]. Additional advances in diagnostic testing have yielded the ability to conduct rapid drug susceptibility testing using either variations on the Xpert MTB/RIF technology, line probe assays for detection of key mutations conferring resistance to first and second line anti-TB drugs, and more recently whole genome sequencing to genotypically test for signature mutations that confer resistance to anti-TB drugs [8, 9]. If the combination of novel rapid tests is successfully implemented in programmatic settings, one can efficiently diagnose active TB within hours dramatically reducing the time to initiation of effective treatment. However, there remain numerous obstacles to the use of newer technologies in resource constrained high TB burden settings, not the least of which are the cost and maintenance of the equipment and supplies, the need for training laboratory staff in their use and clinicians in the interpretation of the results, and the need for strengthening infrastructure and health care systems to use them. With the broad array of newer technology in the diagnostic development pipeline, much more work is needed to determine where and how best to implement newer diagnostic tests.

As of late 2018, there were more than 15 new or novel antimycobacterial compounds in varying stages of development, and a host of phase 1–3 clinical trials underway or planned to evaluate these together with repurposed or existing anti-TB drugs in combination [1]. This, coupled with the recent approval of bedaquiline and delamanid (in Europe) for use in the treatment of drug resistant TB, represents a dramatic change in the landscape for more effective treatment of TB. Efforts at TB treatment shortening for drug-susceptible TB have been mixed. Three phase 3 clinical trials published in 2014, each incorporating the substitution of one or two newer or repurposed drugs (with greater *in vitro* activity against MTB in mouse models) into the induction or continuation phases (or both) of TB treatment in regimens aimed at shortening treatment from 6 to 4 months all failed [10–12]. The RIFAQUIN study reported a 26.9% unfavorable outcome in the 4-month arm compared to 14.4% in the standard arm [10]. The OFLOTUB trial reported a 21% unfavorable outcome rate versus 17% for the standard of care arm, primarily owing to a higher recurrence rate in the shorter course arm [11]. The ReMOX trial reported unfavorable outcomes of 20% and 15% in the two shorter course arms versus 8% in the standard of care arm [12]. While these were disappointing outcomes, it should be noted that in each of these studies 70–80% of participants in the 4-month treatment arms were successfully treated, leading to the further investigation of factors likely to be associated with a favorable outcome that could be used to target shorter course treatment.

More encouraging has been the rapid evolution of shorter course treatment for drug resistant TB. Data from the original Bangladesh regimen studies reported an 84.4% bacteriologically favorable outcome with an intensive 9 to 12-month intensive standardized regimen for the treatment of multidrug resistant TB (MDRTB), with 95% of participants completing the regimen within 12 months [13]. These results have been recapitulated in observational studies among persons with MDRTB in a number of countries in Sub-Saharan Africa and elsewhere, leading to a WHO recommendation in 2016 to treat MDRTB in patients who meet specific criteria with a modified Bangladesh regimen that includes kanamycin, moxifloxacin, prothionamide, clofazimine, pyrazinamide, high dose isoniazid, and ethambutol for 4–6 months followed by moxifloxacin, clofazimine, pyrazinamide and ethambutol for an additional 5 months [14]. Results of two phase 2 trials demonstrating the superiority of bedaquiline (a novel diarylquinoline drug) or delamanid (a novel nitroimidazole drug), respectively, combined with optimized background therapy in the treatment of MDRTB, ultimately led to the availability of two new drugs from novel classes of anti-TB drugs that together with repurposed drugs have substantially improved the successful treatment outcome rates for MDRTB [15–18]. Coupled with the final results of the STREAM-1 randomized controlled trial of a 9-month shorter course MDRTB regimen, the plethora of new data emerging from numerous other studies over the past 2 years reporting more favorable outcomes with shorter courses of combinations of new and existing drugs led to a new WHO rapid communication in August 2018 with key changes to the recommendations for treatment of MDR- and rifampin-resistant (RR-) TB [19, 20]. TB drugs were regrouped into three categories and prioritized based on the evidence supporting their use; Group A includes levofloxacin/moxifloxacin, bedaquiline, and linezolid; Group B includes clofazimine and cycloserine/terizidone; and Group C includes delamanid, ethambutol, pyrazinamide, and other second-line anti-TB drugs. Regimens prioritize Group A, then Group B drugs with Group C drugs reserved for those unable to use one or more of those in the other two groups. The most important changes in the recommendations are that kanamycin and capreomycin are no longer recommended, and all regimens should exclude injectable drugs unless there is a compelling need for them based on drug-susceptibility testing or toxicity management.

Lastly, perhaps the most notable development in the treatment of drug resistant TB has been the interim results reported from the NixTB trial of just three drugs, bedaquiline, pretomanid, and linezolid, used in a 6-month treatment course for extensively drug-resistant TB (XDRTB). Interim results from 75 patients completing treatment as of late 2018 demonstrated durable cure in 88%, with only six deaths and two relapses [21]. Based on final results from 109 participants the U.S. Food and Drug Administration approved pretonamid in August 2019 for persons with highly drug-resistant TB. These results have led to discussion in the field of the possibility of a "universal regimen" for treatment of drugs-susceptible and drug-resistant TB, with clinical trials underway exploring the use of bedaquiline, pretomanid, moxifloxacin, and pyrazinamide as a universal regimen. In addition, there are now more than 20 randomized clinical trials underway worldwide exploring different regimens for treatment shortening for drug resistant TB.

While the next 5 years is likely to see a sea-change in our approach to more effective treatment of TB, there are numerous barriers to the scale-up of new treatment regimens, not the least of which are the need for new and inexpensive rapid diagnostic tests for detecting active TB and for drug-susceptibility testing, assuring that the cost of new drugs and regimens make them accessible, overcoming country level registration and importation barriers, training clinicians in the use of newer drugs and regimens, assuring adequate pharmacovigilance to assess the safety of newer drugs and regimens as they are deployed more broadly and specifically assessing their activity in PLWH who are on antiretroviral agents that may interact with one or more of the anti-TB drugs, and strengthening health systems and infrastructure for public health programs so that the promise of these drugs can be realized.

The past decade has also seen dramatic changes in our armamentarium for the prevention of latent TB infection. Current regimens now include the standard of 9 months of isoniazid, or the alternatives of once weekly isoniazid plus rifapentine for 12 weeks, or daily rifampin for 4 months all of which are similar with regard to efficacy although treatment completion rates are higher with the shorter regimens [22–26]. The most recent trial may be transformative, demonstrating in PLWH the equal efficacy of a short course regimen comprised of 1 month of daily isoniazid and rifapentine compared with daily isoniazid for 9 months [26]. While this has not yet been incorporated into WHO or other treatment guidelines, it will likely be recommended as an alternative, particularly for PLWH. Finally, recent data suggest that a novel TB vaccine construct, M72/ASO1$_E$ might reduce the incidence of progression to active TB by 54% in adults with latent TB infection, a rate of reduction similar to that seen with chemoprevention [27]. The promise of these critically important studies has not yet been realized. Implementation of effective preventive therapy in the settings where it might have the greatest impact has been disappointingly low. Among the principle obstacles to implementation of preventive therapy have been the inability to either identify those at highest risk of TB progression or to convincingly rule out the presence of active TB with currently available diagnostic tests. For example, the positive predictive value of tuberculin skin testing and interferon gamma release assays as diagnostic tests for latent TB have positive predictive values in the range of 2–7% even in the highest risk populations [28]. The focus of research in the field more recently has been on utilizing gene sequencing or key gene signatures to more effectively predict those most likely to progress in the short term. A recent study suggested a single gene pair, *C1QC/TRAV27,* that could successfully predict progression to active TB in household contacts up to 24 months before onset of active disease [29]. Whether this approach will ultimately lead to cost-effective and widely applicable technology that can be implemented in high TB burden settings remains to be established. But without more effective methods of diagnosing those at highest risk of disease progression, and targeting them for intervention, it may not be possible to achieve the rates of reduction in new TB infections resulting from reactivation of latent TB that are required to achieve the new goals set for TB elimination.

In summary, we now have or will have in the near future, a plethora of tools that, if effectively deployed, will allow us to achieve the rates of reduction in TB incidence that will be required to achieve the 2035 TB elimination goals set by the WHO and the United Nations High Level Meeting in 2018. The remaining chapters in this textbook highlight the many elements of TB infection, disease, diagnosis, treatment and prevention that specifically pertain to PLWH and that will challenge our ability to reach these goals in this key patient population.

References

1. World Health Organization. Global Tuberculosis Report 2018
2. Houben RM, Dodd PJ (2016) The global burden of latent tuberculosis infection: a re-estimation using mathematical modeling. PLoS Med 13:e1002152
3. United Nations General Assembly. Political declaration of the UN General Assembly high-level meeting on the fight against tuberculosis. www.un.org/pga/73/event/fight-to-end-tuberculosis/
4. Tornheim JA, Dooley KE (2019) The global landscape of tuberculosis therapeutics. Ann Rev Med 70:105–120
5. Boehme CC, Nabeta P, Hillemann D (2010) Rapid molecular detection of tuberculosis and rifampin resistance. N Engl J Med 363:1005–1015
6. Dorman SE, Schumacher SG, Alland D, Nabeta P, Armstrong DT, King B et al (2018) Xpert MTB/RIF Ultra for detection of Mycobacterium tuberculosis and rifampicin resistance: a prospective multicentre diagnostic accuracy study. Lancet Infect Dis 18:76–84
7. Peter JG, Zijenah LS, Chanda D, Clowes P, Lesosky M, Gina P (2016) Effect on mortality of point-of-care, urine-based lipoarabinomannan testing to guide tuberculosis treatment initiation in HIV-positive hospital inpatients: a pragmatic, parallel-group, multicountry, open-label, randomized controlled trial. Lancet 387:1187–1197
8. Maningi NE, Malinga LA, Antiabong JF, Lekalakala RM, Mbelle NM (2017) Comparison of line probe assay to BACTEC MGIT 960 system for susceptibility testing of first and second-line anti-tuberculosis drugs in a referral laboratory in South Africa. BMC Infect Dis 17:795. (1–8)
9. Gygli SM, Keller PM, Ballif M, Blochliger N, Homke R, Reinhard M et al (2019) Whole-genome sequencing for drug resistance profile prediction in *Mycobacterium tuberculosis*. Antimicrob Agents Chemother 63:e02175–e02118
10. Jindani A, Harrison TS, Nunn AJ, Phillips PP, Churchyard GJ, Charalambous S et al (2014) N Engl J Med 371:1599–1608
11. Merle CS, Fielding K, Sow OB, Gninafon M, Lo MB, Mthiyane T et al (2014) A four-month gatifloxacin-containing regimen for treating tuberculosis. N Engl J Med 371:1588–1598
12. Gillespie SH, Grook AM, McHugh TD, Mendel CM, Meredith SK, Murray SK et al (2014) N Engl J Med 371:1577–1587
13. Aung KJ, Van Deun A, Declercq E, Sarker MR, Das PK, Hossain MA, Rieder HL (2014) Successful '9-month Bangladesh regimen' for multidrug-resistant tuberculosis among consecutive patients. Int J Tuberc Lung Dis 18:1180–1187
14. World Health Organization. WHO treatment guidelines for drug-resistant tuberculosis. 2016 Update

15. Diacon AH, Pym A, Grobusch MP et al (2014) Multidrug-resistant tuberculosis and culture conversion with bedaquiline. NEJM 371:723–732
16. Pym AS, Diacon AH, Tang S-J et al (2016) Bedaquiline in the treatment of multidrug- and extensively drug-resistant tuberculosis. Eur Respir J 47:564–574
17. Gler MT, Skripconoka V, Sanchez-Garavito E et al (2012) Delamanid for multidrug-resistant pulmonary tuberculosis. NEJM 366:2151–2160
18. Skripconoka V, Danilovits M, Pehme L et al (2013) Delamanid improves outcomes and reduces mortality in multidrug-resistant tuberculosis. Eur Respir J 41:1393–1400
19. World Health Organization (2018) WHO treatment guidelines for multi-drug and rifampicin-resistant tuberculosis (MDR/RR-TB), 2018. World Health Organization, Geneva. (https://www.who.int/tb/publications/2018/WHO.2018.MDR-TB.Rx.Guidelines.prefinal.text.pdf)
20. Nunn AJ, Phillips PPJ, Meredith SK, Chiang CY, Conradie F, Dalai D et al (2019) A trial of a shorter regimen for rifampin-resistant tuberculosis. N Engl J Med 380:1201–1213
21. Conradie F, et al. 49th International Union World Conference on Lung Disease, 2018
22. http://www.aidsinfo.nih.gov/guidelines
23. Sterling TR, Villarino ME, Borisov AS et al (2011) Three months of rifapentine and isoniazid for latent tuberculosis infection. N Engl J Med 365:2155–2166
24. Sterling TR, Scott NA, Miro JM et al (2016) Three months of weekly rifapentine plus isoniazid for treatment of *M. tuberculosis* infection in HIV co-infected persons. AIDS 30:1607–1615
25. Menzies D, Adjobimey M, Ruslami R et al (2018) Four months of rifampin or nine months of isoniazid for latent tuberculosis in adults. N Engl J Med 379:440–453
26. Swindells S, Ramchandani R, Gupta A et al (2019) One month of rifapentine plus isoniazid to prevent HIV-related tuberculosis. N Engl J Med 380:1001–1011
27. Van Der Meeren O, Hatherill M, Nduba V, Wilkinson RJ, Muyoyeta M, Van Brakel E et al (2018) Phase 2b controlled trial of M72/$AS01_E$ vaccine to prevent tuberculosis. N Engl J Med 379:1621–1634
28. Diel R, Loddenkemper R, Nienhaus A (2012) Predictive value of interferon-y release assays and tuberculin skin testing for progression from latent TB infection to disease state: a meta-analysis. Chest 142:63–75
29. Suliman S, Thompson E, Sutherland J, Weiner Rd J, Ota MOC, Shankar S et al (2018) Four-gene pan-African blood signature predicts progression to tuberculosis. Am J Resp Crit Care Med. https://doi.org/10.1164/rccm.201711-2340OC. [Epub ahead of print]

Epidemiology of Drug-Susceptible, Drug-Resistant Tuberculosis and HIV in Africa

Kogieleum Naidoo and Nikita Naicker

Abstract In Africa, TB-HIV co-infection rates are estimated to be 72%. Within sub-Saharan Africa, 34 countries recently reported patients with newly diagnosed MDR-TB, representing 14% of all emerging multi-drug resistant reports worldwide. In addition, eight African countries reported patients harboring extensively drug-resistant TB. More importantly, sub-Saharan Africa bears unacceptably high TB incidence rates coupled with the highest HIV prevalence rates globally. Occurrence of TB related death and new TB infections persist, with men disproportionately affected. A gradual reduction in TB incidence rates from 1.4% per annum between 2000 to 2017 to 1.9% per annum between 2015 and 2016 has been observed. Of note, eight out of the sixteen high burden African countries showed a sharp decline in TB incidence rates, surpassing the "End TB Strategy" goal of a 4% per annum decline. Additionally, substantial decreases in TB mortality in HIV infected patients in nine of sixteen high TB burden African countries were also observed. Programmatic uptake and scale up of TB screening, introduction of new TB diagnostic tests, TB preventive therapy and ART have collectively contributed to this success. Regardless of these improvements, TB remains Africa's foremost cause of ill-health and death, with greater efforts required in curbing drug resistant TB incidence, poor drug resistant TB treatment outcomes, and in finding and treating missing TB patients.

Keywords TB incidence · TB prevalence · TB-HIV co-infection · TB mortality · Africa

K. Naidoo (✉)
Centre for the AIDS Programme of Research in South Africa (CAPRISA), 2nd Floor K-RITH Tower Building, Nelson R Mandela School of Medicine, University of KwaZulu-Natal, Durban, South Africa

South African Medical Research Council (SAMRC)-CAPRISA HIV-TB Pathogenesis and Treatment Research Unit, Durban, South Africa
e-mail: Kogie.Naidoo@caprisa.org

N. Naicker
Centre for the AIDS Programme of Research in South Africa (CAPRISA), 2nd Floor K-RITH Tower Building, Nelson R Mandela School of Medicine, University of KwaZulu-Natal, Durban, South Africa

I. Sereti et al. (eds.), *HIV and Tuberculosis*,
https://doi.org/10.1007/978-3-030-29108-2_2

Abbreviations

AIDS	Acquired immunodeficiency syndrome
ART	Antiretroviral therapy
DOTS	Directly observed treatment short course
DR-TB	Drug resistant tuberculosis
HCWs	Healthcare workers
HIV	Human immunodeficiency virus
HTS	HIV testing service
IPT	Isoniazid preventative therapy
MDR-TB	Multi drug resistant tuberculosis
Mtb	Mycobacterium tuberculosis
NSP	National strategic plan
PLHIV	People living with HIV
SA	South Africa
SSA	sub-Saharan Africa
TB	Tuberculosis
TB-IC	Tuberculosis infection control
UN	United Nations
WHO	World Health Organization
XDR-TB	Extensively drug resistant tuberculosis

Role of Human Immunodeficiency Virus (HIV) in Fuelling the Tuberculosis (TB) and Drug-Resistant TB (DR-TB) Epidemic

Approximately 8% of *Mycobacterium tuberculosis* (*Mtb*) infections arise in individuals with HIV, making TB the most significant opportunistic infection in immune compromised patients worldwide. Globally, approximately 9% of newly diagnosed TB patients were living with HIV (72% in Africa) in 2017, were the proportion of known positive TB individuals on ARVs is 78%. More importantly, sub-Saharan Africa (SSA) bears high TB incidence rates as well as the highest HIV prevalence rates globally [1]. Furthermore, SSA represents 14% of all emerging multi-drug resistant (MDR)-TB reports worldwide [1]. Within this region, 34 countries reported on MDR-TB affected patients and 8 countries reported extensively drug-resistant (XDR)-TB affected patients, with approximately 15,000 MDR-TB individuals reported annually in South Africa (SA) alone.

Approximately two billion individuals are presently infected with *Mtb* worldwide, but only 10% of infected immunocompetent patients become symptomatic in their lifetime compared to 50% of immunologically compromised patients [2]. Five up to 15% of the estimated 2 billion individuals infected with *Mtb* will advance to TB disease [1, 3, 4]. However, developing TB disease is greater amongst HIV infected individuals, currently estimated at 10% per annum risk of TB disease in HIV infected individuals not on antiretroviral therapy (ART) [5], and 5% per annum among those on ART [6]. During 2000 and 2016, TB therapy projected 44 million deaths amongst HIV-negative individuals [1]. Amongst HIV-infected individuals, TB therapy supported by ART prevented a further nine million losses [1]. In SSA, the deadly synergy of HIV and TB has accounted for the extremely high TB incidence rates observed over the past 20 years [2]. According to the WHO, highest TB incidence rates were observed in Asian and African regions. Nine SSA countries are currently represented in the 22 high TB burden countries (Fig. 1, Table 1) [1], globally. These countries include DRC, Ethiopia, Kenya, Mozambique, Nigeria, SA, Uganda, Tanzania and Zimbabwe [1]. Of note, known socio-economic drivers of TB including poor housing and working conditions associated with high transmission risk, HIV infection, malnutrition, alcohol abuse, smoking, coupled with delays in presentation for diagnosis and treatment are characteristic of most African settings where TB is endemic [1].

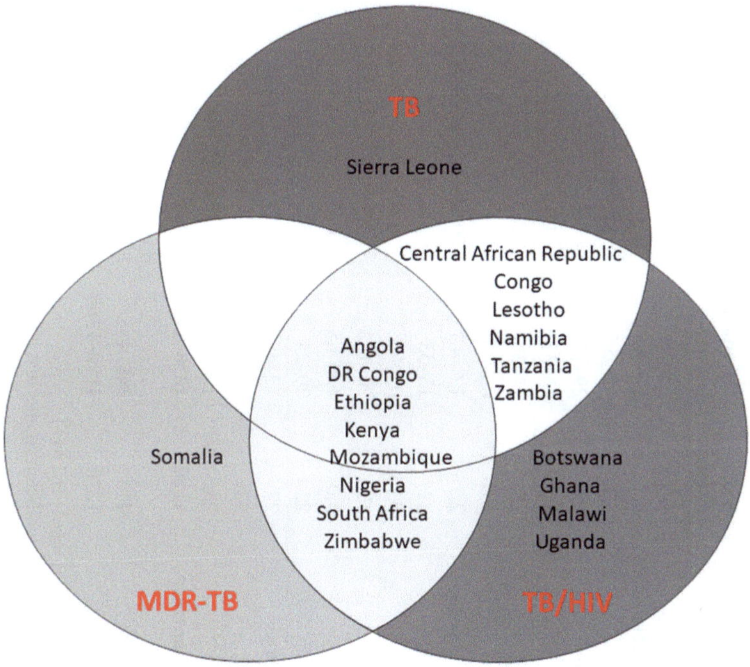

Fig. 1 African Countries with high burden of TB, drug resistant TB and HIV and TB co-infection, including areas of overlap, during 2016–2020. Figure adapted from the WHO global TB report 2018 [1]

Table 1 Projected TB burden in 2017 among 16 high TB burden African countries Numbers in thousands[a]

	HIV-Negative TB Mortality		HIV-Positive TB Mortality		TB Incidence		HIV-Positive TB Incidence	
High TB Burden African Country (total population)	Best estimate	Uncertainty interval	Best estimate	Uncertainty interval	Best estimate	Uncertainty interval	Best estimate	Uncertainty interval
Angola (30,000)	20	12–31	7.8	3.9–13	107	69–153	18	8.5–30
Central African Republic (5000)	3.2	1.8–4.9	2.7	1.4–4.4	20	13–28	6.2	3.3–10
Congo (5000)	3.3	1.9–5.2	2.3	1.2–3.7	20	13–29	5.3	2.7–8.6
DR Congo (81,000)	49	29–74	7.5	3.5–13	262	169–374	20	13–29
Ethiopia (105,000)	25	16–37	4.0	2.5–5.0	172	121–232	12	8.6–17
Kenya (50,000)	25	14–39	18	11–27	158	97–235	45	27–68
Lesotho (5000)	1.0	0.55–1.7	4.6	2.9–6.7	15	9.6–21	11	6.7–15
Liberia (5000)	2.7	1.6–4.1	0.91	0.57–1.3	15	9.4–21	2.2	1.4–3.2
Mozambique (30,000)	22	13–33	27	17–39	163	103–233	66	42–95
Namibia (3000)	0.75	0.48–1.1	0.80	0.55–1.1	11	8.5–14	3.9	2.5–5.5
Nigeria (191,000)	120	70–183	35	21–52	418	273–594	58	37–85
Sierra Leone (8000)	3.0	1.8–4.5	0.78	0.49–1.1	23	15–33	2.8	1.8–4.0
South Africa (57,000)	22	20–24	56	39–77	322	230–428	193	137–258
UR Tanzania (57,000)	27	12–48	22	10–38	154	73–266	48	31–69
Zambia (17,000)	5.0	2.9–7.7	13	8.2–19	62	40–88	36	23–52
Zimbabwe (17,000)	2.0	1.3–2.9	6.3	4.5–8.5	37	27–47	23	15–33
High TB burden countries[b] (4,760,000)	**1110**	**1030–1190**	**247**	**214–282**	**8720**	**7680–9810**	**766**	**680–857**
Africa (1,050,000)	413	348–485	252	219–287	2480	2210–2760	663	585–747
Global (7,520,000)	**1270**	**1190–1360**	**300**	**266–335**	**10,000**	**9000–11,100**	**920**	**832–1010**

Adapted from WHO global TB report 2018 [1]
[a]Less than 100; numbers to two significant figures rest three
[b]Refer to Fig. 1

Incidence of TB in High Burden SSA Countries

Constant with preceding global TB reports, incidence of newly diagnosed TB patients is gradually declining (Fig. 2) from 1.4% per annum between 2000 to 2017 and 1.9% per annum between 2015 to 2016. The rate of decline needs to fast-track to 4–5% yearly by 2020 to attain the End TB Strategy milestones of reduced TB

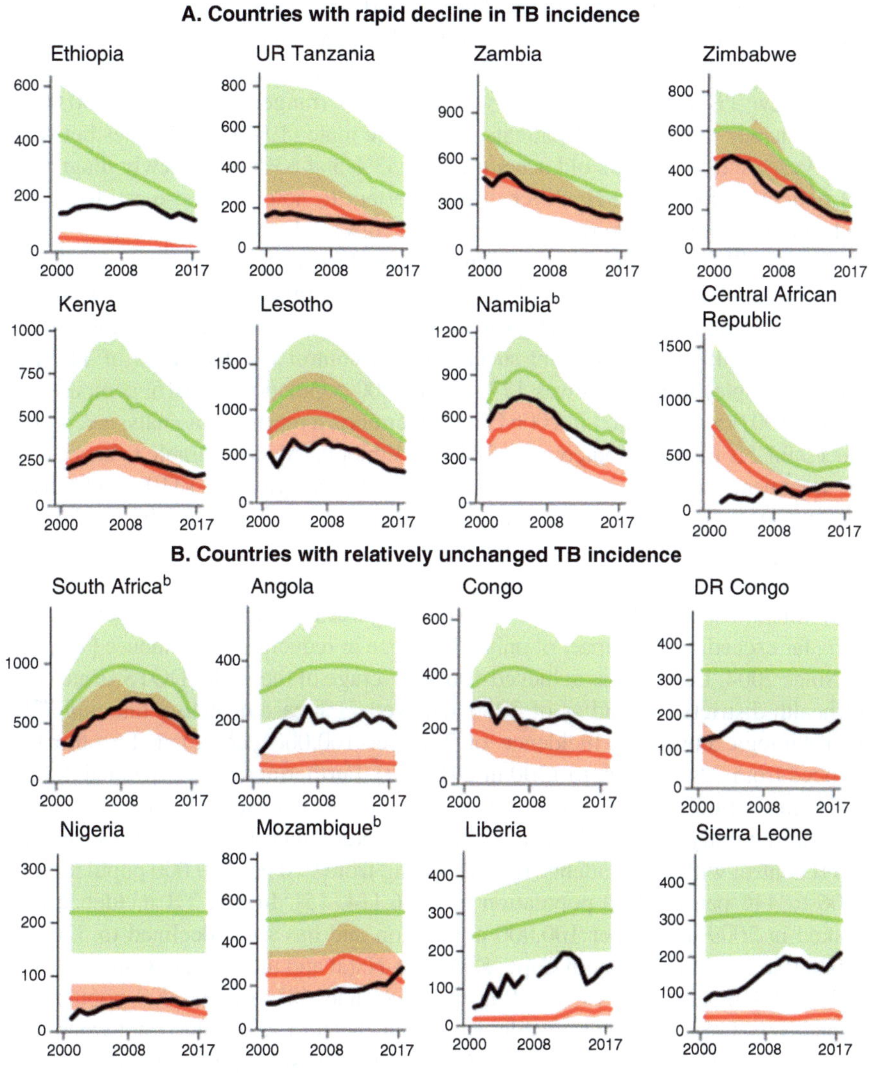

Fig. 2 Projected TB incidence trends in 16 high TB burden African countries; 2000–2017. Incidence rates of HIV-positive TB in red, overall TB incidence rates shown in green, and uncertainty levels shown by shaded areas. Black lines show new and relapse TB patients. A. Eight countries show a sharp decline in TB incidence from 2008–2017. B. Eight countries with minimal TB incidence rate changes [1]

notification and TB related mortality. The projected reduction in incidence rates since 2010 has surpassed 4% yearly in numerous African countries, including Ethiopia (6.9%), Kenya (6.9%), Lesotho (7%), Namibia (6.0%), Tanzania (6.7%), Zimbabwe (11%), and Zambia (4.8%). Data from sixteen African countries show encouraging trends of reduced TB incidence in eleven countries, and incidence rates that appear stable in seven countries (Fig. 2, green lines). Furthermore, in all countries TB incidence in HIV infected patients appears to be declining (Fig. 2, red lines), with rates of reduction of TB incidence reported to be highest in Lesotho and SA. New and relapse TB cases increased in Liberia, Mozambique and Sierra Leone, despite the decline in other African countries. Furthermore, in 2017, projected number of new TB patients in Africa was 2,480,000 (range: 2,210,000–2,760,000). Among newly diagnosed TB patients, 62% were male (1,540,000). Males between 0–14 years and > 15 years old accounted for 53% and 63% of all newly diagnosed TB patients respectively. However, it is important to note that these estimates are wide and may not be truly representative of the disease burden.

Various TB control strategies implemented by African countries have contributed to the declining trends in TB incidence and mortality. Ethiopia has implemented the WHO-recommended DOTS strategy to improve TB treatment success and case detection rates resulting in significant success in TB control [7]. Incidence of TB (per 100,000 people) in Ethiopia was 177 per 100,000 population in 2016, compared to 340 per 100,000 population in 2006 [7]. Furthermore, a recent meta-analysis reported successful TB treatment outcomes of 84% in Ethiopia [8], comparable to successful TB treatment outcomes reported in SA (82%) [9], Kenya (82%) [10], and Ghana (88%) [11]. A multi-centred retrospective cohort study conducted in Ethiopia demonstrated significantly reduced TB incidence rates of 96% among exposed to isoniazid (INH) preventative therapy (IPT) compared to non-IPT exposed patients. Moreover concomitant use of ART with IPT showed a significant decline in TB incidence of 94%, far exceeding the impact of only ART alone in reducing TB incidence [12].

Since 2004, Lesotho has achieved 100% coverage of the WHO DOTS strategy in all health districts [7], and as of 2016 achieved 71% case detection rate and 73% microbiologic coverage. Incidence of TB (per 100,000 people) in Lesotho has reduced from 724 in 2006, to 1300 in 2016 [53]. Furthermore, the expanded decentralisation of public-access nurse-driven ART services in Lesotho has likely contributed substantially to TB incidence decline [13]. Namibia has seen similar successes in TB control with TB incidence rates dropping from 910 per 100,000 population in 2006 to 446 per 100,000 population in 2016 [14, 15]. In SA, TB incidence rate peaked in 2009 at 832 per 100,000 population and has since declined to 520 per 100,000 in 2016. The Eastern Cape, KwaZulu-Natal and the Western Cape have the highest TB incidence rates in SA at 692, 685 and 681 per 100,000 population, respectively [16]. The most notable decline has been in KwaZulu-Natal where the incidence has decreased from 1185 to 685 per 100,000 since 2013. The reduction in TB incidence seen in SA is due to multiple factors. People living with HIV remain vulnerable to TB disease acquisition, and this vulnerability is enhanced among those not on ART and those with advanced immunosuppression [17]. The large public access ART programme has enrolled almost two-thirds of the country's HIV

Table 2 Estimated TB burden (in thousands) in WHO regions among adults and children, in Africa and globally, in 2017 [1]

	Total		Male		Female	
	Best estimate	Uncertainty interval	Best estimate	Uncertainty interval	Best estimate	Uncertainty interval
Africa	2480	2210–2760	1540	1310–1770	941	798–1080
Global	10,000	9000–11,100	6360	5440–7290	3680	3140–4210
	Total ≥ 15 years		Male ≥15 years		Female ≥15 years	
Africa	2180	1910–2450	1380	1150–1620	800	665–936
Global	9030	7980–10,100	5830	4900–6760	3200	2690–3710
	Total 0–14 years		Male 0–14 years		Female 0–14 years	
Africa	296	260–333	156	129–182	141	117–164
Global	1010	888–1120	529	445–613	478	401–554

positive people onto ART. Furthermore, since September 2016, SA has implemented the Universal HIV test and treat policy resulting in larger numbers of patients accessing ART, including a higher proportion of patients initiating ART at CD4 counts >500 [18]. In addition, programmatic implementation of IPT in HIV positive individuals has also contributed to reduced TB incidence. Other measures such as universal TB symptom screening for all patients presenting to health services, as well as roll-out of the GeneXpert test for TB diagnosis has also assisted with early detection and possibly reduced TB transmission. Zambia has a policy of ensuring the availability of quality first line anti-TB drugs in all public health facilities, driving the county's attainment of a TB treatment success rate of 86%, surpassing the WHO target of 85% (Table 2).

The Risk of TB in HIV Infected Individuals

Tuberculosis is the most common opportunistic infection among people with HIV infection. An estimated 1.2 million HIV positive incident TB cases and 430,000 deaths from HIV associated TB, occurred globally in 2015 despite widespread availability of effective treatment and prevention [7]. In 2015, worldwide estimations of people with HIV was 36.7 million, and 2–3 billion TB infected, with 10.4 million new patients with TB disease. In patients with new or latent *Mtb* infection [7, 19–21], HIV remains the sturdiest risk factor for TB disease and TB associated death. The risk of TB disease is high soon after HIV sero-conversion, doubles within the first year of HIV acquisition [22–24], and is highest with advancing immunosuppression at CD4 counts <100 cells/μL [25, 26]. Tuberculosis risk in HIV positive individuals is 20–37% higher compared to non-HIV infected individuals, and in parts of SSA, HIV-TB co-infection rates are as high as 80% [27]. The burden of TB disease is 450,000 individuals per annum in SA, among mainly HIV infected patients—the country bearing the highest TB incidence rate globally, where it occurs mainly in the HIV uninfected [7].

TB-HIV Integration: When to Start ART During TB Treatment

WHO recommends that HIV-TB infected patients ought to begin ART regardless of their CD4 count. This will potentially reduce mortality. ART must be administered within 8 weeks of initiation of anti-TB treatment. Randomized studies underpinning this guidance included large numbers of African patients and showed strong survival benefit with integrated ART in TB therapy. The South African Starting Antiretroviral Therapy at Three Points in Tuberculosis (SAPIT) trial demonstrated a 56% rise in survival rate once TB and ART treatment were combined, when compared to deferred ART following TB treatment [28, 29]. Additional investigation of the study data displayed no change in incidence rate of acquired immune deficiency syndrome (AIDS) or mortality compared to ART initiation amongst patients randomised to receive ART within 4 weeks of TB treatment start (early arm) versus those randomized to receive ART during the first 4 weeks of the continuation phase of TB treatment (late integrated arm). Parallel results were published from additional randomised clinical trials conducted in other settings, the STRIDE (AIDS Clinical Trials Group Study 5221), and the CAMELIA (Cambodian Early versus Late Introduction of Antiretrovirals study) [30, 31] studies. The STRIDE study enrolled 554/809 (69%) of TB-HIV co-infected patients from seven African countries, found that immediate ART, i.e. ART initiated within 2 weeks of TB treatment start compared to early ART, defined as ART initiated 8–12 weeks after TB treatment start, did not reduce AIDS-defining illness and death. Furthermore, the STRIDE study showed that among patients with CD4+ counts <50 cells/mm^3, there was 42% lower incidence of AIDS defining illness and mortality in immediate compared to initial ART. The vast majority of deaths were attributed to HIV related disease, including progression of TB [31].

Since findings from these three landmark studies became available, several other research groups investigating optimal timing of ART in TB patients have published findings from systematic reviews and metanalyses [32], modelling studies [32], cohort and clinical trials [33, 34], which uniformly conclude that early ART in TB therapy is associated with decreased death, however, the death advantage from early ART was most pronounced amongst individuals with CD4+ counts <50 cells/mm^3. The recent WHO guidelines echo these results, recommending that TB therapy should be initiated followed by ART immediately within the first 8 weeks of therapy regardless of CD4+ count. Individuals with severe immunosuppression (CD4+ <50 cells/mm^3), should begin ART in the first 2 weeks of TB therapy. HIV infected individuals with TB meningitis remain exempt to this recommendation. In-country ART guidelines from high TB burden African countries reflect the WHO policy on ART timing in TB.

Impact of ART in Patients with DR-TB

WHO estimates that between 36,000 and 44,000 MDR-TB cases occurred in the African region in 2016. The proportion of DR-TB cases co-infected with HIV differs by background burden of HIV, and ranges from 0.4–28.8% in low HIV burden

African settings to approximately 80% in settings with a high background burden of HIV [35, 36]. Patients with DR-TB and HIV co-infection are characterized by excessively high mortality rates of 50–80% [37–39]. Empiric evidence for the survival benefit of early ART initiation in patients with MDR-TB was provided in a sub-group secondary analysis of 23 MDR-TB patients that were enrolled into the SAPIT study. This study showed an 86% reduction in mortality among MDR-TB patients initiating ART early in DR-TB treatment [40]. This study together with other published literature observed that HIV and ART status significantly impacted DR-TB treatment outcomes including patient survival [37, 41, 42]. Another study concluded that ART is a noteworthy determinant of treatment success in individuals co-infected with HIV [52]. This SA study reported 114/748 DR-TB and HIV co-infected patients (15%) were not on ART. Furthermore 67 (59%) of HIV-positive individuals not on ART had failed treatment outcomes [52]. Those 116 (35%) HIV-positive patients on ART represented patients who had successful treatment outcomes [52]. It is anticipated that the introduction of the test and treat approach will further reduce the number of HIV positive DR-TB patients not on ART. Another study by Brust et al. [3] also reported that therapeutic and mortality outcomes among MDR-TB patients receiving simultaneous ART was similar to HIV-uninfected patients. Results displayed 191 participants with MDR-TB outcome, 130 were cured or completed therapy, which did not differ by HIV status ($P = 0.50$). HIV-infected and HIV-uninfected individuals had greater survival rates (86% and 94%, respectively; $P = 0.34$). The sturdiest mortality risk factor among DR-TB and HIV co-infected patients was a CD4 count ≤100 cells/mm^3 [3]. These findings taken collectively, support early concurrent treatment of MDR-TB and HIV in co-infected patients. While studies consistently show no association between HIV and mortality in patients with XDR-TB despite receiving XDR-TB therapy, two studies independently showed that use of ART in XDR-TB-HIV co-infected patients reduced mortality [37, 43]. Furthermore, a substantially lower risk of death was found even among patients initiating ART at CD4 cell counts >200/mm^3, (HR 0.094, 95% CI 0.007–1.22) [37].

Universal HIV and TB Case Finding and Universal ART for HIV-Infected TB Patients

WHO commends that all HIV-TB infected patients should be initiated on ART regardless of CD4 count creating huge potential for further reductions to TB associated mortality in HIV infected patients. Most African countries with a high burden of TB-HIV have specific policies supporting HIV counselling and testing for those with presumptive or confirmed TB and recommend ART for all TB cases regardless of CD4+ cell count [44].

In SA, separate reports have confirmed the high national-level uptake of HIV testing services (HTS), demonstrating HTS uptake of >90% can be attained in primary healthcare facilities irrespective of the level of TB and HIV service integration

[45], and in DR-TB services regardless of whether these are centralized or decentralized [46]. Despite calls to ensure that all persons investigated for TB are also routinely offered HIV testing [47, 48], there is a shortage of data to show to what extent programmes are achieving this within different African countries. There is also limited available data describing HIV testing for children undergoing investigation and treatment for TB disease. National-level data in SA shows that the proportion of child TB cases (under 15 years) reported to have unknown HIV status declined from 77% to 25% between 2008 and 2012 [49].

WHO recommends the Xpert® MTB/RIF test for early diagnostic testing for all children and adults with symptoms and signs of TB, including testing of extrapulmonary samples e.g. tissue specimens, lymph nodes and cerebrospinal fluid. Amid 2010 and 2016, 6659 GeneXpert devices were procured by 130 of 145 countries suitable for concessional pricing. Of the 6.9 million test cartridges obtained by qualified countries in 2016, 35% (2.4 million) went to SA. Nevertheless numerous functioning challenges in the scale-up and implementation of GeneXpert were observed in five SSA countries: low coverage, poor laboratory infrastructure, limited access, poor associations to treatment, inadequate data on outcomes, difficulties with specimen transport and analytic algorithms that weren't associated with updated WHO recommendations on target patient groups and sponsoring challenges [50]. The GeneXpert strategy should poise the need to advance its application with general health systems establishment. To achieve the full impact of innovative diagnostics, quality of health service delivery and quality of care must improve. Furthermore obligation of resources is required to take new diagnostics into the field including rearrangement of laboratory services and enhanced access to technical expertise to support implementation [50] (Fig. 3).

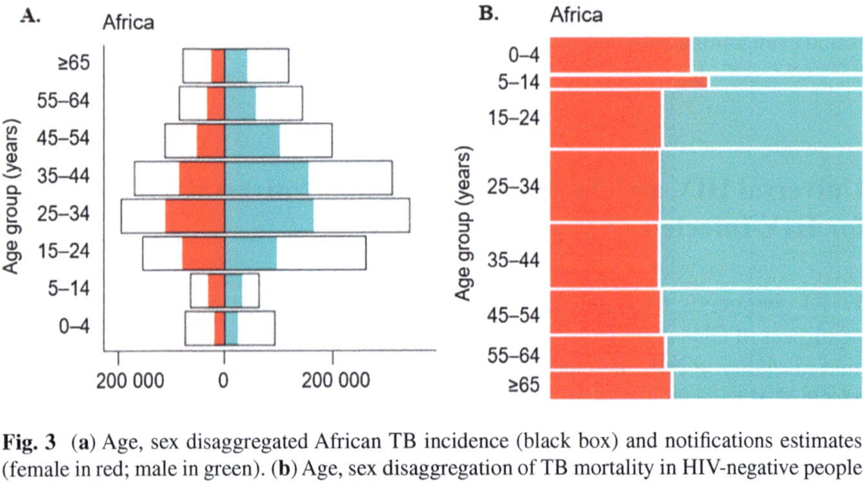

Fig. 3 (**a**) Age, sex disaggregated African TB incidence (black box) and notifications estimates (female in red; male in green). (**b**) Age, sex disaggregation of TB mortality in HIV-negative people in Africa (female in red; male in green) [1]

Tuberculosis Related Mortality in High Burden SSA Countries

Despite more than a decade of public access ART, TB treatment and preventive therapy, TB contributes most to morbidity and mortality from an infectious disease, globally. This remains starkly apparent in many SSA settings, including SA. About 1.5 million TB deaths occurred worldwide in 2016, with approximately 90% of all TB related deaths occurring in SSA and South East Asia [51]. ART scale-up reduces HIV-associated morbidity and mortality. TB case-fatality rates of 16–35% have been observed in HIV infected patients not on ART compared to 4–9% in HIV-negative patients [51]. Trends from 16 high TB burden countries in SSA (Fig. 4) show substantial decrease in TB mortality rates in HIV-positive individuals living in 9 of 16 high TB burden settings in the past few years: CAR, DRC, Ethiopia, Kenya, Lesotho, Mozambique, Tanzania, Namibia, and Nigeria. Despite huge programmatic investments in HIV and TB, TB mortality rates in HIV infected patients remain unchanged in SA, Angola, Sierra Leone and Liberia. Comorbidities in these settings, such as opportunistic infections, anaemia, TB drug resistance coupled with high initial loss-to-follow-up in ART and TB programmes are commonly identified factors contributing to the ongoing high TB mortality rates.

Tuberculosis remains an international emergency accounting for 1.7 million deaths annually. Africa is disproportionately affected by TB. One quarter of the global TB burden resides in Africa, making this continent a key geographical area for health TB related interventions. The widespread mismanagement of INH and rifampicin over three decades has caused the emergence of potentially untreatable forms of TB. These forms of TB undermine clinical and programmatic outcomes in disease endemic settings like SSA. Regardless of improvements in prevention and care, TB is still one of the world's foremost causes of ill-health and death. The present rate of reduction in TB incidence and TB mortality is insufficient to reach targets set in the SDGs and in the End TB Strategy. Failure to respond timely and appropriately to the dual escalation of HIV and TB incidence between 1992–2005 in SSA resulted in unacceptably high rates of TB-associated morbidity and mortality, in HIV infected patients with devastating consequences within communities. Since then, the rapid expansion of ART has resulted in a dramatic reduction in HIV-associated TB incidence and mortality in SSA. Several factors threaten realisation of the 95–95-95 End TB strategy targets, key among these being increasing DR-TB incidence, and failure to find and treat missing TB patients. Robust high-quality implementation strategies suitable for disease endemic resource limited settings that appropriately use new technologies in TB prevention, diagnosis and treatment is essential for the 2035 TB elimination targets to be met in Africa.

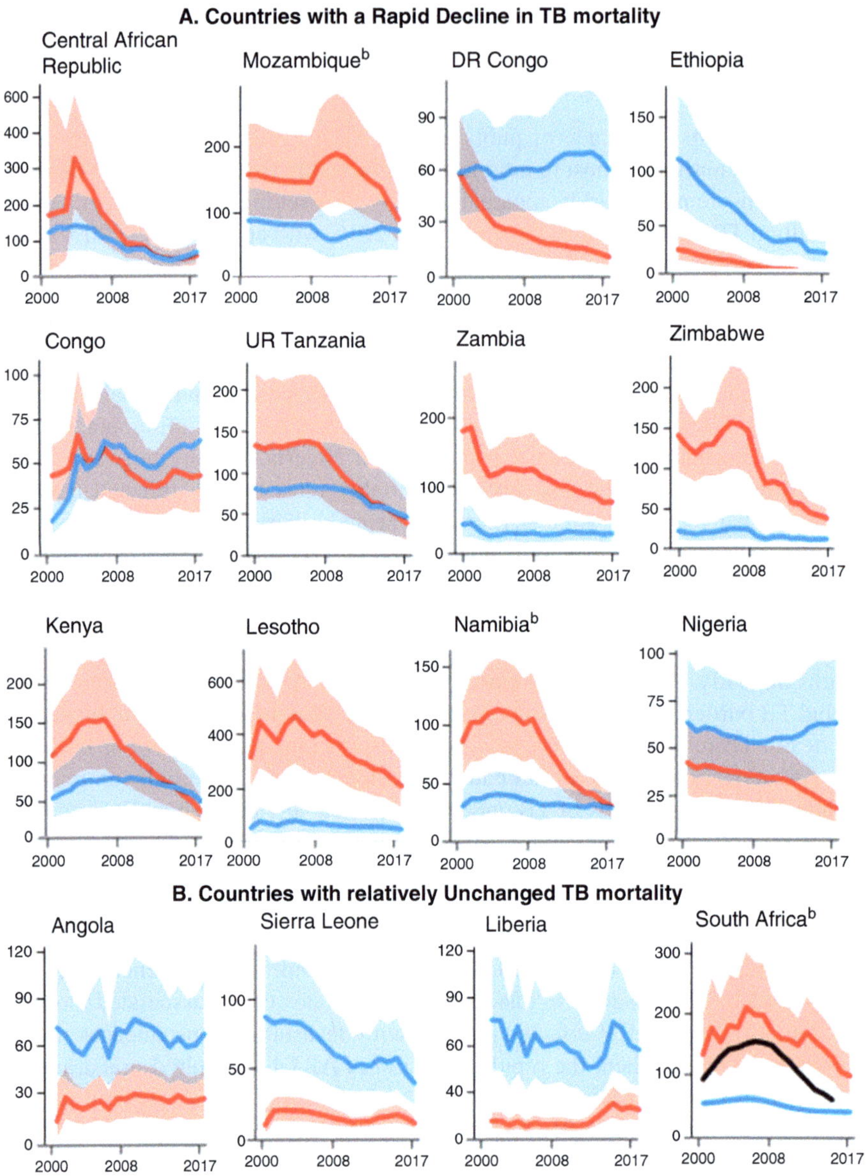

Fig. 4 Estimated TB mortality rate trends within 16 TB high burden African countries, 2000–2016. TB mortality rates of HIV-positive TB in blue, mortality rates of HIV-positive, and shaded areas represent uncertainty intervals

References

1. World Health Organization (2018). Global Tuberculosis Report. from http://apps.who.int/iris/bitstream/10665/250441/1/9789241565394-eng.pdf?ua=1
2. Adeiza M, Abba A, Okpapi J (2014) HIV-Associated tuberculosis: a sub-Saharan African perspective. Sub-Saharan Afr J Med 1(1):1–14
3. Brust JCM, Shah NS, Mlisana K, Moodley P, Allana S, Campbell A, Johnson BA, Master I, Mthiyane T, Lachman S, Larkan L-M, Ning Y, Malik A, Smith JP, Gandhi NR (2018) Improved survival and cure rates with concurrent treatment for multidrug-resistant tuberculosis–human immunodeficiency virus coinfection in South Africa. Clin Infect Dis 66(8):1246–1253
4. World Health Organization (2017). Global Tuberculosis Report: 1–147
5. Corbett EL, Watt CJ, Walker N, Maher D, Williams BG, Raviglione MC, Dye C (2003) The growing burden of tuberculosis: global trends and interactions with the HIV epidemic. Arch Intern Med 163(9):1009–1021
6. Lawn SD, Harries AD, Williams BG, Chaisson RE, Losina E, De Cock KM, Wood R (2011) Antiretroviral therapy and the control of HIV-associated tuberculosis. Will ART do it? [Unresolved issues]. Int J Tuberc Lung Dis 15(5):571–581
7. World Health Organization. (2016). Global tuberculosis report 2016., from http://apps.who.int/iris/bitstream/10665/250441/1/9789241565394-eng.pdf]
8. Eshetie S, Gizachew M, Alebel A, van Soolingen D (2018) Tuberculosis treatment outcomes in Ethiopia from 2003 to 2016, and impact of HIV co-infection and prior drug exposure: a systematic review and meta-analysis. PLoS One 13(3):e0194675
9. Jacobson KB, Moll AP, Friedland GH, Shenoi SV (2015) Successful tuberculosis treatment outcomes among HIV/TB coinfected patients down-referred from a district hospital to primary health clinics in rural South Africa. PLoS One 10(5):e0127024
10. Mibei DJ, Kiarie JW, Wairia A, Kamene M, Okumu ME (2016) Treatment outcomes of drug-resistant tuberculosis patients in Kenya. Int J Tuberc Lung Dis 20(11):1477–1482
11. Amo-Adjei J, Awusabo-Asare K (2013) Reflections on tuberculosis diagnosis and treatment outcomes in Ghana. Arch Public Health 71(1):22
12. Semu M, Fenta TG, Medhin G, Assefa D (2017) Effectiveness of isoniazid preventative therapy in reducing incidence of active tuberculosis among people living with HIV/AIDS in public health facilities of Addis Ababa, Ethiopia: a historical cohort study. BMC Infect Dis 17(1):5
13. Government of Lesotho (2017). Final Report for a Joint Review of HIV/Tuberculosis and Hepatitis Programmes
14. Hayes-Larson E, Hirsch-Moverman Y, Saito S, Frederix K, Pitt B, Maama BL, Howard AA (2017) Prevalence, patterns, and correlates of HIV disclosure among TB-HIV patients initiating antiretroviral therapy in Lesotho. AIDS Care 29(8):978–984
15. Meyer-Rath G, McGillen JB, Cuadros DF, Hallett TB, Bhatt S, Wabiri N, Tanser F, Rehle T (2018) Targeting the right interventions to the right people and places: the role of geospatial analysis in HIV program planning. AIDS (London, England) 32(8):957–963
16. Stats South Africa. (2016). Mortality and cause of death report 2016 from http://www.statssa.gov.za/publications/P03093/P030932016.pdf
17. Lawn SD, Myer L, Orrell C, Bekker L-G, Wood R (2005) Early mortality among adults accessing a community-based antiretroviral service in South Africa: implications for programme design. AIDS 19(18):2141–2148
18. Coetzee LM, Cassim N, Glencross DK (2017) Analysis of HIV disease burden by calculating the percentages of patients with CD4 counts <100 cells/μL across 52 districts reveals hot spots for intensified commitment to programmatic support. S Afr Med J 107(6):507–513
19. Dye C, Scheele S, Dolin P, Pathania V, Raviglione MC (1999) Consensus statement. Global burden of tuberculosis: estimated incidence, prevalence, and mortality by country. WHO Global Surveillance and Monitoring Project. JAMA 282(7):677–686
20. Getahun H, Gunneberg C, Granich R, Nunn P (2010) HIV infection-associated tuberculosis: the epidemiology and the response. Clin Infect Dis 50(Suppl 3):S201–S207

21. Woldehanna S, Volmink J (2004) Treatment of latent tuberculosis infection in HIV infected persons. Cochrane Database Syst Rev 1:CD000171
22. Murray J, Sonnenberg P, Shearer SC, Godfrey-Faussett P (1999) Human immunodeficiency virus and the outcome of treatment for new and recurrent pulmonary tuberculosis in African patients. Am J Respir Crit Care Med 159(3):733–740
23. Sonnenberg P, Glynn JR, Fielding K, Murray J, Godfrey-Faussett P, Shearer S (2005) How soon after infection with HIV does the risk of tuberculosis start to increase? A retrospective cohort study in South African gold miners. J Infect Dis 191(2):150–158
24. Whalen C, Okwera A, Johnson J, Vjecha M, Hom D, Wallis R, Huebner R, Mugerwa R, Ellner J (1996) Predictors of survival in human immunodeficiency virus-infected patients with pulmonary tuberculosis. The Makerere University-Case Western Reserve University Research Collaboration. Am J Respir Crit Care Med 153(6 Pt 1):1977–1981
25. Antonucci G, Girardi E, Raviglione MC, Ippolito G (1995) Risk factors for tuberculosis in HIV-infected persons. A prospective cohort study. The Gruppo Italiano di Studio Tubercolosi e AIDS (GISTA). JAMA 274(2):143–148
26. Crampin AC, Floyd S, Glynn JR, Sibande F, Mulawa D, Nyondo A, Broadbent P, Bliss L, Ngwira B, Fine PE (2002) Long-term follow-up of HIV-positive and HIV-negative individuals in rural Malawi. AIDS 16(11):1545–1550
27. Granich R, Akolo C, Gunneberg C, Getahun H, Williams P, Williams B (2010) Prevention of tuberculosis in people living with HIV. Clin Infect Dis 50(Suppl 3):S215–S222
28. Abdool Karim SS, Naidoo K, Grobler A, Padayatchi N, Baxter C, Gray A, Gengiah T, Nair G, Bamber S, Singh A, Khan M, Pienaar J, El-Sadr W, Friedland G, Abdool Karim Q (2010) Timing of initiation of antiretroviral drugs during tuberculosis therapy. N Engl J Med 362(8):697–706
29. Abdool Karim SS, Naidoo K, Grobler A, Padayatchi N, Baxter C, Gray AL, Gengiah T, Gengiah S, Naidoo A, Jithoo N, Nair G, El-Sadr WM, Friedland G, Abdool Karim Q (2011) Integration of antiretroviral therapy with tuberculosis treatment. N Engl J Med 365(16):1492–1501
30. Blanc FX, Sok T, Laureillard D, Borand L, Rekacewicz C, Nerrienet E, Madec Y, Marcy O, Chan S, Prak N, Kim C, Lak KK, Hak C, Dim B, Sin CI, Sun S, Guillard B, Sar B, Vong S, Fernandez M, Fox L, Delfraissy JF, Goldfeld AE, Team CS (2011) Earlier versus later start of antiretroviral therapy in HIV-infected adults with tuberculosis. N Engl J Med 365(16):1471–1481
31. Havlir DV, Kendall MA, Ive P, Kumwenda J, Swindells S, Qasba SS, Luetkemeyer AF, Hogg E, Rooney JF, Wu X, Hosseinipour MC, Lalloo U, Veloso VG, Some FF, Kumarasamy N, Padayatchi N, Santos BR, Reid S, Hakim J, Mohapi L, Mugyenyi P, Sanchez J, Lama JR, Pape JW, Sanchez A, Asmelash A, Moko E, Sawe F, Andersen J, Sanne I, AIDS Clinical Trials Group Study A5221 (2011) Timing of antiretroviral therapy for HIV-1 infection and tuberculosis. N Engl J Med 365(16):1482–1491
32. Franke MF, Robins JM, Mugabo J, Kaigamba F, Cain LE, Fleming JG, Murray MB (2011) Effectiveness of early antiretroviral therapy initiation to improve survival among HIV-infected adults with tuberculosis: a retrospective cohort study. PLoS Med 8(5):e1001029
33. Manosuthi W, Mankatitham W, Lueangniyomkul A, Thongyen S, Likanonsakul S, Suwanvattana P, Thawornwan U, Suntisuklappon B, Nilkamhang S, Sungkanuparph S, Team TS (2012) Time to initiate antiretroviral therapy between 4 weeks and 12 weeks of tuberculosis treatment in HIV-infected patients: results from the TIME study. J Acquir Immune Defic Syndr 60(4):377–383
34. Sinha S, Shekhar RC, Singh G, Shah N, Ahmad H, Kumar N, Sharma SK, Samantaray JC, Ranjan S, Ekka M, Sreenivas V, Mitsuyasu RT (2012) Early versus delayed initiation of antiretroviral therapy for Indian HIV-Infected individuals with tuberculosis on antituberculosis treatment. BMC Infect Dis 12:168
35. Daftary A, Padayatchi N, O'Donnell M (2014) Preferential adherence to antiretroviral therapy over tuberculosis treatment: a qualitative study of drug-resistant TB/HIV co-infected patients in South Africa. Glob Public Health 9(9):1107–1116

36. Isaakidis P, Das M, Kumar AM, Peskett C, Khetarpal M, Bamne A, Adsul B, Manglani M, Sachdeva KS, Parmar M (2014) Alarming levels of drug-resistant tuberculosis in HIV-infected patients in metropolitan Mumbai, India. PLoS One 9(10):e110461
37. O'Donnell MR, Padayatchi N, Kvasnovsky C, Werner L, Master I, Horsburgh CR (2013) Treatment outcomes for extensively drug-resistant tuberculosis and HIV co-infection. Emerg Infect Dis 19(3):416–424
38. Padayatchi N, Mahomed S, O'Donnell M, Conradie F, Naidoo K (2017) The World Health Organization excludes Mycobacterium tuberculosis from the 2017 priority pathogens list. South African Med J 107(6):466–466
39. Pietersen E, Ignatius E, Streicher EM, Mastrapa B, Padanilam X, Pooran A, Badri M, Lesosky M, van Helden P, Sirgel FA (2014) Long-term outcomes of patients with extensively drug-resistant tuberculosis in South Africa: a cohort study. Lancet 383(9924):1230–1239
40. Padayatchi N, Abdool Karim S, Naidoo K, Grobler A, Friedland G (2014) Improved survival in multidrug-resistant tuberculosis patients receiving integrated tuberculosis and antiretroviral treatment in the SAPiT Trial. Int J Tuberc Lung Dis 18(2):147–154
41. O'Donnell MR, Zelnick J, Werner L, Master I, Loveday M, Horsburgh CR, Padayatchi N (2011) Extensively drug-resistant tuberculosis in women, KwaZulu-Natal, South Africa. Emerg Infect Dis 17(10):1942–1945
42. Shenoi S, Heysell S, Moll A, Friedland G (2009) Multidrug-resistant and extensively drug-resistant tuberculosis: consequences for the global HIV community. Curr Opin Infect Dis 22(1):11–17
43. Dheda K, Limberis JD, Pietersen E, Phelan J, Esmail A, Lesosky M, Fennelly KP, te Riele J, Mastrapa B, Streicher EM, Dolby T, Abdallah AM, Ben-Rached F, Simpson J, Smith L, Gumbo T, van Helden P, Sirgel FA, McNerney R, Theron G, Pain A, Clark TG, Warren RM (2017) Outcomes, infectiousness, and transmission dynamics of patients with extensively drug-resistant tuberculosis and home-discharged patients with programmatically incurable tuberculosis: a prospective cohort study. Lancet Respir Med 5(4):269–281
44. Gupta S, Granich R, Date A, Lepere P, Hersh B, Gouws E, Samb B (2014) Review of policy and status of implementation of collaborative HIV-TB activities in 23 high-burden countries. Int J Tuberc Lung Dis 18(10):1149–1158
45. Kaplan R, Caldwell J, Bekker LG, Jennings K, Lombard C, Enarson DA, Wood R, Beyers N (2014) Integration of TB and ART services fails to improve TB treatment outcomes: comparison of ART/TB primary healthcare services in Cape Town, South Africa. S Afr Med J 104(3):204–209
46. Loveday M, Wallengren K, Brust J, Roberts J, Voce A, Margot B, Ngozo J, Master I, Cassell G, Padayatchi N (2015) Community-based care vs. centralised hospitalisation for MDR-TB patients, KwaZulu-Natal, South Africa. Int J Tuberc Lung Dis 19(2):163–171
47. Kumar AM, Gupta D, Gupta RS, Satyanarayana S, Wilson N, Zachariah R, Lawn SD, Harries AD (2013) HIV testing in people with presumptive tuberculosis: time for implementation. Lancet Respir Med 1(1):7–9
48. Word Health Organization. (2012). WHO policy on collaborative TB/HIV activities: guidelines for national programmes and other stakeholders
49. Smith J, Moyo S, Day C (2014) A review of TB in children and adolescents in South Africa 2008–2012. In: Padarath A, English R (eds) South African Health Review 2013/14. Health Systems Trust, Durban
50. Umubyeyi AN, Bonsu F, Chimzizi R, Jemal S, Melese M, Ruttoh E, Mundy C (2016) The role of technical assistance in expanding access to Xpert(®) MTB/RIF: experience in sub-Saharan Africa. Public Health Action 6(1):32–34
51. Nagu TJ, Aboud S, Mwiru R, Matee MI, Rao M, Fawzi WW, Zumla A, Maeurer MJ, Mugusi F (2017) Tuberculosis associated mortality in a prospective cohort in Sub Saharan Africa: association with HIV and antiretroviral therapy. Int J Infect Dis 56:39–44
52. Loveday M, Wallengren K, Reddy T, Besada D, Brust JCM, Voce A, Desai H, Ngozo J, Radebe Z, Master I, Padayatchi N, Daviaud E, Caylà JA (2018) MDR-TB patients in KwaZulu-Natal, South Africa: Cost-effectiveness of 5 models of care. PLOS ONE 13(4):e0196003
53. World Health Organization (2007). Kingdom of Lesotho: National Tuberculosis Programme

Modelling the HIV-Associated TB Epidemic and the Impact of Interventions Aimed at Epidemic Control

P. J. Dodd, C. Pretorius, and B. G. Williams

Abstract In this chapter, we focus on mathematical models of tuberculosis epidemiology (TB) that include interactions with HIV and an explicit representation of transmission. We review the natural history of TB and illustrate how its features are simplified and incorporated in mathematical models. We then review the ways HIV influences the natural history of TB, the interventions that have been considered in models, and the way these individual-level effects are represented in models. We then go on to consider population-level effects, reviewing the TB/HIV modelling literature. We first review studies whose focus was on purely epidemiological modelling, and then studies whose focus was on modelling the impact of interventions. We conclude with a summary of the uses and achievements of TB/HIV modelling and some suggested future directions.

Keywords Tuberculosis · Human immunodeficiency virus · Mathematical modelling · Transmission modelling · TB/HIV · Epidemiological modelling · Economic evaluation · Natural history · Epidemiology

---

The original version of this chapter was revised. The correction to this chapter is available at https://doi.org/10.1007/978-3-030-29108-2_14

P. J. Dodd (✉)
School of Health and Related Research, University of Sheffield, Sheffield, UK
e-mail: p.j.dodd@sheffield.ac.uk

C. Pretorius
Avenir Health, Glastonbury, CT, USA

B. G. Williams
South African Centre for Epidemiological Modelling and Analysis,
University of Stellenbosch, Stellenbosch, South Africa

I. Sereti et al. (eds.), *HIV and Tuberculosis*,
https://doi.org/10.1007/978-3-030-29108-2_3

Introduction

The emergence of the HIV pandemic in sub-Saharan Africa and the simultaneous rise in the incidence of tuberculosis (TB) in this region is a stark reminder of the importance of HIV-epidemiology for modelling TB (Fig. 1). As HIV prevalence increased during the 1990s, the per-capita TB notification rate increased several-fold in some countries. As the HIV epidemics matured and the mean duration of HIV-infection grew longer, the mean level of population immunocompetence declined, and the strength of association increased. By the late 1990s for example, over 70% of TB notifications in Zimbabwe were in people living with HIV (PLHIV) [1]. Comparing this proportion to the HIV infection prevalence in the population without TB yields an odds ratio of around 10, which can also be interpreted as an average incidence rate ratio for developing TB if infected with HIV [2].

With the wider roll-out of antiretroviral therapy for HIV (ART) beginning around 2004, initially provided to those meeting certain CD4 count thresholds and now recommended for all PLHIV, these population-level associations between HIV and TB incidence began to weaken. As the epidemic declined in some countries, including Zimbabwe and Ethiopia, and ART was rolled out very aggressively, as in Botswana and South Africa, the TB notification rates declined but not to the pre-HIV levels. Today, around 10% of TB is thought to be HIV-associated globally (74% of this in sub-Saharan Africa) [3]. A higher, but more uncertain, proportion of TB deaths are HIV-associated, but vital registration cause-of-death coding rules and frequent comorbidities make such estimates more difficult. However, the fairly constant HIV prevalence and still-elevated risks of developing TB even among PLHIV on ART means that HIV will continue to affect TB epidemiology for decades to come and remain a crucial aspect of modelling in high-HIV burden settings.

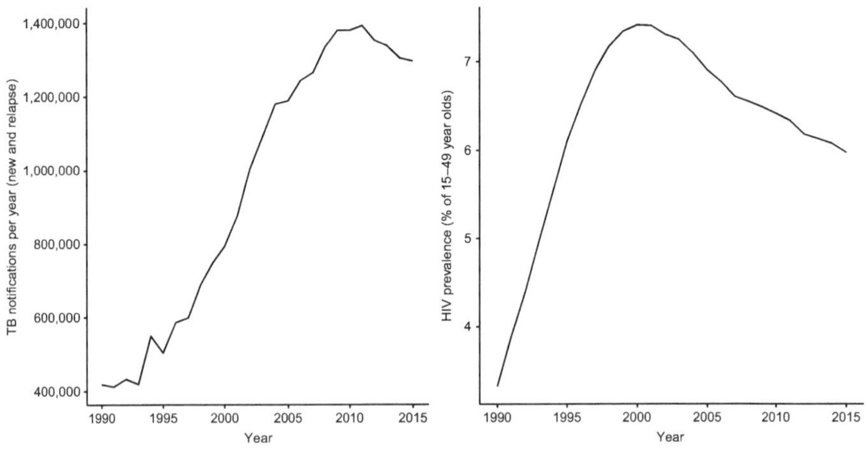

Fig. 1 TB and HIV in sub-Saharan Africa between 1990 and 2015: total tuberculosis notifications reported to the World Health Organisation (WHO) by countries in the WHO African region (left); UNAIDS estimates of HIV prevalence in 15–49 year olds (right)

Focus of This Chapter

The use of mathematical modelling as a tool in public health for understanding the behaviour of complex systems, and of assessing the impact and cost-effectiveness of interventions has grown. For an infectious disease like TB, making quantitative predictions about the population-level impact of interventions requires models that explicitly account for the transmission process. Without including transmission, there is no way of assessing the indirect benefits interventions may have through onward cases averted. For interventions whose benefit accrues mainly or wholly at the individual-patient level, the standard static modelling approaches of health economics are likely to be appropriate, but for interventions that are likely to reduce transmission, static models may substantially underestimate benefits from indirect effects. In this chapter, we focus explicitly on dynamic models, that is population-level transmission models where the risks of infection depend on the prevalence of infectious cases and may change in response to interventions.

We first review the natural history and transmission epidemiology of TB, and describe in nontechnical terms the way these features are approximated and abstractly represented in transmission models. We describe the way HIV affects individuals' natural history of TB and the approaches that have been used to model these effects. We highlight a number of interventions and describe how these have been allowed for in TB models, and we review the literature on modelling the impact of these interventions on TB at a population-level. We conclude by critically reflecting on the achievements and limitations of the work to date and suggest future directions.

TB Modelling Without HIV

Before going on to consider how TB/HIV is modelled, it is important to understand how TB is usually represented in transmission models and what modelling techniques are used. In this section, we introduce compartmental models of TB and step through various aspects of TB natural history, discussing how modellers approach incorporating these into mathematical models.

Modelling Approaches

We will centre our description of TB modelling on Fig. 2, which is largely based on the early work of Dye and colleagues [4] (see also the book by Dye [5]), which, with variations largely reflecting different model purpose, has become the canonical basis for TB transmission models. Different disciplines have brought different techniques, focus and approaches to the analysis, with TB/HIV modelling papers appearing in journals ranging from mathematical biology focussed on formally characterizing properties of specific models with abstract applications, through population biology/ecology studies aimed at exploring actual or potential behaviour

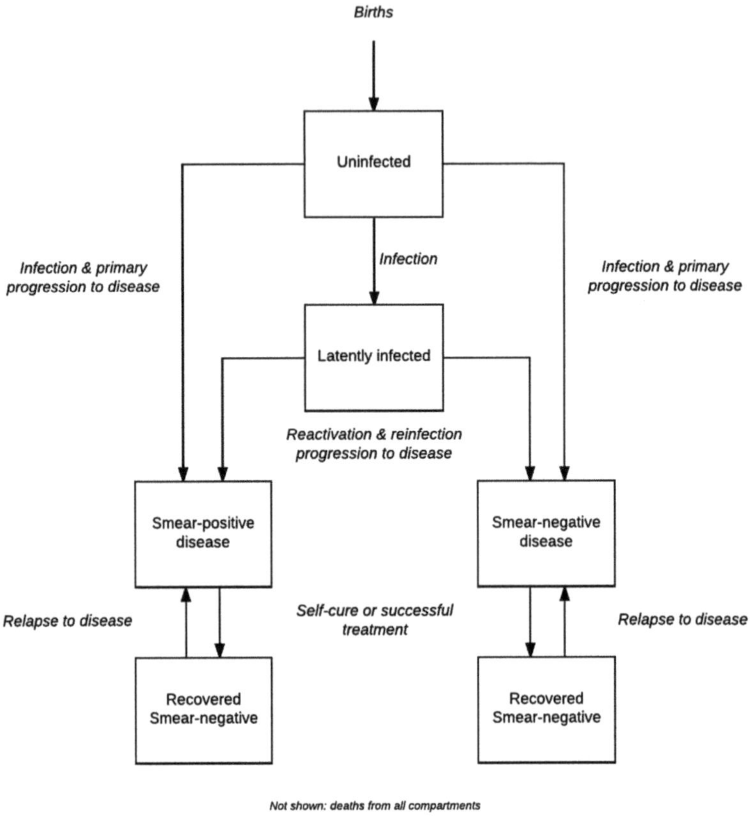

Fig. 2 A typical TB transmission model structure

in more realistically-specified particular populations, through to very detailed models tightly tied to the epidemiology and costs in specific settings, and aimed at policy evaluation and decision support. A list of modeling studies evaluating HIV and TB is given in Table 1.

It should be noted that appropriate model structure is determined not only by the natural history or epidemiology, but also by the question at hand, and the published models vary widely in their structure. However, it is also the case that some models have appeared in the literature whose structure or parametrization means that they seriously misrepresent the natural history of TB. Menzies et al. [6] in a systematic review of TB modelling literature found that in 40% of published TB models there was an implied cumulative risk of progression to active disease that was substantially at odds with empirical data.

The model diagram in Fig. 2 maps precisely to simple ‘compartmental models’. In these models, a set of mutually exclusive states that exhaustively classify individuals in the population are chosen. Each person in the population can then be thought of as being in one of these ‘boxes’ or ‘compartments’. The model state at a given time is then given by the population number in each compartment and the population dynamics are determined by quantitative rules for the rates of flow

Table 1 Summary of topics considered by TB modelling papers for high HIV prevalence settings

Focus on intervention	Topic	First author, year and reference
No	HIV on TB incidence	Bermejo 1992 [67], Schulzer 1992 [68], Schulzer 1994 [70], Dolin 1994 [69]
	TB dynamics with HIV	Kapitanov 2015 [74], Massad 1993 [71], Naresh 2009 [72], Roeger 2009 [73]
	HIV on TB evolution & DR	Basu 2009 [76], Basu 2008 [75], Sergeev 2012 [77]
	HIV on TB outbreaks	Murray 2002 [79], Porco 2001 [78], Pretorius 2011 [80]
	HIV on TB transmission	Andrews 2014 [84], Andrews 2013 [85], Dodd 2016 [86], Escombe 2008 [81], Uys 2011 [83], Wood 2010 [82]
	Interpreting TB/HIV epidemiology	Blaser 2016 [89], Hughes 2006 [87], Pretorius 2014 [90], Sánchez 2009 [88]
Yes	TB/HIV dynamics with interventions	Agusto 2014 [91], Bacaër 2008 [94], Kaur 2014 [92], Sharomi 2008 [93]
	Control of DR-TB in TB/HIV epidemics	Basu 2007 [63], Basu 2009 [96], Basu 2011 [97]
	Preventive therapy	Basu 2009 [101], Cohen 2006 [100], Guwatudde 2004 [99], Heymann 1993 [98], Kunkel 2016 [103], Mills 2011 [102]
	Case finding	Azman 2014 [110], Baltussen 2005 [104], Dodd 2011 [42], Dowdy 2009 [107], Dye 1998 [4], Laxminarayan 2009 [105], Mellor 2011 [108], Sánchez 2008 [106], Yaesoubi 2013 [109]
	Diagnostics	Dowdy 2014 [62], Dowdy 2006 [111], Langley 2012 [114], Langley 2014 [61], Lin 2011 [113], Menzies 2012 [112]
	ART	Bhunu 2009 [117], Dodd 2013 [118], Pretorius 2014 [119], Williams 2005 [115], Williams 2003 [35], Williams 2010 [1]
	Combined interventions	Chindelevitch 2015 [122], Currie 2005 [120], Currie 2003 [121], Gilbert 2015 [124], Gilbert 2016 [125], Houben 2016 [127], Houben 2016 [129], Knight 2015 [123], Trauer 2017 [128]

Topics are in order of discussion in text. DR = drug resistant. The topic word 'dynamics' implies a more mathematically focussed study

between boxes, which may depend on the state of the system and on time. Certain flows are sometimes omitted from such model diagrams, e.g. the inflow of births into the uninfected class; the outflow of death from all states (at higher rates from disease states). Model diagrams, such as in Fig. 2 may suppress some detail for the sake of simplicity. An example is where models include a discrete number of age categories, each one of which would require a duplication of the infection/disease logic of Fig. 2. Model diagrams can also be used to describe more approximately the logic of modelling approaches that are not compartmental, e.g. individual- or agent-based models, which may include much finer subdivisions of the states and more flexible rules governing transition between states. Individual-based models (IBMs) explicitly track *in silico* representations of people, and can therefore deal with effects that are hard to represent when only dealing with groups of a certain type of person (e.g. a detailed dependence on the history that led to the current state rather than simply depending on the current state alone).

Modelling TB Natural History and Epidemiology

Infection and Progression

One of the features of TB natural history that sets it aside from other infectious diseases and therefore distinguishes TB models from models of other infectious diseases is its long latent state and the complexity of this latent state's relationship to active TB disease. At birth, individuals are assumed to be uninfected and subject to infection with *Mycobacterium tuberculosis* (*M. tuberculosis*) with a force of infection (FOI) that depends on the prevalence of active TB disease at that time (see below). Individuals infected with *M. tuberculosis* for the first time are subject to an age-dependent risk of progressing to TB disease that decreases during childhood ages, remains low until adolescence and then increases as people reach adulthood [7]. This is normally represented as an elevated risk of 'primary progression' during the first 2–5 years following infection, with a subsequent lower risk of 'endogenous reactivation' thereafter [8]. These rates of progression compete with mortality, such that only a minority of those infected will develop active disease as a result: the canonical rule-of-thumb is that once infected with TB people have a 10% lifetime risk of disease [9], evenly divided between primary progression and endogenous reactivation, and more modern data broadly corroborate this picture [10]. In models, this is most often modelled as a certain fraction of infections progressing directly to active disease (an approximation to primary progression) and other individuals entering a latent category. Another modelling approach is to introduce two latent categories, with newly infected individuals moving initially into a fast-progressing latent category, before moving into a lower TB risk slow-latent category over a 2–5 year time scale [11].

Reinfection and Protection

Another feature that is different from many other infectious diseases is the potential for those already carrying a latent *M. tuberculosis* infection (LTBI) to be reinfected. Evidence from modelling of population trends and more directly from cohorts of healthcare workers suggests that an existing latent infection conveys some protection against infection and progression to disease [8, 12, 13]. However, it is not possible to distinguish whether this protection applies against re-infection itself or progression following reinfection. If the latter, it is not clear whether there is any difference in the way this protection applies to progression in the initial years following reinfection or more distant re-activation. Re-infection is conventionally modelled as a partially effective 'degree' vaccine: i.e. infections in the latent (or slow-latent) category act like initial infections in the uninfected category, but occur at a lower rate with the FOI multiplied by some hazard ratio representing protection. This provides a contribution to incident TB disease from the latent category or via another route back into the fast-latent compartment. For modelling approaches with a single latent category, this approach covers either protection via reduced possibility of infection or via reduced progression. For modelling approaches with fast- and slow-latent compartments, this approach represents protection via reduced susceptibility to infection.

Infectiousness and Risks of Infection

Upon development of active disease, many models assign individuals to either sputum smear-positive or smear-negative disease. This definition has been made operationally as it affects the changes of diagnosis by particular algorithms and represents a group historically emphasised in TB control approaches, but also because smear-positive TB is on average more infectious per unit time than smear-negative TB disease [14, 15]. Some models introduce a low rate of progression from smear-negative to smear-positive disease [4], but this is poorly evidenced and considered small. While incipient smear-positive TB disease must at some point transition from smear-negative to smear-positive, the conventional modelling approach assumes this happens early on if at all, with around half of TB disease being smear-positive. It should be noted that new insights suggesting that the dichotomy between LTBI and clinical disease is an overly-simplistic representation of the dynamic spectrum existing in biological reality [16], and epidemiological data from prevalence surveys that have found substantial amounts of asymptomatic TB [17] have yet to be explored in terms of their implications for models. Typically, the force-of-infection in TB models is taken as being proportional to the per capita TB prevalence with a discounting factor for smear-negative TB prevalence. The coefficient of proportionality between per capita smear-positive TB prevalence and the force-of-infection is known in the epidemiological literature as Styblo's ratio: the number of infections per year generated by 1 smear-positive TB case. Historically this was estimated to be in the region of 10 infections per year, but may be closer to 6 infections per year in the modern era [18, 19]. It is this dependence of force-of-infection on the current prevalence of active disease that makes a TB model 'dynamic' as opposed to 'static'. Changes in TB prevalence due to interventions here are allowed to accrue indirect benefits, influencing the future incidence of TB disease via a reduced force-of-infection.

Mortality, Self-Cure and Detection

The best data on the natural history of TB disease in the absence of treatment comes from the pre-chemotherapy literature [20]. A substantial proportion of those with TB disease would be expected to die: around 70% of those with smear-positive disease and around 30% of those with smear-negative disease, giving an average case-fatality ratio of around 50%. The remaining cases are said to 'self-cure', represented as a return to latent infection compartments in TB models. The course of TB disease was found to run over an average of around 3 years (independent of smear-status), though data from the early 20th century suggest that some individuals could have active TB for over a decade [21].

The era of chemotherapy for TB has added a third class of outcome to TB disease: that a case is diagnosed and put onto treatment. This detection process is often abstracted into a single rate that captures both the efforts of a patient to seek care, and the sensitivity of the clinical algorithms used for diagnosis. Modelled in this way, rates of death, self-cure and detection can be considered as a competing hazard framework, with the rate representing detection competing against the rates of self-cure and death to determine the fraction of incident TB cases that are detected. In practice, this logic is usually inverted: information about the fraction of cases detected is used to determine the corresponding rate of detection.

Treatment and Recovery

Detection is usually modelled as synonymous with treatment initiation and those in the treatment category are normally assumed not to be infectious. This is based on the rapid decrease in both bacillary load and coughing frequency of TB patients on effective treatment [22, 23]. Exceptions include models that are used to investigate the significance of delays to treatment, and work including drug-resistant TB which may still be infectious on inappropriate treatment. Often, this low infectiousness on treatment and the small fraction of the population on TB treatment at any point mean that models do not include an explicit compartment for those on TB treatment. If an explicit TB treatment compartment is included it is normally modelled as a non-infectious compartment with a mean duration of 6 months (i.e. a total per capita exit rate of 2 per year).

The outcomes of treatment are often modelled as simply death or successful treatment, neglecting treatment failure or loss-to-follow-up (LTFU), with the total out rate split in proportion to the fraction experiencing each outcome. Surveillance data reported to WHO suggest that the treatment success for the 2015 cohort of new TB cases was 83%, and 78% in PLHIV [3]. Death, LTFU, and unevaluated outcomes heterogeneously account for most of the non-successful outcomes, with treatment failure rare in settings without very high levels of drug resistance. Globally, around 11% of TB treatments for new and relapse cases in PLHIV reported death as an outcome, compared with 4% of those without HIV [3]. Pre-treatment LTFU (i.e. LTFU between diagnosis and treatment initiation) is often incorporated into diagnostic rates, which may misrepresent the complexity in subsequent care-seeking by those who have at some point been diagnosed with TB.

Successful treatment outcomes and often self-cure of disease, are frequently modelled by a transition to a 'recovered' compartment. In contrast to the 'recovered' designation in most infectious disease models, this compartment is used to capture the elevated risk of recurrent TB disease in those previously treated for TB. TB treatment is not now thought to be sterilizing, and even after a documented cure, the risk of TB infection is greater than in people who have never developed the disease [24]. In many settings, approaching 10% of notified TB cases will be individuals who have been previously treated for TB, sometimes after LTFU, but more often after treatment completion. Recurrence is a complex phenomenon which includes disease following reinfection, reactivation, and mis-classified episodes of ongoing TB disease. Individuals with previous TB may be at elevated risk of TB infection and subsequent progression due to constant social or biological risk factors, or potentially due to changed risk factors resulting from their initial TB episode (lung scarring, exacerbated poverty). Many of these factors also increase the risk of reactivation disease from the original *M. tuberculosis* infection.

This has important implications for TB control suggesting that previously treated patients should be followed-up regularly, possibly until the end of their lives. More recently a series of studies by Marx et al. have shown that, in the Western Cape Province of South Africa, previously treated patients have a relapse rate and a recurrence rate both of which are about 3000/100,000 in the first year after successful treatment which has important implications for the control of TB, especially in high burden settings [25, 26].

Drug Resistance

Recurrence may be linked with the local epidemiology of drug-resistant TB (DR-TB), which varies hugely around the globe, with proportions of multidrug-resistant or rifampicin-resistant TB (MDR/RR) among TB cases ranging from over 30% in some former Soviet republics to less rates typically below 3% in much of sub-Saharan Africa [3]. In all settings, the rates of drug resistance in cases with a history of TB treatment are higher than treatment-naive patients; treatment-naive patients are a barometer of transmitted resistance whereas cases previously treated reflect both acquired resistance during treatment and undiagnosed primary resistance that has been empirically determined to need second line treatment. MDR/RR TB is significant due to its worse outcomes and the much higher cost of treatment. Other resistance patterns, such as extensively drug-resistant (XDR) TB include resistance to common second-line compounds. Models including drug-resistant TB must replicate much of the structure to allow infection and transmission by strains of *M. tuberculosis* with different resistance patterns. The different outcomes, durations, relevant diagnostic algorithms and mechanisms for becoming a drug-resistant TB case must all be included in transmission models of DR-TB.

Risk Factors Other Than HIV

HIV is the single strongest risk factor for TB and the way its influence on the natural history of TB is typically incorporated in models of TB will be discussed in the following section. However, it is worth noting that other factors also influence the risk and characteristics of TB disease. Risks of progression to disease change rapidly during childhood, as does the spectrum of disease, requiring specific modelling approaches [27, 28]. The proportion of disease that is smear positive may increase through adult ages [8]. Biological and social risk factors mean that there are typically around twice as many TB cases in men as in women [29]. Diabetes mellitus is a moderate risk factor for TB incidence, but forecasts of increasing diabetes prevalence suggest a potentially important future role in global TB epidemiology [30]. Smoking, indoor air pollution, and silicosis are all risk factors that may make substantial contributions to TB incidence in some locales [31, 32]. A curious, but still unexplained, observation is the extremely strong association of TB with body-mass index (BMI) [33]. Across six longitudinal follow-up studies of navy recruits in the

USA (2 studies), male smokers in Finland, Mass Radiography in Norway, Health Centres for the Elderly in Hong Kong and participants in an NHANES study in the USA, there was a 13.8%±0.4% reduction in TB risk per unit increase in BMI over a range of BMI from 17 kg/m^2 to 34 kg/m^2, making BMI the single best predictor of an individual's relative risk of tuberculosis. Finally, through a variety of mechanisms, socioeconomic status has associations with TB incidence and outcomes and may well be a strong influence on global TB epidemiology during a period of rapid economic development in many high-TB prevalence regions [34].

Modelling TB in Individuals with HIV

Data for Modelling

In this section, we review the ways that the individual-level effects of HIV have been incorporated in TB models. It is worth noting that epidemiological evidence often needs particular interpretations, sometimes requiring additional assumptions, for use in mechanistic models. This is because mechanistic models are frequently more specific in their representations of causality than the statistical models conventionally used for analysing epidemiological data. Therefore, when incorporating effects which may act on one (or several) different pathways to effect, a more nuanced account of effect is required.

Progression, Infection and Protection

The main way TB models include the association between HIV and TB is through an increased rate of progression to TB disease for individuals with HIV infections. However, increased rates of progression could apply solely to 'primary progression' in the first couple of years following an infection, or to subsequent 'endogenous reactivation', or to both these processes (potentially in different ways). Data on TB incidence in cohorts of PLHIV suggest that the incidence rate ratio for developing TB increases immediately after infection by a factor of around 2, and remains higher, increasing as immunocompetence declines to over 30. Most models in the literature account for an increased rate of primary TB progression in PLHIV. Individual-based models and partial differential equation (PDE) models may include rates of progression to disease that vary continuously by time-since-infection. In these cases, without an enforced distinction between primary progression and endogenous reactivation, the most natural and commonly used approach is application of a single incidence rate ratio (IRR), which therefore elevates of both processes. The division between primary progression and endogenous reactivation, which is in any case a somewhat arbitrarily drawn distinction in practice, is further blurred in PLHIV when large IRRs may mean reactivation timescales are on a par with primary progression rates in HIV-uninfected individuals. Taken together, averaging across levels of immunocompetence, a rate of progression to TB disease in those with *M. tuberculosis* infections of 10% per year has been suggested among PLHIV, compared with a similar lifetime risk of progression among HIV-uninfected individuals [2].

Determining *M. tuberculosis* infection in PLHIV is particularly problematic, as sensitivity of both traditional tuberculin skin tests (TSTs) and newer interferon gamma release assays (IGRAs) is typically lower in this group due to immune dysfunction. It is therefore not possible to reliably determine whether HIV infection increases susceptibility to infection (as opposed to increasing progression to disease following infection). In models, it is usually assumed that HIV does not affect the risk of *M. tuberculosis* infection, and is a risk factor for higher TB incidence due to higher rates of progression to disease. The effect of HIV infection on the protection against reinfection disease conferred by LTBI is also poorly defined. Most models assume that HIV infection reduces or completely removes the protection against reinfection disease due to previous infection.

Influence of CD4 Count

There are a number of motivations for including a more detailed representation of HIV-related immunosuppression in TB models and for formulating this in terms of CD4 cell count. As mentioned above, there is more than an order of magnitude variation in IRR among PLHIV depending on their degree of immunosuppression. This is substantial enough to generate secular trends in the association between HIV and TB at a population level as HIV epidemics have aged. The count of CD4-positive lymphocytes in peripheral whole blood samples (CD4 cell count) has been widely studied as a marker of immune suppression in PLHIV, and a guide for clinical decisions. Many guidelines on when to start ART have historically been based on CD4 counts, and there is much surveillance and survey data on CD4 counts at various stages of accessing care. This means that models including CD4 count could examine CD4-based ART policies and draw upon a wide range of data in their parametrization.

A number of analyses have identified a consistent relationship between CD4 count decline and increasing risk of TB incidence [1, 35, 36]. This relationship is exponential, that is a fixed decrement of 100 cells/mm^3 CD4 count results in a 1.4-fold factor increase in the IRR for TB [36]. Thus the roughly 20% drop in CD4 cell count during the acute phase of HIV infection roughly doubles the risk of TB [35]. After the acute phase of HIV infection a person's CD4 cell count drops linearly (without ART) until death implying the risk of TB increases exponentially [35].

The IRR for TB given HIV varies widely on a population level: for example, the incidence rate ratio for TB in those with and without HIV ranges from 7.5 in Lesotho to 17.5 in Zimbabwe [37]. This may partly be due to the relative importance of socio-epidemiological factors, e.g. IRRs for TB given HIV are typically much lower for populations with generalized HIV epidemics than for populations with HIV epidemics driven by specific risk groups, whose members may be at elevated risk of TB due to other correlated risk factors amplified by assortative mixing. But this is likely only part of the story, and there is also substantial variation in baseline CD4-cell counts within and between populations [38]. Furthermore, for reasons still not understood, it seems that while the decline of CD4 cell counts is linear in time after HIV infection, the survival after infection is independent of the initial CD4 cell count, implying that people with a high CD4 cell count have a correspondingly faster rate of decline. However, this consistent exponential relationship with CD4 decrement across a number of settings and background TB risks means the strong biological component of increased risk from HIV can be separated from factors that do not depend on CD4 count.

Using CD4 count as a measure of immunosuppression for the IRR for TB has several advantages. Statistical analyses of data from the CASCADE cohorts provide models of CD4 count progression and their determinants, as well as HIV-related mortality [39, 40]. Modelling increases in TB risk through CD4 count means existing work on HIV natural history can be used directly in HIV/TB models, and allows TB models to be built on top of existing HIV model structures. There is an advantage in terms of parsimony: a single parameter can be used to capture the increased risk of TB in a variety of CD4 count compartments, by using an average implied IRR from the exponential model rather than separately assessing the risk for each CD4 category, and for models which continuously track time-since-HIV-infection or CD4 (IBMs, PDEs) this parametrization avoids binning into categories altogether. Incorporating several CD4 categories and parametrizing their TB risks via this exponential model is the approach most commonly used in applied models aimed at informing country policy. An example is the Spectrum HIV model developed by Avenir Health, which includes 7 CD4 categories, and the TB Impact Model and Estimates (TIME) modelling module that is part of the Spectrum tool and makes use of this underlying HIV model and the exponential model of TB risk with respect to CD4 count.

The natural history of HIV and TB are both different in children, who have much higher absolute CD4 cell counts that vary rapidly with age. The CD4 percentage is often used as an age-adjusted measure of immunosuppression. Systematic review and meta-analysis suggests that, as with adults, there is an exponential increase in the IRR for TB as CD4 percentage declines [41]. At a population level, an average over different levels of immunosuppression yields an IRR of around 8.

TB Disease and Outcomes

HIV infection shifts the spectrum of TB disease away from cavitary disease towards smear negative and extrapulmonary disease. One consequence of this is that the average HIV-positive TB case is usually assumed to be less infectious per unit time than the average HIV-negative TB case. This is often modelled as a lower proportion of disease that is classified as smear-positive. The natural history of TB/HIV without

TB treatment is uncertain because HIV emerged after TB treatment was available, however, outcomes in individuals with DR-TB that was inappropriately treated, expert clinical opinion and other information do paint a consistent picture of a very much shorter duration of disease (of the order a few months rather than the few years of TB in HIV-negative individuals), with very limited or no self-cure (i.e. a case fatality rate of close to 100%). Thus, although HIV increases the risk of TB by a factor of the order ten, TB disease progresses about ten times faster than in HIV-positive than in HIV-negative people. This means that while incidence may increase by up to ten times prevalence remains more or less constant as first shown by Corbett et al. [31]

Similarly, there is little evidence on TB case-detection rates in PLHIV. One could quantify rates of detection by HIV-status by examining HIV-stratified prevalence:notification ratios from using TB prevalence survey data, however TB prevalence surveys are not powered for conclusions by HIV status. The usual assumption in modelling is that TB has the same probability of being detected whether HIV positive or negative, although some studies have explored differential case detection [42]. This reflects much higher rates of care-seeking or screening for TB among PLHIV to compete with their higher rates of mortality. While evidence to directly inform this assumption is lacking, given population HIV prevalence and an IRR for TB among PLHIV, this assumption does generate HIV-prevalence among notified TB cases in keeping with observation.

TB treatment outcomes for PLHIV are somewhat worse than those of HIV-negative TB patients. In reality, outcomes are worse for more immunocompromised patients. All other characteristics of TB disease and detection are also likely to vary with level of immunosuppression, though this is rarely modelled. An exception is Williams et al. [1], which modelled the duration of untreated TB disease as being shorter for lower CD4 count.

Effects of TB on HIV

TB is one of the leading opportunistic infections listed as cause of death in PLHIV. Even if not fatal, it has been shown that TB disease can affect CD4 count in PLHIV and worsen HIV progression [43]. Finally, during some periods in some locales, TB disease has been a major route of HIV diagnosis and entry into HIV care. However, few studies have investigated the potential for bidirectional cross-talk between the HIV and TB epidemics. Most TB-focussed modelling exercises in high HIV-prevalence settings have used HIV incidence or prevalence as an input. This is no doubt partly because this requires dynamic HIV and TB models as well as specification of all their sources of interaction, which is challenging given the scarcity of data. It also potentially reflects the sense that the level of TB in a population may not greatly affect survival of untreated HIV; that if the primary cause of death at some level of immunosuppression had not been TB, it would have been something else. However, this assumption is unlikely to be true, especially at higher CD4 counts.

Modelling Interventions

Antiretroviral Therapy (ART)

Coverage of ART among PLHIV globally has increased rapidly since 2004, and has potent benefits in reducing mortality almost to levels of HIV-negative individuals [44]. ART also benefits as a population-level HIV preventive tool [45], via reduced probability of transmission [46]. WHO guidelines now suggest initiation of ART in PLHIV regardless of CD4 count [47], though historically and still in some national guidelines, ART initiation is recommended only once CD4 count has fallen below some threshold.

Given its power to reduce HIV incidence and HIV-related mortality, ART provision has the potential to influence TB incidence through reductions in HIV prevalence in populations. However, ART also has direct individual level effects on TB.

Most notably, ART reduces the incidence of TB in adults, with a hazard ratio of around 0.35 across all CD4 counts [48], and to a similar degree in children [41]. The proportionate reduction in TB incidence may be higher for adults initiated at the lowest CD4 counts, whose rates of TB would otherwise be highest [48]. The protection from TB appears to increase over the first few years on ART in line with markers of immune status [49, 50], and somewhat more rapidly in children [41]. However, TB rates in PLHIV on ART compared to HIV-uninfected adults appear to remain at least 4 fold higher, even after 5 years or more on ART [51].

Being on ART may shift the spectrum of TB disease back in the direction of the HIV-negative spectrum, likely increasing infectiousness, decreasing case fatality rate, and potentially allowing some degree of protection stemming from LTBI and lengthening disease duration. Most models have treated TB disease in PLHIV on ART as very similar or identical to TB disease in PLHIV not on ART, and transmission models have not considered unmasking, immune reconstitution inflammatory syndrome (IRIS), or drug-drug interactions [52]. TB treatment outcomes are improved in PLHIV by being on ART, and are comparable with those for HIV-uninfected TB patients [53].

However, regular attendance of ART clinics may well also have implications for TB detection rates. Aside from acting as a marker for those able and motivated to access care, it also signals a heightened index of suspicion for TB among both patients and clinicians. Conversely, ART clinics have also been suggested as the source of a potentially heightened exposure to TB: the Tugela Ferry outbreak of XDR TB centred on an ART clinic, and high rates of active TB have been found among ART clinic attendees [54]. These features, which may have national-level implications in high HIV burden settings, remain largely unexplored in modelling studies.

Isoniazid Preventive Therapy (IPT)

Isoniazid preventive therapy (IPT) to lower the rate of progression to TB disease is currently recommended for 36 months in PLHIV [55]. Shorter courses of IPT have been shown to lower TB incidence by around 60% in PLHIV with a positive skin test [56], and data comparing TB incidence for individuals on ART with that in those on both ART and IPT [57], and comparing TB incidence or mortality reductions from IPT in cohorts with or without ART [58] suggest IPT has an incremental benefit while taking ART.

While IPT has sometimes been regarded historically as protecting by clearing a latent infection, trial data evaluating extended durations of IPT in PLHIV (including those on ART) [59] and mathematical modelling fitting to data from a set of IPT studies in PLHIV not on ART with post-prophylaxis follow-up [60] both suggest that IPT is unlikely to clear *M. tuberculosis* infection in most PLHIV. This means that the risk of TB incidence after treatment cessation is likely to return to its level before prophylaxis, and has implications for the way IPT should be modelled in PLHIV (i.e. as a temporary reduction in risk). Houben et al.'s modelling also suggested a higher curative effect of rifamycin-containing prophylactic regimens among PLHIV [60], tallying with biological understanding of mechanism.

Improved TB Detection

Transmission models that have focussed on evaluating diagnostic algorithms within clinics tend to have relatively simple representations of either HIV or of the diagnosis process, with the focus of detail being on the complementary aspect. Most frequently, the effects of changed algorithm are reflected in different rates into treatment states in a compartmental model. This may neglect the true complexity of repeated attempts to obtain a diagnosis, diagnosis of previously treated TB and similar details. Sometimes these changes in transmission parameters may be derived from more detailed operational models of diagnostic procedures [61], or calculated by approximations to decision trees giving mean sensitivity, specificity and delay for patient types represented in the transmission model [62]. Applying simple changes in a detection rate (e.g. in proportion to changes in test sensitivity) without considering additional compartments representing pre-care-seeking infectiousness may exaggerate the potential for improvements in passive detection to affect transmission.

More active approaches to case-detection are variously-termed and understood (e.g. active case-finding, enhanced case-finding, systematic screening), and have been the subject of transmission modelling analyses in high-HIV burden settings (intensified case-finding, conventionally referring to screening among PLHIV, seems to have received less attention from transmission modellers). Periodic rounds of active case-finding have been modelled as mechanistically removing a proportion of prevalent cases each round. Often, active or enhanced case-finding is conceptualized more nebulously as improved case-detection from direct efforts and potentially improved community awareness, and represented as increases in detection rate parameters. Understanding when transmission occurs on average with respect to care-seeking and symptoms is a crucial uncertainty that affects the relative impact of active case-finding interventions.

Household contact-tracing has also been considered in these settings, typically using individual-based models that can naturally represent households of index cases.

Infection Control

Improvements in ventilation for preventing nosocomial transmission of TB have been modelled by using the Wells-Riley equation and its variants, which quantify the reduction in infectiousness achieved by changes in ventilation rates [63]. More sophisticated specific approaches have been used to model the impact on ventilation of specific changes to buildings [64], and of upper room ultraviolet germicidal irradiation [65], but this has yet to be combined with transmission modelling in high-HIV burden settings.

Population-Level Impacts

In this section, we review the mathematical modelling literature that has included explicit representation of TB transmission and of the effect of HIV on TB epidemiology. We have divided the literature into work describing the influence of HIV on TB epidemics that does not evaluate the impact of interventions, and work whose main focus is in evaluating single or multiple interventions that reduce TB burden. In focussing only on modelling work that includes transmission, we have excluded a substantial body of work evaluating the individual-level impact and cost-effectiveness of TB interventions among PLHIV (e.g. some of the work on IPT, diagnostic approaches). We have also included some work focussed on modelling the transmission process in high-HIV burden settings (e.g. using Wells-Riley approaches), but which does not use a transmission model (in the usual sense) to propagate the consequences of this transmission. In writing this section, we have drawn on a systematic review of the TB/HIV modelling literature up to 2012 [66], and updated it using the same search terms. See Fig. 3 for temporal trends in number and topic of these publications.

There are several key points that determine the way in which one models HIV-related TB. First of all, standard TB models for modelling HIV-negative TB can be used even in the presence of a substantial HIV epidemic TB because HIV has a much larger impact on TB incidence than prevalence, and therefore transmission. Secondly, the time-scales for HIV are of the order of years but for TB are of the order of decades which allows the separation of the models. Given a model of TB that fits the data before the epidemic of HIV starts, simple approaches need only a few additional parameters to include TB/HIV: a parameter (around 2), which gives the increase in the risk of TB immediately after infection with HIV; a parameter which gives the relative risk of TB when HIV-positive people are on ART; and a

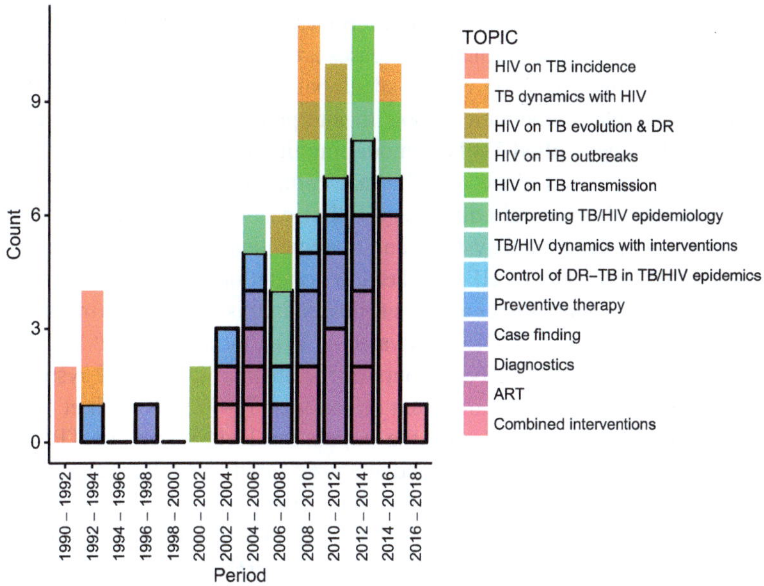

Fig. 3 Temporal evolution of topics considered by TB modelling papers for high HIV prevalence settings. Bars with solid borders represent studies with a focus on modelling interventions. Bar colours are in the same order as in the legend. DR = drug resistant. The topic word 'dynamics' implies a more mathematically focussed study

parameter specifying the exponential rate of increase in TB incidence as people progress to successive clinical stages of HIV, which may differ between populations to account for the substantial variation in the incidence rate ratio in different populations. A detailed discussion of these observations is given in the supporting information of Williams et al. [37]

Population-Level Impact of HIV on TB Epidemics

The earliest papers using modelling tools to understand TB/HIV epidemics focussed on drawing attention to the threat posed by HIV to TB control and making estimates of the increases in TB incidence due to HIV. Bermejo et al. [67] in 1992 used a simple mathematical model and ecological data on associations between HIV and TB to conclude that TB incidence would double going forward when adult HIV prevalence hit 13%. Schulzer et al. [68] in the same year concluded that the TB incidence in 15–45 year olds was likely to increase by factors of between 4 and 12 by the year 2000 compared to 1980, depending on the baseline annual risks of *M. tuberculosis* infection, and projected HIV prevalence. Dolin et al. [69] in 1994 generated TB burden forecasts for 1990–1999, projecting 88 million new TB cases for this period with 10% of them HIV associated. The same year, Schulzer et al. [70] introduced a more sophisticated actuarial back calculation approach, and projected

2–6 fold increases in TB incidence in sub-Saharan African populations by 2000 compared with 1980. These predicted increases are in line with those observed over the period. Later, in their 2003 review of global TB/HIV epidemiology, Corbett et al. [2] used simple mathematical tools to interpret data that were subsequently very influential in the applied TB/HIV modelling literature.

The first article to introduce a traditional compartmental transmission model based on ordinary differential equations (ODEs) was Massad et al. [71] in 1993. This work was in a dynamical systems and mathematical biology tradition, but did conclude that the influence of HIV on TB at a population level was much stronger than the other way around. Other authors followed in this vein, exploring the stability properties of systems of ODEs motivated as TB/HIV models, but with a focus more on mapping out possible behaviours of the abstract dynamical systems than closely tying the models to a real setting or settings [71–73]. (Naresh et al. [72] is notable and unusual for considering the population-level impact of TB on HIV, however.) Kapitanov [74] introduces and analyses a PDE model, separately including time-since-infection for HIV and *M. tuberculosis*.

Authors beginning with Basu et al. [75] in 2008 have followed a population biology tradition and focussed on the potential impact of HIV for the evolution of TB, especially DR-TB. This first article considered cross-immunity between TB strains in a multi-strain HIV/TB model, and highlighted the potential for HIV to promote the emergence of DR-TB at a population-level by compensating for reduced fitness in DR-TB strains. In a later paper Basu et al. [76] consider the evolution of TB virulence, again concluding that HIV may facilitate increased virulence. Sergeev et al. [77] agree that HIV can facilitate DR-TB epidemics, but explore the dynamics of the relationship between HIV and DR-TB, predicting a lower proportion of DR-TB among PLHIV early on in DR-TB epidemics due to enhanced reactivation of older strains in this group.

Models have also been used to explore the influence of HIV on stochastic aspects of TB incidence in small communities or lower burden settings. In 2001, Porco et al. [78] used discrete event simulation (DES) to study the influence of HIV on the frequency and severity of TB outbreaks. Murray et al. [79] included HIV in their analysis of the determinants of molecular cluster size. More recently, Pretorius et al. [80] used a van Kampen expansion to explore the effects of stochasticity in a TB model applied to a medium size high-HIV community, and examined the temporal correlations for active TB cases.

Another strand of work has made use of the classical Wells-Riley model of indoor transmission and ventilation to interpret experimental work on the relative infectiousness of TB in PLHIV [81], to consider the role of household transmission in generating observed acute respiratory illnesses (ARI) in South Africa [82], and arguing implications at the population-level and a necessity for more-than-proportionately intense interventions in intense transmission settings (again focussed on South Africa) [83]. Andrews et al. [84] have modelled the implications of data on social contact patterns, and CO_2-derived measures of ventilation, for the location of TB transmission in South Africa, singling out the role of public transport in another study [85]. Dodd et al. [86] also make use of social contact data (from Zambia and South Africa), together with data on TB prevalence and infection rates, to model the age- and sex-specific rates of TB infection, arguing ARI measured in children under-estimates that experienced by adults.

Lastly, transmission models have been used to interpret the patterns and trends in epidemiologic data from high-HIV burden settings. In 2006, Hughes et al. [87] used a DES model calibrated to Zimbabwe, to understand the impact of HIV on the TB epidemic there. In an intriguing article, Sanchez et al. [88] fitted a difference equation model of HIV and TB in sub-Saharan African settings, notably Kenya, highlighting the difficulties of matching the observed trends and exploring potential reasons for discrepancies. Blaser et al. [89] developed an age-structured TB/HIV model calibrated to Cape Town in order to understand the epidemiology. They conclude that protection from LTBI and higher progression rates in previously treated individuals are key in being able to reproduce age-patterns of TB. Finally, Pretorius et al. [90] developed a regression methodology to model the influence of population CD4 changes on TB incidence and is used to disaggregate TB incidence by CD4 stratum in the TIME model (part of the Spectrum model suite for country-level estimation of TB/HIV mortality burden).

Population-Level Impact of Interventions

A number of studies considering the impact of interventions on TB/HIV have appeared in the mathematical literature. Agusto et al. [91] considered optimal control theory applied to a two-(TB)strain TB/HIV ODE model. Kaur et al. [92] studied an ODE TB/HIV model that includes abstractly defined screening and treatment interventions from a dynamical systems perspective. Sharomi et al. [93] also applied dynamical systems analysis to a TB/HIV ODE model, but with more explicit representation of intervention strategies including ART. The study of Bacaër et al. [94] is more realistic, and calibrates an TB/HIV ODE model to a specific (South African) setting. Interventions for HIV including ART and condom use promotion, as well as IPT for TB are considered: ART was predicted to strongly reduce TB notifications.

Motivated by the 2006 outbreak of XDR-TB in Tugela Ferry, South Africa, in which ART clinics appeared to play a key role, Basu et al. [63] use an TB/HIV ODE model including drug-resistant TB to model intervention strategies to reduce nosocomial transmission of TB, concluding that a combination of strategies could prevent around half of XDR cases. It was shown that the outbreak of XDR-TB in Tugela Ferry was largely due to nosocomial transmission resulting from poor infection control in a clinic with very high rates of both HIV and TB [95]. Basu et al. [96] went on to consider this with a stochastic model appropriate to the relatively small numbers of this outbreak situation, concluding that community-based interventions would be needed to curtail the outbreak. Later Basu et al. [97] considered the role of institutions such as prisons in amplifying TB epidemics, and argue that reducing the inflow to these institutions would have impacts on both TB incidence and the propagation of drug-resistant TB.

A number of studies have considered IPT. In keeping with our scope, we discuss only those including transmission; it should be noted that IPT in PLHIV has been the focus of a similar number of static cost-effectiveness modelling analyses. In 1993, Heyman [98] considered the impact of HIV on TB in HIV-hyperendemic settings with low TB treatment coverage, and the impact of preventive therapy (assumed curative) aimed at PLHIV, suggesting a large reduction in prevalence could result over a decade. In 2004, Guwatudde et al. [99] developed an TB/HIV ODE model and concluded that IPT for PLHIV would have a limited impact on the TB epidemics of sub-Saharan Africa. In 2006, Cohen et al. [100] also modelled IPT for PLHIV in sub-Saharan Africa and found a greater potential for this intervention, with up to around 20% of cumulative HIV deaths preventable through this strategy over a 5-year period. However, they also found a potential for increases in drug-resistant TB, and argued that IPT should therefore be coupled with improved diagnostic and treatment options. Basu et al. [101] specifically consider IPT delivered through ART clinics in Botswana, found that increases in resistance were small and more than counterbalanced by reduced TB incidence and mortality. Mills et al. [102] introduced a dual-network model - one network for contacts relevant to HIV transmission, a second network with contacts relevant to TB transmission - and investigated the role of correlations between these structures in generating heterogeneity in the impact of IPT. IPT was found to be effective at a population-level, but networks with clustering of HIV and TB transmission in intense foci had reduced local impact from IPT due to high rates of TB reinfection. Kunkel et al. [103] calibrated a multi-strain TB model with HIV to data from Botswana and consider the health benefits and resistance dynamics resulting from continuous IPT for PLHIV. They find health benefits outweigh concerns around increased resistance, so long as sufficient control is maintained for the overall TB epidemic.

Various studies have used modelling to evaluate TB case finding in high-HIV burden settings, including both improvements in passive case detection and cure under the DOTS strategy, and through active case finding. The early paper of Dye et al. [4] projected the global and regional impact of the DOTS strategy (focussed on case detection and treatment success) and included HIV. They concluded that DOTS has a greater impact on mortality than incidence and that this difference is heightened in settings where HIV is prevalent. Baltussen et al. [104] built on the work of Dye et al. [4] to evaluate the cost-effectiveness of DOTS and DOTS-plus on a regional and global level. Laxminarayan et al. [105] undertook a later global country-level economic evaluation of sustaining DOTS, finding that the benefits exceeded costs in all of the (then-designated) 22 high TB-burden countries. Sanchez et al. [106] considered increased detection as well as decreased LTFU and higher cure of shorter regimens for PLHIV in a transmission model calibrated to the epidemiology of Kenya, projecting a 20% reduction in TB incidence and mortality by 2030 for these interventions in combination. Dowdy et al. [107] considered the effects of improvements in case detection on TB incidence, finding that declines in TB incidence reduce over time since an improvement in case-detection. Dodd et al. [42] considered what features affect periodic active case finding for TB in high-HIV burden settings using a PDE model, and allowing different baseline detection characteristics for HIV-positive and HIV-negative TB cases. For a given case-finding round efficacy, the impact is higher in settings where a higher proportion of TB incidence is due to recent infection. Mellor et al. [108] used an individual-based household-structured model based on data from Zimbabwe to consider the impact of different household contact tracing (HHCT) interventions to screen for TB disease and offer IPT, and found HHCT for late-stage HIV-positive individuals was particularly effective in reducing mortality. Yaesoubi et al. [109] considered optimal dynamic rules for switching on and off active or enhanced TB case finding efforts in addition to passive case finding given a decision rule based on a policy-make willingness-to-pay threshold, concluding that dynamic strategies are more efficient and therefore potentially more feasible and sustainable in practice than fixed case-finding strategies. Lastly, Azman et al. [110] consider the cost-effectiveness of active case-finding in three countries, including South Africa, concluding that the modelled intervention would be cost-effective in South Africa, and that active case-finding strategies have greater cost-effectiveness when considered over longer time horizons.

The impact of different diagnostic strategies in HIV-driven TB epidemics has also been considered. Dowdy et al. [111] in 2006 considered the impact of rapid molecular testing or culture compared to active case-finding or wider ART use; finding that while improved diagnostic strategies only moderately reduce TB incidence (by around 10%), their effect on TB mortality is larger (around 20%). Dowdy et al. [62] introduce a transmission modelling framework focussed on diagnostic strategies, which did include a simple aggregate representation of HIV. Menzies et al. [112] undertook a rigorous economic evaluation of Xpert MTB/RIF in five southern African countries, using a TB transmission model that included the effects of HIV. This analysis too found limited impacts on TB incidence, but benefits in terms of TB mortality and morbidity. Longer time horizons favoured the intervention because they allowed indirect benefits to accrue. Another stream of modelling work has sought to integrate operational research models of practical implementation details and workflows with transmission models for evaluating new diagnostics in high TB (and high HIV) burden settings [61, 113, 114]. For example, Langley et al. [61] also evaluated Xpert (together with fluorescence LED microscopy) in Tanzania, finding Xpert to have the greatest cost and impact, and allowing estimation of quantities such as reductions in patient visits to health facilities.

ART for HIV is one of the key interventions whose impact on TB in high-HIV burden settings has been considered. In 2003, Williams et al. [35] developed a cohort model and brought evidence together on the efficacy of ART for preventing TB and the incidence rate ratio for TB at different CD4 counts, highlighting the importance of starting ART early and achieving high coverage and compliance in reducing cumulative TB incidence. In Williams et al. [115], a transmission model with a 4-stage HIV structure was used to project the impact of HIV on the Indian TB epidemic, finding that continued progress with TB control activities should contain the impact of HIV, but recommending ART to PLHIV who develop TB. The evidence on the individual impact of CD4 cell count and ART on TB risk was updated in Williams et al. [1] and incorporated in a TB/HIV model applied to 9 sub-Saharan African countries to investigate the impact of a 'universal test-and-treat' (UTT) HIV control strategy on TB. UTT comprises frequent HIV testing at high coverage followed by immediate ART initiation, and has been projected to have the potential to reduce HIV incidence by reducing the infectiousness of PLHIV [116]. This study included the impact of aggressive ART beginning in 2010 on HIV incidence, and predicted a potential reduction in TB/HIV by more than 50% by 2015; strategies achieving ART initiation within 2 years of HIV infection led to more than 95% reductions in TB/HIV by 2050. The approaches to modelling the relationship between TB and HIV introduced in these papers, confirmed by more recent evidence synthesis [36], have been very influential on other modelling approaches to TB/HIV.

Some work on the impact of ART on TB has included work in a mathematical tradition such as Bhunu et al. [117], who pursue a dynamical systems analysis of an TB/HIV model including ART, and Bacaër et al. [94], discussed above. Dodd et al. [118] caution that the still-high relative risks of TB on ART and the longer life-expectancies imply that long-term reduction in TB from ART must be mediated by reductions in HIV incidence or background population risks of TB. Pretorius et al. [119] report results from three independent models assessing the impact of ART policy changes in South Africa over the period 2014–2033, finding expanded coverage and universal eligibility could reduce cumulative TB incidence and mortality by around 20% and 30%, respectively over this period, with one TB case averted for every person-decade or so on ART.

In addition to the studies focussed on ART discussed here, many of the models considering combinations of interventions (discussed below) have included ART as a component of their policy options, alongside other interventions. Currie et al. [120] consider three interventions to prevent TB in Uganda, Kenya and South Africa - namely IPT, ART, and reduced HIV transmission - with curative interventions (improved TB case detection and treatment success). They found the largest impacts on TB from curative interventions, but cautioned that alone they were able to contain but not reverse TB epidemics. Currie et al. [121] built on this work to consider the cost-effectiveness of interventions in Kenya, finding that improvements in TB detection and cure were extremely cost-effective, but noting that ART policies had the largest potential for general health gains, while being the most expensive and relatively expensive measured as a TB-prevention strategy. Chindelevitch et al. [122] compared wider ART provision and improvements to TB control for improving TB control in South Africa, finding that expanded coverage of the TB programme had the greatest potential for impact on TB, but suggesting a potential 22% reduction in cumulative TB incidence over 5 years from expanded ART eligibility. Knight et al. [123] also considered South Africa, using an individual-based model to evaluate portfolios of interventions including expanded ART, long-duration IPT, and active case finding for achieving the national targets to find that the most aggressive combination considered could result in TB incidence and mortality rates that were 70% and 86% lower than those in 2012. The most effective single intervention was general-population active case finding for TB, whereas improvements in ART had more modest impacts: UTT alone generated around a 20% reduction in TB incidence. Gilbert et al. [124] considered a combination of intervention strategies in rural South African settings, including ART, IPT, and a community-based integrated intensified case finding strategy for HIV and TB. The intensified case finding was found to be the single most effective intervention for TB incidence, with around a 25% reduction over 10 years and a comparable impact on combined TB and HIV mortality as expanded ART policies. Gilbert et al. [125] built on this to evaluate the projected cost-effectiveness of their intensified case finding strategy, finding it cost-effective in rural South Africa.

A number of studies recently have gathered multiple modelling teams together to address the same question. The study of Pretorius et al. [119], discussed above, was a result from a multi-modelling study focussed on ART policies for HIV [126]. In 2016, Houben et al. [127] undertook a study with 11 different models focussed on 3 countries to assess approaches to meeting the End TB strategy goals. South Africa was one of the countries considered by the 8 models including the necessary HIV structure for this context, and it was found that a combination of continuous IPT for PLHIV on ART, TB screening at health facilities, and improved TB care could reduce TB incidence and mortality in 2025 by 55% and 72%, respectively, compared with 2015.

Finally, a number of country-level TB modelling tools are being developed to undertake national policy analyses, often for informing applications for support from the Global Fund to Fight AIDS, Malaria and Tuberculosis [128, 129]. For example, the TIME model is built into the widely-used Spectrum suite of models, which have a heavy emphasis on HIV policy, and has been applied to a number of high-HIV burden countries. Most of the outputs from this work have yet to be described in peer-reviewed literature.

Discussion

Uses and Achievements of Modelling

One of the key uses of TB/HIV epidemic modelling has been as a tool to understand the drivers of epidemiological patterns and project the effects of HIV on TB epidemics. Some of the early studies were based on simple models but provided important insights. The first projections of the effect of HIV on TB incidence proved broadly correct, and the understanding that the association between TB and HIV is a dynamic one, driven by mean population immune status is also borne out by observation. Early fears that TB in PLHIV might be more infectious than in HIV-uninfected individuals proved wrong, and together with the much shorter duration of TB/HIV compared with TB in HIV-uninfected individuals means that even in settings where the majority of incident TB is HIV-associated, it will still often be the case that TB in the HIV-negative population drives the majority of transmission. This implies that efforts to strengthen TB control in these settings must also encompass the HIV-negative population, while recognising the special challenges of TB/HIV from increased risks of mortality and different clinical presentation.

Modelling work has also established the potential for HIV to have a facilitating effect on the development and establishment of DR-TB in populations; the impaired host immune responses compensating for reductions in pathogen transmission fitness or virulence. Modelling has also explored the role of ART clinics in facilitating TB, and particular DR-TB transmission in the wider community.

Understanding the impact of new or existing interventions and prioritizing their future development has also been an important area where models have been useful. The application of modelling to IPT in PLHIV is a good example of understanding existing interventions: data from randomized studies are often analyzed at an individual level; sometimes transmission effects and the potential for reinfection with *M. tuberculosis* may have a bearing on the interpretation of results from individual-level trials. Modelling has been particularly influential in investigating the potential for impact from innovative strategies that have not yet been trialled at population level. HIV modelling provided some of the evidence supporting the policy shifts to universal ART provision, and related TB modelling has studied the potential for additional benefits from these policies effects on the TB epidemic. Modelling studies are able to explore more interventions, intervention variants, and combinations of interventions than would ever be possible to trial in empirical studies, thus helping to prioritize and design interventions to take forward for empirical evaluation.

Models of TB/HIV are also increasingly being used to guide policy and investment decisions, including both epidemiological impact from changes in TB transmission and the costs and benefits needed to generate health economic evaluations of policy options. Multiple models have been applied to harmonized scenarios in specific settings, particularly with the aim of establishing a consensus about ingredients needed to move towards TB control targets using current tools. Practical use of country-level modelling of TB in high-HIV burden settings to support policy design and applications for donor funding has also been increasing. This is an important area of work that is as yet less documented in the academic literature.

Future Directions

For a combination of serious conditions that afflicts in the region of 1 million people in the world each year, TB/HIV has not received the attention from epidemiological modellers that it should have. This is perhaps due to the genuine complexity of dealing of the interactions between two conditions which each have their own complexities in terms of epidemiology, natural history, and control policies. However, it means that there are many areas where modelling could be usefully applied going forwards.

More could be done to use modelling as a tool to understand epidemiology and the performance of control efforts. ART has scaled up hugely over the last decade; modelling using routine and study data could be used more to understand the long-term impact of ART on TB/HIV epidemiology and chart its likely future course. This may require increasing attention to interactions with other societal features that are rapidly changing in high-HIV and TB burden settings, such as urbanization, improvements in nutritional status, and increasing prevalence of diabetes mellitus. Modelling has been used in conjunction with social contact data and ventilation data to assess likely contributions of particular locales to TB transmission, and the impact of infection control interventions aimed at particular hot-spots; more work in this area would help improve our knowledge of an important but poorly-understood aspect of TB epidemiology and potentially suggest relatively easily-implemented interventions.

Finally, some workers have begun to include the operational details of health systems and care provision in models, and there is an increasing demand for these details that mesh with the level where design decisions are taken and where costs-accrue. Developing such models presents challenges in terms of the range of expertise required by the teams involved, but also in terms of modelling techniques for combining models that may have very different emphases and ideal approaches to implementation separately. Increasing availability of electronic health records may also make easier the job of parametrizing the health system aspects of such models.

Conclusion

Population-level modelling of TB in high-HIV burden settings has helped bring together a wealth of understanding around TB/HIV natural history, epidemiology and interventions to provide important insights into the implications for TB transmission and control. Increasingly, models are being used to guide policy and investment decisions at a country and supranational level. TB/HIV remains a substantial global health concern, and important questions and challenges remain to be addressed by transmission modelling.

Acknowledgments This publication was produced by TREATS which is part of the EDCTP2 programme supported by the European Union (grant number RIA2016S- 1632-TREATS). The views and opinions of authors expressed herein do not necessarily state or reflect those of EDCTP.

References

1. Williams BG, Granich R, De Cock KM, Glaziou P, Sharma A, Dye C (2010) Antiretroviral therapy for tuberculosis control in nine African countries. Proc Natl Acad Sci U S A 107:19485–19489
2. Corbett EL, Watt CJ, Walker N et al (2003) The growing burden of tuberculosis. Arch Intern Med 163:1009
3. World Health Organisation (2017) Global tuberculosis report 2017
4. Dye C, Garnett GP, Sleeman K, Williams BG (1998) Prospects for worldwide tuberculosis control under the WHO DOTS strategy. Directly observed short-course therapy. Lancet 352:1886–1891
5. Dye C. The population biology of tuberculosis. 2017.
6. Menzies NA, Wolf E, Connors D, Cohen T, Hill AN, Yaesoubi R, Galer K, White PJ, Abubakar I, Salomon JA (2018) Progression from latent infection to active disease in dynamic TB transmission models: a systematic review. Lancet Infect Dis 18(8):e228–e238

7. Marais BJ, Gie RP, Schaaf HS et al (2004) The natural history of childhood intra-thoracic tuberculosis: a critical review of literature from the pre-chemotherapy era. Int J Tuberc Lung Dis 8:392–402
8. Vynnycky E, Fine PE (1997) The natural history of tuberculosis: the implications of age-dependent risks of disease and the role of reinfection. Epidemiol Infect 119:183–201
9. Comstock GW, Livesay VT, Woolpert SF (1974) The prognosis of a positive tuberculin reaction in childhood and adolescence. Am J Epidemiol 99:131–138
10. Sloot R, van der Loeff MF S, Kouw PM, Borgdorff MW (2014) Risk of tuberculosis after recent exposure. A 10-year follow-up study of contacts in Amsterdam. Am J Respir Crit Care Med 190:1044–1052
11. Ragonnet R, Trauer JM, Scott N, Meehan MT, Denholm JT, McBryde ES. Optimally capturing latency dynamics in models of tuberculosis transmission. Epidemics 2017. https://doi.org/10.1016/j.epidem.2017.06.002
12. Andrews JR, Noubary F, Walensky RP, Cerda R, Losina E, Horsburgh CR (2012) Risk of progression to active tuberculosis following reinfection with Mycobacterium tuberculosis. Clin Infect Dis 54:784–791
13. Sutherland I, Svandova E, Radhakrishna S (1976) Alternative models for the development of tuberculosis disease following infection with tubercle bacilli. Bull Int Union Tuberc 51:171–179
14. Tostmann A, Kik SV, Kalisvaart NA et al (2008) Tuberculosis transmission by patients with smear-negative pulmonary tuberculosis in a large cohort in The Netherlands. Clin Infect Dis 47:1135–1142
15. Hernandez-Garduno E (2004) Transmission of tuberculosis from smear negative patients: a molecular epidemiology study. Thorax 59:286–290
16. Esmail H, Barry CE 3rd, Young DB, Wilkinson RJ (2014) The ongoing challenge of latent tuberculosis. Philos Trans R Soc Lond Ser B Biol Sci 369:20130437
17. Onozaki I, Law I, Sismanidis C, Zignol M, Glaziou P, Floyd K (2015) National tuberculosis prevalence surveys in Asia, 1990–2012: an overview of results and lessons learned. Trop Med Int Health 20:1128–1145
18. Trunz BB, Bourdin Trunz B, Fine P, Dye C (2006) Effect of BCG vaccination on childhood tuberculous meningitis and miliary tuberculosis worldwide: a meta-analysis and assessment of cost-effectiveness. Lancet 367:1173–1180
19. van Leth F, van der Werf MJ, Borgdorff MW (2008) Prevalence of tuberculous infection and incidence of tuberculosis: a re-assessment of the Styblo rule. Bull World Health Organ 86:20–26
20. Tiemersma EW, van der Werf MJ, Borgdorff MW, Williams BG, Nagelkerke NJD (2011) Natural history of tuberculosis: duration and fatality of untreated pulmonary tuberculosis in HIV negative patients: a systematic review. PLoS One 6:e17601
21. Thompson BC (1943) Survival rates in pulmonary tuberculosis. Br Med J 2:721–721
22. Jindani A, Aber VR, Edwards EA, Mitchison DA (1980) The early bactericidal activity of drugs in patients with pulmonary tuberculosis. Am Rev Respir Dis 121:939–949
23. Loudon RG, Spohn SK (1969) Cough frequency and infectivity in patients with pulmonary tuberculosis. Am Rev Respir Dis 99:109–111
24. Davies C. The eradication of tuberculosis in Rhodesia: with particular reference to the Midlands and South Eastern Provinces. 1966.
25. Marx FM, Dunbar R, Enarson DA et al (2014) The temporal dynamics of relapse and reinfection tuberculosis after successful treatment: a retrospective cohort study. Clin Infect Dis 58:1676–1683
26. Marx FM, Floyd S, Ayles H, Godfrey-Faussett P, Beyers N, Cohen T (2016) High burden of prevalent tuberculosis among previously treated people in Southern Africa suggests potential for targeted control interventions. Eur Respir J 48:1227–1230
27. Dodd PJ, Gardiner E, Coghlan R, Seddon JA (2014) Burden of childhood tuberculosis in 22 high-burden countries: a mathematical modelling study. Lancet Glob Health 2:e453–e459
28. Dodd PJ, Sismanidis C, Seddon JA (2016) Global burden of drug-resistant tuberculosis in children: a mathematical modelling study. Lancet Infect Dis 16:1193–1201

29. Horton KC, MacPherson P, Houben RMGJ, White RG, Corbett EL (2016) Sex differences in tuberculosis burden and notifications in low- and middle-income countries: a systematic review and meta-analysis. PLoS Med 13:e1002119
30. Stevenson CR, Forouhi NG, Roglic G et al (2007) Diabetes and tuberculosis: the impact of the diabetes epidemic on tuberculosis incidence. BMC Public Health 7:234
31. Corbett EL, Churchyard GJ, Clayton TC et al (2000) HIV infection and silicosis: the impact of two potent risk factors on the incidence of mycobacterial disease in South African miners. AIDS 14:2759–2768
32. Lin H-H, Ezzati M, Murray M (2007) Tobacco smoke, indoor air pollution and tuberculosis: a systematic review and meta-analysis. PLoS Med 4:e20
33. Lönnroth K, Williams BG, Cegielski P, Dye C (2010) A consistent log-linear relationship between tuberculosis incidence and body mass index. Int J Epidemiol 39:149–155
34. Lönnroth K, Jaramillo E, Williams BG, Dye C, Raviglione M (2009) Drivers of tuberculosis epidemics: the role of risk factors and social determinants. Soc Sci Med 68:2240–2246
35. Williams BG, Dye C (2003) Antiretroviral drugs for tuberculosis control in the era of HIV/AIDS. Science 301:1535–1537
36. Ellis PK, Martin WJ, Dodd PJ. CD4 count and tuberculosis risk in HIV-positive adults not on ART: a systematic review and meta-analysis. PeerJ 2017; 5. https://doi.org/10.7717/peerj.4165
37. Williams BG, Gouws E, Somse P et al (2015) Epidemiological trends for HIV in Southern Africa: implications for reaching the elimination targets. Curr HIV/AIDS Rep 12:196–206
38. Williams BG, Korenromp EL, Gouws E, Schmid GP, Auvert B, Dye C (2006) HIV infection, antiretroviral therapy, and CD4+ cell count distributions in African populations. J Infect Dis 194:1450–1458
39. Wolbers M, Babiker A, Sabin C et al (2010) Pretreatment CD4 Cell slope and progression to AIDS or death in HIV-infected patients initiating antiretroviral therapy—The CASCADE Collaboration: a collaboration of 23 cohort studies. PLoS Med 7:e1000239
40. Touloumi G, Pantazis N, Pillay D et al (2013) Impact of HIV-1 subtype on CD4 count at HIV seroconversion, rate of decline, and viral load set point in European seroconverter cohorts. Clin Infect Dis 56:888–897
41. Dodd PJ, Prendergast AJ, Beecroft C, Kampmann B, Seddon JA (2017) The impact of HIV and antiretroviral therapy on TB risk in children: a systematic review and meta-analysis. Thorax 72:559–575
42. Dodd PJ, White RG, Corbett EL (2011) Periodic active case finding for TB: when to look? PLoS One 6:e29130
43. Toossi Z, Mayanja-Kizza H, Hirsch CS et al (2001) Impact of tuberculosis (TB) on HIV-1 activity in dually infected patients. Clin Exp Immunol 123:233–238
44. Samji H, Cescon A, Hogg RS et al (2013) Closing the gap: increases in life expectancy among treated HIV-positive individuals in the United States and Canada. PLoS One 8:e81355
45. Tanser F, Bärnighausen T, Grapsa E, Zaidi J, Newell M-L (2013) High coverage of ART associated with decline in risk of HIV acquisition in rural KwaZulu-Natal, South Africa. Science 339:966–971
46. Cohen MS, Chen YQ, McCauley M et al (2016) Antiretroviral therapy for the prevention of HIV-1 transmission. N Engl J Med 375:830–839
47. Guideline on when to start antiretroviral therapy and on pre-exposure prophylaxis for HIV. Geneva: World Health Organization; 2015.
48. Suthar AB, Lawn SD, del Amo J et al (2012) Antiretroviral therapy for prevention of tuberculosis in adults with HIV: a systematic review and meta-analysis. PLoS Med 9:e1001270
49. Lawn SD, Myer L, Edwards D, Bekker L-G, Wood R (2009) Short-term and long-term risk of tuberculosis associated with CD4 cell recovery during antiretroviral therapy in South Africa. AIDS 23:1717–1725
50. Nicholas S, Sabapathy K, Ferreyra C, Varaine F, Pujades-Rodríguez M, AIDS Working Group of Médecins Sans Frontières (2011) Incidence of tuberculosis in HIV-infected patients before and after starting combined antiretroviral therapy in 8 sub-Saharan African HIV programs. J Acquir Immune Defic Syndr 57:311–318

51. Gupta A, Wood R, Kaplan R, Bekker L-G, Lawn SD (2012) Tuberculosis incidence rates during 8 years of follow-up of an antiretroviral treatment cohort in South Africa: comparison with rates in the community. PLoS One 7:e34156
52. McIlleron H, Meintjes G, Burman WJ, Maartens G (2007) Complications of antiretroviral therapy in patients with tuberculosis: drug interactions, toxicity, and immune reconstitution inflammatory syndrome. J Infect Dis 196:S63–S75
53. Nglazi MD, Bekker L-G, Wood R, Kaplan R (2015) The impact of HIV status and antiretroviral treatment on TB treatment outcomes of new tuberculosis patients attending co-located TB and ART services in South Africa: a retrospective cohort study. BMC Infect Dis 15:536
54. Giri PA, Deshpande JD, Phalke DB (2013) Prevalence of pulmonary tuberculosis among HIV positive patients attending antiretroviral therapy clinic. N Am J Med Sci 5:367–370
55. Recommendation on 36 months isoniazid preventive therapy to adults and adolescents living with HIV in resource-constrained and high TB- and HIV-prevalence settings: 2015 Update. World Health Organization, Geneva, p 2015
56. Akolo C, Adetifa I, Shepperd S, Volmink J (2010) Treatment of latent tuberculosis infection in HIV infected persons. Cochrane Database Syst Rev:CD000171
57. Golub JE, Saraceni V, Cavalcante SC et al (2007) The impact of antiretroviral therapy and isoniazid preventive therapy on tuberculosis incidence in HIV-infected patients in Rio de Janeiro, Brazil. AIDS 21:1441–1448
58. Ayele HT, van Mourik MSM, Debray TPA, Bonten MJM (2015) Isoniazid prophylactic therapy for the prevention of tuberculosis in HIV infected adults: a systematic review and meta-analysis of randomized trials. PLoS One 10:e0142290
59. Samandari T, Agizew TB, Nyirenda S et al (2011) 6-month versus 36-month isoniazid preventive treatment for tuberculosis in adults with HIV infection in Botswana: a randomised, double-blind, placebo-controlled trial. Lancet 377:1588–1598
60. Houben RMGJ, Sumner T, Grant AD, White RG (2014) Ability of preventive therapy to cure latent Mycobacterium tuberculosis infection in HIV-infected individuals in high-burden settings. Proc Natl Acad Sci U S A 111:5325–5330
61. Langley I, Lin H-H, Egwaga S et al (2014) Assessment of the patient, health system, and population effects of Xpert MTB/RIF and alternative diagnostics for tuberculosis in Tanzania: an integrated modelling approach. Lancet Glob Health 2:e581–e591
62. Dowdy DW, Andrews JR, Dodd PJ, Gilman RH. A user-friendly, open-source tool to project impact and cost of diagnostic tests for tuberculosis. elife 2014; 3. https://doi.org/10.7554/elife.02565
63. Basu S, Andrews JR, Poolman EM et al (2007) Prevention of nosocomial transmission of extensively drug-resistant tuberculosis in rural South African district hospitals: an epidemiological modelling study. Lancet 370:1500–1507
64. Taylor JG, Yates TA, Mthethwa M, Tanser F, Abubakar I, Altamirano H (2016) Measuring ventilation and modelling M. tuberculosis transmission in indoor congregate settings, rural KwaZulu-Natal. Int J Tuberc Lung Dis 20:1155–1161
65. Noakes CJ, Beggs CB, Sleigh PA (2004) Modelling the Performance of Upper Room Ultraviolet Germicidal Irradiation Devices in Ventilated Rooms: Comparison of Analytical and CFD Methods. Indoor Built Environ 13:477–488
66. RMGJ H, Dowdy DW, Vassall A et al (2014) How can mathematical models advance tuberculosis control in high HIV prevalence settings? Int J Tuberc Lung Dis 18:509–514
67. Bermejo A, Veeken H, Berra A (1992) Tuberculosis incidence in developing countries with high prevalence of HIV infection. AIDS 6:1203–1206
68. Schulzer M (1992) An estimate of the future size of the tuberculosis problem in sub-Saharan Africa resulting from HIV infection. Tuber Lung Dis 73:52–58
69. Dolin PJ, Raviglione MC, Kochi A (1994) Global tuberculosis incidence and mortality during 1990-2000. Bull World Health Organ 72:213–220
70. Schulzer M, Radhamani MP, Grzybowski S, Mak E, Fitzgerald JM (1994) A mathematical model for the prediction of the impact of HIV infection on tuberculosis. Int J Epidemiol 23:400–407

71. Massad E, Burattini MN, Coutinho FAB, Yang HM, Raimundo SM (1993) Modeling the interaction between aids and tuberculosis. Math Comput Model 17:7–21
72. Naresh R, Sharma D, Tripathi A (2009) Modelling the effect of tuberculosis on the spread of HIV infection in a population with density-dependent birth and death rate. Math Comput Model 50:1154–1166
73. Roeger L-IW, Feng Z, Castillo-Chavez C (2009) Modeling TB and HIV co-infections. Math Biosci Eng 6:815–837
74. Kapitanov G (2015) A double age-structured model of the co-infection of tuberculosis and HIV. Math Biosci Eng 12:23–40
75. Basu S, Orenstein E, Galvani AP (2008) The theoretical influence of immunity between strain groups on the progression of drug-resistant tuberculosis epidemics. J Infect Dis 198:1502–1513
76. Basu S, Galvani AP (2009) The evolution of tuberculosis virulence. Bull Math Biol 71:1073–1088
77. Sergeev R, Colijn C, Murray M, Cohen T (2012) Modeling the dynamic relationship between HIV and the risk of drug-resistant tuberculosis. Sci Transl Med 4:135ra67
78. Porco TC, Small PM, Blower SM (2001) Amplification dynamics: predicting the effect of HIV on tuberculosis outbreaks. J Acquir Immune Defic Syndr 28:437–444
79. Murray M (2002) Determinants of cluster distribution in the molecular epidemiology of tuberculosis. Proc Natl Acad Sci U S A 99:1538–1543
80. Pretorius C, Dodd P, Wood R (2011) An investigation into the statistical properties of TB episodes in a South African community with high HIV prevalence. J Theor Biol 270:154–163
81. Escombe AR, DAJ M, Gilman RH et al (2008) The infectiousness of tuberculosis patients coinfected with HIV. PLoS Med 5:e188
82. Wood R, Johnstone-Robertson S, Uys P et al (2010) Tuberculosis transmission to young children in a South African community: modeling household and community infection risks. Clin Infect Dis 51:401–408
83. Uys P, Marais BJ, Johnstone-Robertson S, Hargrove J, Wood R (2011) Transmission elasticity in communities hyperendemic for tuberculosis. Clin Infect Dis 52:1399–1404
84. Andrews JR, Morrow C, Walensky RP, Wood R (2014) Integrating social contact and environmental data in evaluating tuberculosis transmission in a South African township. J Infect Dis 210:597–603
85. Andrews JR, Morrow C, Wood R (2013) Modeling the role of public transportation in sustaining tuberculosis transmission in South Africa. Am J Epidemiol 177:556–561
86. Dodd PJ, Looker C, Plumb ID et al (2016) Age- and sex-specific social contact patterns and incidence of Mycobacterium tuberculosis infection. Am J Epidemiol 183:156–166
87. Hughes G, Currie C, Corbett E. Modeling tuberculosis in areas of High HIV prevalence. In: Proceedings of the 2006 Winter Simulation Conference. 2006. https://doi.org/10.1109/wsc.2006.323116
88. Sánchez MS, Lloyd-Smith JO, Williams BG et al (2009) Incongruent HIV and tuberculosis co-dynamics in Kenya: interacting epidemics monitor each other. Epidemics 1:14–20
89. Blaser N, Zahnd C, Hermans S et al (2016) Tuberculosis in Cape Town: An age-structured transmission model. Epidemics 14:54–61
90. Pretorius C, Glaziou P, Dodd PJ, White R, Houben R (2014) Using the TIME model in Spectrum to estimate tuberculosis-HIV incidence and mortality. AIDS (28 Suppl 4):S477–S487
91. Agusto FB, Adekunle AI (2014) Optimal control of a two-strain tuberculosis-HIV/AIDS co-infection model. Biosystems 119:20–44
92. Kaur N, Ghosh M, Bhatia SS (2014) The role of screening and treatment in the transmission dynamics of HIV/AIDS and tuberculosis co-infection: a mathematical study. J Biol Phys 40:139–166
93. Sharomi O, Podder CN, Gumel AB, Song B (2008) Mathematical analysis of the transmission dynamics of HIV/TB coinfection in the presence of treatment. Math Biosci Eng 5:145–174

94. Bacaër N, Ouifki R, Pretorius C, Wood R, Williams B (2008) Modeling the joint epidemics of TB and HIV in a South African township. J Math Biol 57:557–593
95. Wallengren K, Scano F, Nunn P, Margot B, Buthelezi S, Williams B, Pym A, Samuel EY, Mirzayev F, Nkhoma W, Mvusi L, Pillay Y. Resistance to TB drugs in KwaZulu-Natal: causes and prospects for control. 2011. https://arxiv.org/pdf/1107.1800.pdf.
96. Basu S, Friedland GH, Medlock J et al (2009) Averting epidemics of extensively drug-resistant tuberculosis. Proc Natl Acad Sci U S A 106:7672–7677
97. Basu S, Stuckler D, McKee M (2011) Addressing institutional amplifiers in the dynamics and control of tuberculosis epidemics. Am J Trop Med Hyg 84:30–37
98. Heymann SJ (1993) Modelling the efficacy of prophylactic and curative therapies for preventing the spread of tuberculosis in Africa. Trans R Soc Trop Med Hyg 87:406–411
99. Guwatudde D, Debanne SM, Diaz M, King C, Whalen CC (2004) A re-examination of the potential impact of preventive therapy on the public health problem of tuberculosis in contemporary sub-Saharan Africa. Prev Med 39:1036–1046
100. Cohen T, Lipsitch M, Walensky RP, Murray M (2006) Beneficial and perverse effects of isoniazid preventive therapy for latent tuberculosis infection in HIV-tuberculosis coinfected populations. Proc Natl Acad Sci U S A 103:7042–7047
101. Basu S, Maru D, Poolman E, Galvani A (2009) Primary and secondary tuberculosis preventive treatment in HIV clinics: simulating alternative strategies. Int J Tuberc Lung Dis 13:652–658
102. Mills HL, Cohen T, Colijn C (2011) Modelling the performance of isoniazid preventive therapy for reducing tuberculosis in HIV endemic settings: the effects of network structure. J R Soc Interface 8:1510–1520
103. Kunkel A, Crawford FW, Shepherd J, Cohen T (2016) Benefits of continuous isoniazid preventive therapy may outweigh resistance risks in a declining tuberculosis/HIV coepidemic. AIDS 30:2715–2723
104. Baltussen R, Floyd K, Dye C (2005) Cost effectiveness analysis of strategies for tuberculosis control in developing countries. BMJ 331:1364
105. Laxminarayan R, Klein EY, Darley S, Adeyi O (2009) Global investments in TB control: economic benefits. Health Aff 28:w730–w742
106. Sánchez MS, Lloyd-Smith JO, Porco TC et al (2008) Impact of HIV on novel therapies for tuberculosis control. AIDS 22:963–972
107. Dowdy DW, Chaisson RE (2009) The persistence of tuberculosis in the age of DOTS: reassessing the effect of case detection. Bull World Health Organ 87:296–304
108. Mellor GR, Currie CSM, Corbett EL (2011) Incorporating household structure into a discrete-event simulation model of tuberculosis and HIV. ACM Trans Model Comput Simul 21:1–17
109. Yaesoubi R, Cohen T (2013) Identifying dynamic tuberculosis case-finding policies for HIV/TB coepidemics. Proc Natl Acad Sci U S A 110:9457–9462
110. Azman AS, Golub JE, Dowdy DW. How much is tuberculosis screening worth? Estimating the value of active case finding for tuberculosis in South Africa, China, and India. BMC Med 2014; 12 https://doi.org/10.1186/s12916-014-0216-0
111. Dowdy DW, Chaisson RE, Moulton LH, Dorman SE (2006) The potential impact of enhanced diagnostic techniques for tuberculosis driven by HIV: a mathematical model. AIDS 20:751–762
112. Menzies NA, Cohen T, Lin H-H, Murray M, Salomon JA (2012) Population health impact and cost-effectiveness of tuberculosis diagnosis with Xpert MTB/RIF: a dynamic simulation and economic evaluation. PLoS Med 9:e1001347
113. Lin H-H, Langley I, Mwenda R et al (2011) A modelling framework to support the selection and implementation of new tuberculosis diagnostic tools. Int J Tuberc Lung Dis 15:996–1004
114. Langley I, Doulla B, Lin H-H, Millington K, Squire B (2012) Modelling the impacts of new diagnostic tools for tuberculosis in developing countries to enhance policy decisions. Health Care Manag Sci 15:239–253
115. Williams BG, Granich R, Chauhan LS, Dharmshaktu NS, Dye C (2005) The impact of HIV/AIDS on the control of tuberculosis in India. Proc Natl Acad Sci U S A 102:9619–9624

116. Granich RM, Gilks CF, Dye C, De Cock KM, Williams BG (2009) Universal voluntary HIV testing with immediate antiretroviral therapy as a strategy for elimination of HIV transmission: a mathematical model. Lancet 373:48–57
117. Bhunu CP, Garira W, Mukandavire Z (2009) Modeling HIV/AIDS and tuberculosis coinfection. Bull Math Biol 71:1745–1780
118. Dodd PJ, Knight GM, Lawn SD, Corbett EL, White RG (2013) Predicting the long-term impact of antiretroviral therapy scale-up on population incidence of tuberculosis. PLoS One 8:e75466
119. Pretorius C, Menzies NA, Chindelevitch L et al (2014) The potential effects of changing HIV treatment policy on tuberculosis outcomes in South Africa: results from three tuberculosis-HIV transmission models. AIDS (28 Suppl 1):S25–S34
120. Currie CSM, Williams BG, Cheng RCH, Dye C (2003) Tuberculosis epidemics driven by HIV: is prevention better than cure? AIDS 17:2501–2508
121. Currie CSM, Floyd K, Williams BG, Dye C (2005) Cost, affordability and cost-effectiveness of strategies to control tuberculosis in countries with high HIV prevalence. BMC Public Health 5:130
122. Chindelevitch L, Menzies NA, Pretorius C, Stover J, Salomon JA, Cohen T. Evaluating the potential impact of enhancing HIV treatment and tuberculosis control programmes on the burden of tuberculosis. J R Soc Interface 2015; 12 https://doi.org/10.1098/rsif.2015.0146
123. Knight GM, Dodd PJ, Grant AD, Fielding KL, Churchyard GJ, White RG (2015) Tuberculosis prevention in South Africa. PLoS One 10:e0122514
124. Gilbert JA, Long EF, Brooks RP et al (2015) Integrating community-based interventions to reverse the convergent TB/HIV epidemics in rural South Africa. PLoS One 10:e0126267
125. Gilbert JA, Shenoi SV, Moll AP, Friedland GH, Paltiel AD, Galvani AP (2016) Cost-Effectiveness of Community-Based TB/HIV Screening and Linkage to Care in Rural South Africa. PLoS One 11:e0165614
126. Eaton JW, Menzies NA, Stover J et al (2014) Health benefits, costs, and cost-effectiveness of earlier eligibility for adult antiretroviral therapy and expanded treatment coverage: a combined analysis of 12 mathematical models. Lancet Glob Health 2:e23–e34
127. RMGJ H, Menzies NA, Sumner T et al (2016) Feasibility of achieving the 2025 WHO global tuberculosis targets in South Africa, China, and India: a combined analysis of 11 mathematical models. Lancet Glob Health 4:e806–e815
128. Trauer JM, Ragonnet R, Doan TN, McBryde ES (2017) Modular programming for tuberculosis control, the 'AuTuMN' platform. BMC Infect Dis 17:546
129. RMGJ H, Lalli M, Sumner T et al (2016) TIME Impact - a new user-friendly tuberculosis (TB) model to inform TB policy decisions. BMC Med 14:56

Immune Responses to *Mycobacterium tuberculosis* and the Impact of HIV Infection

Catherine Riou, Cari Stek, and Elsa Du Bruyn

Abstract *Mycobacterium tuberculosis* control relies on a well-orchestrated immune response, where a complex array of innate and adaptive immune cells responses act synergistically to restrict *Mycobacterium tuberculosis* growth. While different immune cell subsets have been associated with protection in experimental models of TB, it is still unclear exactly what type of immune responses are required to confer protection in humans.

People living with HIV are around 20 times more likely to develop active TB. The clearest immune defect caused by HIV is a progressive reduction in absolute CD4 T cell numbers that correlates with increasing risk of active TB. However, shortly after HIV acquisition or when CD4 T cell numbers improve upon HIV treatment, the risk of active TB remains heightened. This indicates that, independently of the overall CD4 T cell depletion, HIV infection also induces qualitative changes weakening protective TB immune responses.

This chapter section covers the human immune response to *Mycobacterium tuberculosis* and describes the impact of HIV infection.

Keywords *Mycobacterium tuberculosis* · HIV · Immunology · T cells · Macrophages · Dendritic cells · Neutrophils · inflammation · Cytokines

C. Riou (✉)
Wellcome Centre for Infectious Diseases Research in Africa, Institute of Infectious Disease and Molecular Medicine, University of Cape Town, Cape Town, South Africa

Division of Medical Virology, Department of Pathology, University of Cape Town, Cape Town, South Africa
e-mail: cr.riou@uct.ac.za

C. Stek · E. Du Bruyn
Wellcome Centre for Infectious Diseases Research in Africa, Institute of Infectious Disease and Molecular Medicine, University of Cape Town, Cape Town, South Africa
e-mail: cari.stek@uct.ac.za; elsa.dubruyn@uct.ac.za

I. Sereti et al. (eds.), *HIV and Tuberculosis*,
https://doi.org/10.1007/978-3-030-29108-2_4

Mycobacterium tuberculosis infection leads to a highly orchestrated immune response, in which both innate and adaptive immune cells are required to contain the infection [1]. HIV disease is associated with generalized immunodeficiency and systemic chronic immune activation, leading to a progressive deterioration of both innate and adaptive immune responses [2] that together cripple the host's ability to mount and/or maintain effective responses against *M. tuberculosis*.

First Line of Defence Against *M. tuberculosis*: The Innate Immune System

Infection with *M. tuberculosis* occurs via the aerosol route, and bacilli are first recognized by phagocytic cells including macrophages, monocytes, neutrophils and dendritic cells, with alveolar macrophages being the primary targets for *M. tuberculosis* infection. The recognition of *M. tuberculosis* by phagocytic cells occurs via the interaction of Pathogen-Associated Molecular Patterns (PAMPs, present at the surface of bacteria) with Pattern Recognition Receptors (PRRs) expressed by phagocytes. PRRs are capable of binding to conserved molecular structures (such as lipoproteins or peptido-glycans) that are expressed by a large variety of microbes, including *M. tuberculosis*. PRRs include toll like receptors (TLR), C-type lectin receptor, scavenger receptors and the intracellular nucleotide oligomerization domain (NOD)-like receptors (NLRs) [3, 4].

Once *M. tuberculosis* has been sensed by PRRs, it is engulfed by the cell, creating intracellular phagosomes. The maturation of these phagosomal compartments promotes their acidification, which is required for the optimal activity of antimycobacterial digestive enzymes and reactive oxygen species, triggering intracellular bacterial elimination. *M. tuberculosis*-infected cells also produce an array of inflammatory cytokines and chemokines (including TNFα, IL-1β, IL-12, MCP1 and IL-8). These cytokines promote the recruitment of additional macrophages, neutrophils and dendritic cells to the site of infection enhancing the innate response and initiating the formation of granulomas to contain *M. tuberculosis*. Neutrophils play an important role in the innate response, contributing to *M. tuberculosis* clearance through the production of antimicrobial peptides, but they also participate in the dissemination of viable bacteria in established disease, exacerbating pathology. Elevated peripheral blood neutrophil count are associated with death in tuberculosis patients [5]. Dendritic cells are a link between the innate and adaptive immune response through their significant role in capturing, processing and presenting antigens. Infection of dendritic cells with *M. tuberculosis* induces their maturation and migration to the secondary draining lymph nodes, where the adaptive immune response is initiated by priming of naïve T lymphocytes by the dendritic cells.

Additional players, at the interface of the innate and adaptive immune system, are unconventional donor-unrestricted T (DURT) cells. These cell subsets interact with dendritic cells and macrophages, through highly conserved molecules (unlike the classical HLA- class I or HLA class II restricted antigen presentation). DURT cells include: CD1-restricted Natural Killer T cells (NKT cells) capable of recognizing lipid and glycolipid moieties present on *M. tuberculosis* cell walls; MR1-restricted mucosal associated invariant T cells (MAIT cells) that can bind and recognize riboflavin (vitamin B2) or folate (vitamin B9) derivatives; and γδ T cells which exhibit a restricted TCR repertoire and recognize small metabolites (called "phospho-antigens" structurally related to isopentenyl pyrophosphate) produced by mammalian cells and intracellular pathogens including *M. tuberculosis* [6]. Our understanding of the role of DURT cells in anti-bacterial immunity is only partial, given that this is a young field of investigation, but increasing evidence suggests that these cell subsets may contribute to protection against *M. tuberculosis* infection and/or to *M. tuberculosis* containment. Indeed, a rapid expansion of MAIT and γδ T cells is observed in healthy individuals recently exposed to *M. tuberculosis* [7]. Moreover, human studies have shown that the frequency of NKT cells and MAIT cell populations are reduced in quantity and functionally impaired during active TB compare to latent infection [8].

However, *M. tuberculosis* is endowed with a wide range of strategies to counteract innate defences, thereby preventing its eradication. Once engulfed by the macrophage, *M. tuberculosis* interferes with macrophage functions, inhibiting phagosomal acidification and maturation and preventing cell apoptosis; thus allowing *M. tuberculosis* to survive within the macrophage [9, 10]. Moreover, *M. tuberculosis* uses virulence mechanisms to disseminate by inducing necrotic death of infected cells. This results in the release of bacteria that are then taken up by freshly recruited phagocytes and this promotes bacterial population expansion as well as tissue inflammation and necrosis.

In most cases, due to the ability of *M. tuberculosis* to "highjack" the innate immune system, innate responses are not capable of eradicating *M. tuberculosis* and appear to only moderately restrict bacterial growth during the initial phase of infection. The initial innate response does, however, create an inflammatory environment that promotes the recruitment of additional innate cells, and the priming and recruitment of the adaptive immune response. These events lead to the formation of granulomas that are capable of containing *M. tuberculosis* [11].

The Impact of HIV Infection on Innate Immune Responses to *M. tuberculosis*

The hallmark of HIV infection is progressive destruction of CD4+ T cells (HIV primary target cells) but macrophages and dendritic cells are also permissive to HIV infection, leading to significant dysfunction of these cells [12, 13]. While HIV infects only 1–10% of alveolar macrophages in vivo (with limited cytopathic effect), it alters key aspects of macrophage functions such as receptor-mediated phagocytosis and cell apoptosis; thereby impairing their capacity to eliminate intracellular pathogens. In acute HIV infection, dendritic cell numbers (especially plasmacytoid dendritic cells) decrease markedly. Moreover, HIV also interferes with the processing and/or presentation of *M. tuberculosis* antigens by dendritic cells, thereby disrupting important cellular functions linking innate and adaptive immune response and impairing immune responses to *M. tuberculosis*. Additionally, HIV infection induces a systemic and chronic state of immune activation, affecting the functions and trafficking of uninfected cells. The aberrant immune activation during HIV infection has been attributed to several mechanisms including HIV viral replication, the loss of gut mucosal integrity (leading to the translocation of gut bacteria and bacterial products into the peripheral blood stream), increased concentration of soluble pro-inflammatory molecules (such as IP-10, IL-6, TNFα) and increased homeostatic proliferation in response to HIV-induced lymphopenia. All these factors probably play a role in inducing a hyper-inflammatory environment, thereby eliciting innate cell activation and maturation. Indeed, in the context of HIV infection, neutrophils exhibit a hyperactivated phenotype and are more susceptible to necrotic cell death, affecting their ability to restrict *M. tuberculosis* growth. HIV also has an impact on DURT cells. The frequency of MAIT cells, CD1-restricted NKT cells and γδ T cells are markedly decreased in the blood of HIV-infected patients, with poor recovery after initiating ART [14–16]. Moreover, further characterization of the residual peripheral iNKT and MAIT cell populations suggests that HIV infection also alter their functional potential, impairing their cytokine production capacity and proliferation potential [17]. Overall, via direct and indirect mechanisms, HIV impairs the capacity of the innate immune system to contain and clear *M. tuberculosis* infection.

The Second Line of Defence: The Adaptive Immune System

Once primed by antigen presenting cells in the lymph nodes, naïve *M. tuberculosis*-specific T cells mature, proliferate and migrate to the site of infection where they participate in the formation of granulomas to contain *M. tuberculosis* infection. As a facultative intracellular pathogen, *M. tuberculosis* is preferentially recognized by the major histocompatibility complex (MHC) class II processing pathway, leading to the development of a predominant CD4 response. *M. tuberculosis*-specific CD4+ T cell responses are diverse, including a vast array of T helper (Th) subsets endowed with distinct effector or regulatory functions. Th1 CD4+ T cells, producing interferon-gamma (IFNγ), contribute to the recruitment of monocytes and granulocytes and activate the antimicrobial functions of macrophages. These cells are necessary but not sufficient to control TB infection. Of note, it is this IFNγ response that is measured in the diagnostic tests for *M. tuberculosis* infection such as the interferon-gamma release assay (IGRA) including QuantiFERON-TB gold-in-tube or T-SPOT TB test, detecting recent or remote *M. tuberculosis* exposure that elicited an adaptive immune response. Besides Th1 responses, other T helper CD4 subsets such as IL-17 producing Th17 and IL-10 producing regulatory CD4+ T cells (Treg) (which exhibit pro-inflammatory and suppressive functions, respectively) also participate in *M. tuberculosis* containment. Overall, the clinical outcome of

M. tuberculosis infection relies on the capacity of the adaptive immune system to generate a balanced CD4+ T cell response, reaching equilibrium between effector and regulatory T helper subsets. Pro-inflammatory responses enhance bacterial killing required to control *M. tuberculosis*; while anti-inflammatory responses limit pathology and inflammation during initial infection and latency. The disruption of this pro- vs anti-inflammatory equilibrium (tipping the balance one way or the other) appears to impair *M. tuberculosis* containment. While immune activation and inflammatory reactions are essential for host protection (as demonstrated by increased TB susceptibility in humans treated with TNFα blocking agents [18] or with genetic disorders disrupting IFNγ signalling pathways [19]), there is mounting evidence that excessive inflammation is also detrimental for *M. tuberculosis* control. For example, increasing the per capita production of IFNγ in CD4+ T cells led to premature death in *M. tuberculosis* infected mice [20]. Moreover, in the mouse model, deleting PD-1 (an inhibitory receptor regulating T cell responsiveness) in CD4+ T cells exacerbated TB disease, where PD-1-knock out mice developed large necrotic lesions with high bacterial load and succumbed faster compared to wild-type animals [21]. While the mechanisms underlying IFNγ-driven TB disease exacerbation are yet undefined, it is plausible that excessive inflammatory signals could lead to tissue injury, promoting cell necrosis and enhancing TB pathology. Interestingly, a few cases of tuberculosis reactivation have been reported in patients on PD-1-blocking therapy [22], reinforcing the idea that *M. tuberculosis* control requires a balanced inflammatory response.

While *M. tuberculosis* is preferentially recognized by the MHC class II processing pathway, secreted immunodominant *M. tuberculosis* antigens can escape from the phagosome to the cytosol of the macrophage, be processed by cytosolic pathways, and be presented by MHC class I molecules and elicit CD8 responses. Alternatively, *M. tuberculosis* antigens can also be degraded into peptides in the phagosome, where they are then loaded on MHC class I molecules (vacuolar pathway), inducing *M. tuberculosis*-specific CD8 responses. *M. tuberculosis*-specific CD8 responses are more frequently detected in patients with TB active disease (~70%) compared to latently infected individuals (~20%). But the precise role of CD8 cell responses in *M. tuberculosis* control is still unclear. However, these cells have been shown to be recruited to the lung and can be found in granulomas. Like CD4+ T cells, CD8+ T cells can produce IFNγ, and TNFα, cytokines playing a role in orchestrating immune responses against *M. tuberculosis*. Moreover, in vitro studies indicate *M. tuberculosis*-specific CD8+ T cells can directly kill *M. tuberculosis*-infected macrophages, preferentially targeting cells that are heavily infected. Mechanistically, upon recognition of *M. tuberculosis*-infected cells, CD8+ T cells release the contents of cytotoxic granules containing perforin, granzyme B, and granulysin. Perforin establishes pores in the cytoplasmic membrane of the targeted cell, allowing entry of granzyme B, which eliminates infected cells via apoptosis. These observations support the hypothesis that *M. tuberculosis*-specific CD8+ T cell responses may also contribute to *M. tuberculosis* control.

Although cell-mediated immunity (predominantly based on T cells and mononuclear phagocytes) is the cornerstone in the defense against *M. tuberculosis*, increasing evidences suggest that humoral immunity to TB (i.e. antibody response against *M. tuberculosis*) could also participate in *M. tuberculosis* containment. While it is clear, based on animal models and human data, that antibodies alone are not sufficient to ensure protection, they could be necessary for an optimal immune response to *M. tuberculosis* infection, limit TB disease severity and/or participate in the prevention of initial *M. tuberculosis* infection [23]. For example, children without detectable antibodies against lipoarabinomannan (LAM), a glycolipid of *M. tuberculosis* cell wall, are at greater risk of disseminated TB [24]. Moreover, it has been shown that a significant minority of healthcare workers who are exposed to high doses of *M. tuberculosis* make *M. tuberculosis*-specific antibodies. Some of these individuals had no prior evidence of latent TB infection, suggesting that they may represent a subset of "restrictors" who can resist infection by *M. tuberculosis* [25] and it is plausible that antibodies play a role in protection against infection in these individuals.

The Impact of HIV Infection on Adaptive Immune Responses to *M. tuberculosis*

The depletion of CD4+ T cells, which is the major immunological feature of HIV infection, is the main contributor to the increased risk of reactivation of latent TB and susceptibility to progression of new *M. tuberculosis* infection seen in HIV-infected patients. During the initial stages of HIV infection mucosal CD4+ T cells are the prime target cells for HIV viral entry and replication [26]. This is particularly true for the effector memory subset of CD4+ T cells owing to their abundant expression of the CC-chemokine receptor 5 (CCR5) [27]. HIV envelope glycoproteins (gp120) bind to the CD4 receptor and to a co-receptor to gain cellular entry. During early infection with M-tropic HIV strains CCR5 is the predominant co-receptor used by HIV. It is reasonable to assume that profound, rapid mucosal memory CD4+ T cell loss ensues early after HIV infection, as non-human primate (NHP) studies of SIV infection have shown that 60% of gut mucosal memory CD4+ T cells are infected at peak viremia and 80% of these cells are destroyed by 4 days post infection [28]. This profound depletion of the mucosal memory CD4+ T cells is not confined to the gut, but also extends to lung interstitium [29], and this has been thought to explain the phenomenon of increased TB risk being apparent even with a relatively preserved peripheral CD4 count [30]. Moreover, this increased TB susceptibility during the early phase of HIV infection may also be explained by the finding that peripheral *M. tuberculosis*-specific CD4+ T cells are preferentially depleted upon HIV infection compared to memory T cells targeting other pathogens (such as cytomegalovirus, CMV). Mechanistically, the vulnerability of *M. tuberculosis*-specific $CD4^+$ T cells to HIV was linked to their ability to produce elevated levels of IL-2 and their poor capacity to secrete Macrophage Inflammatory Protein 1β (MIP-1β), a ligand for CCR5 acting as a natural antagonist of HIV entry

[31]. Furthermore, TB disease has been associated with increased expression of CCR5 on CD4+ T cells, increasing the pool of potential target cells for HIV. This HIV-induced depletion of *M. tuberculosis*-specific CD4+ T cells explains the reduced sensitivity of *M. tuberculosis* infection assays (i.e. QuantiFERON and T-SPOT TB) [32]. Few studies have focused on the impact of HIV infection on *M. tuberculosis*-specific CD4+ T cells at the site of disease. Studies of broncho-alveolar lavage (BAL) samples from *M. tuberculosis*-sensitized individuals with and without HIV have shown that although there is a decrease in frequency of *M. tuberculosis*-specific CD4+ T cells in HIV infected individuals, absolute numbers may be maintained in early HIV infection [33]. This can possibly be explained by the HIV induced influx of lymphocytes to the lung known as lymphocytic alveolitis, a phenomenon observed in ART-naive HIV infected individuals [34]. However, BAL sampling only reflects the alveolar CD4+ T cell compartment and may not adequately reflect the cellular composition of the lung as a whole (i.e. including the interstitium). There is evidence that lung interstitial CD4+ T cells with a resident memory-like phenotype (CD3+ CD4+ CD45RO+ TCRαβ+ CD25- CD62L-CD69+) are early targets of HIV owing to their abundant expression of CCR5 and their susceptibility to productive HIV infection. Although it is not known to what degree these interstitial CD4+ T cells are *M. tuberculosis*-specific, their depletion in co-infected NHPs was associated with dissemination of *M. tuberculosis* to distant organs [29].

In addition to the numerical depletion of CD4+ T cells, HIV also impairs the functional properties of the remaining *M. tuberculosis*-specific CD4+ T cells. Indeed, in HIV infection, *M. tuberculosis*-specific Th1 CD4 responses are skewed towards a monofunctional TNFα response; and *M. tuberculosis*-specific CD4+ T cells exhibit decreased ability to produce IL-2 that is associated with level of HIV viremia [35]. As IL-2 is important for T cell proliferation, this could limit the expansion of antigen-specific T cells upon antigen recall. HIV infection also impairs tuberculosis immunity by distorting the spectrum of *M. tuberculosis*-specific T helper responses. HIV is known to preferentially eliminate Th17 subsets from the periphery and the gut, and a reduced proportion of Th17 cells was also observed in pleural effusions from HIV-coinfected patients compared to HIV-uninfected patients. In the context of latent *M. tuberculosis* infection, HIV co-infected persons with relatively maintained CD4 counts exhibited preserved Th1 *M. tuberculosis* responses, but showed deficient IL-10-inducible responses, suggesting that by impairing anti-inflammatory regulatory pathways, HIV may shift *M. tuberculosis*-specific responses toward a more pathogenic/inflammatory profile [36].

The impact of HIV coinfection on *M. tuberculosis*-specific CD8+ T cells is relatively understudied. One study of participants with latent tuberculosis showed that *M. tuberculosis*- specific CD8+ T cells from HIV-coinfected people exhibited decreased cytotoxic potential and impaired proliferative capacity.

As previously mentioned, chronic immune activation is a hallmark of HIV disease progression and alters T cell phenotype, affecting the CD8 compartment even more severely than the CD4 compartment. During HIV infection, T cells have upregulation of activation markers such as HLA-DR and CD38, cell memory profiles shift towards a more differentiated profile with the accumulation of effector cells, and there is a significant increase in the expression of markers of senescence (such as CD57 and PD1) [37]. Together, these alterations result in T cell exhaustion and accelerate their turnover impairing their ability to mount an effective immune response upon antigen recall.

Overall, there is gradual but progressive impairment of *M. tuberculosis* immune responses during the course of untreated HIV infection. In the early phase of HIV infection, when CD4 counts are relatively maintained, sustained systemic inflammation may promote the generation of *M. tuberculosis*-specific CD4 responses of a suboptimal T helper type with increased cell turnover, weakening their ability to control *M. tuberculosis*. In more advanced HIV, severe reduction of absolute numbers of CD4+ T cells further increases tuberculosis risk by eliminating *M. tuberculosis*-specific CD4+ T cells, which are key players for *M. tuberculosis* control. Although the early depletion of *M. tuberculosis*-specific CD4 T cells following HIV infection is thought to be a major contributor to the early increased risk of TB disease [31], the impact of HIV on the innate arm of the immune system likely also contributes to the increased risk of progression to active tuberculosis disease.

The Impact of HIV Infection on Granuloma Formation, *M. tuberculosis* Containment and Dissemination Beyond the Lung

The granuloma is a classic pathologic feature of *M. tuberculosis* infection. It is constituted by many different immune cell types including primarily *M. tuberculosis*-infected macrophages, highly differentiated cells such as multinucleated giant cells (also known as Langerhans giant cells), foamy cells and epithelioid cells; all these cells are surrounded by a rim of lymphocytes, giving it a solid structure. Unlike other granulomatous diseases such as sarcoidosis, *M. tuberculosis* granulomas are often characterized by the presence of a caseous necrotic center but *M. tuberculosis* granulomas can also be non-necrotic or fibrotic. The granuloma creates an immunologic microenvironment that should facilitate the control *M. tuberculosis* growth and prevent its dissemination. However, due to the ability of *M. tuberculosis* to evade immune responses, it also provides bacilli a niche in which to potentially survive for decades in a latent form with ongoing potential for re-activation and spread [38]. T cells are instrumental in the maintenance and functional capacity of granulomas. Indeed, the arrival of *M. tuberculosis*-specific T cells at the site of disease coincides with the curbing of bacterial proliferation by producing IFNγ and TNFα that enhance macrophage microbicidal activity [39]. In HIV-infected patients with relatively preserved CD4 counts, typical granuloma architecture is observed. By contrast, TB patients with advanced HIV infection present with diffuse lesions which are multibacillary with ill-formed or absent granulomas, and can be necrotic. It is most likely that the disruption of granuloma structure is linked to HIV-induced immune dysregulation. HIV-induced killing of resident CD4+ T cells probably results in a direct disruption of granuloma structure; and combined with the alterations of T cells and macrophage function observed during HIV infection, this weakens the capacity of granulomas to contain *M. tuberculosis*, enhancing susceptibility to active disease and promoting *M. tuberculosis* dissemination. Consistent with this, disseminated TB and mycobacteremia are common in HIV co-infected individuals, especially in those with lower peripheral CD4 counts [40]. There is evidence from phylogenetic analysis of *M. tuberculosis* samples collected post mortem from lung and extrapulmonary biopsies that dissemination from the lung to distant organs occurs as frequently as within the lung itself in HIV-infected individuals [41]. Furthermore, dissemination of *M. tuberculosis* may take place soon after HIV infection. In the study by Corleis et al. [29] CCR5+ interstitial lung CD4+ T cells became productively infected with SIV and were preferentially depleted in NHPs with latent TB infection early after infection with SIV, which was in turn associated with dissemination of *M. tuberculosis* to the liver, spleen and kidney [29]. Furthermore, it has been shown that the *M. tuberculosis* virulence factor ESAT-6 is upregulated in mycobacteremic blood from HIV infected individuals and has been implicated in enhancing HIV replication [42]. From animal TB models we know that *M. tuberculosis*-specific T cells are required to halt *M. tuberculosis* infected dendritic cell migration in infected tissues, and this may to some extent explain why, with the depletion of *M. tuberculosis*-specific T cells in HIV infection, there is greater dissemination of *M. tuberculosis* [43]. In *M. tuberculosis* infected zebrafish the pharmacological inhibition of vascular endothelial growth factor (VEGF) lead to reduction in *M. tuberculosis* induced-granuloma angiogenesis, with concomitant decrease in *M. tuberculosis* burden and dissemination [44]. It is unknown whether there is abnormal granuloma angiogenesis in *M. tuberculosis*/HIV co-infected granulomas, but it is well recognised that HIV disrupts normal *M. tuberculosis* granuloma architecture and this may in turn give rise to a loss in containment of *M. tuberculosis* bacilli and consequent dissemination [45, 46].

Lung Damage in TB and the Impact of HIV

The lung is the primary target organ of *M. tuberculosis* and lung damage is common in TB, with cavities being one of TB's characteristic features. The development of cavities in TB has been studied extensively in rabbits. These studies show cavities develop from liquefied caseating granulomas, which contain large numbers of bacteria. The release of high amounts of *M. tuberculosis* antigens triggers a tissue-damaging delayed-type hypersensitivity reaction. The granuloma ruptures and spills its contents into a bronchus, leaving behind a cavity [47]. Pathologic studies in humans, which became rare after the 1950s, show a different pathway to cavity formation: cavities did not appear to develop from liquefied caseating granulomas, but from a caseous pneumonia, in which host lipids and mycobacterial antigens—but relatively few bacteria—accumulate in the alveoli. Similar to the rabbit model, necrosis occurs related to a delayed-type hypersensitivity reaction against mycobacterial antigens. The necrotic tissue either fragments to produce a cavity or hardens to develop fibrocaseous disease [48].

In both scenario's, the formation of cavities requires degradation of the lung extracellular matrix (ECM). The ECM is comprised of the interstitium, which forms the parenchyma of the lung, surrounding cells and providing structural scaffolding, and the basement membrane, which separates the epithelium or endothelium from the surrounding stroma. The ECM of the lung is mainly made up of type I collagen and elastin, and type III and IV collagen are important components of the alveolar wall and basement membrane. All these large fibers are connected by smaller fibrils. To destroy the ECM, cleavage of both small fibrils and large fibers is necessary. Collagens are highly resistant to cleavage by proteolytic enzymes; only matrix metalloproteinases (MMP's) are capable of completely degrading the ECM. Both immune cells like macrophages/monocytes and tissue cells like lung epithelial cells can secrete MMP's. Their generation is tightly regulated by tissue inhibitors of metalloproteinases (TIMPs). MMPs are not stored but require gene transcription before secretion; exemptions are MMP-8 and -9, which are stored in neutrophils. Prostaglandin and several cytokines (IL-1β, TNFα, IFNγ, IL-4, and IL-10) play a role in regulating MMP expression [49]. Data from animal models and human studies suggest that MMPs play a central role in mediating lung damage in TB. For example, several MMPs, primarily MMP-1, -3, -7, -8, and -9, are upregulated in blood, sputum and BAL fluid of patients with active TB, and increased levels of MMPs correlate with extent pulmonary disease—both cavities and infiltrates—seen on chest radiographs [50].

Neutrophils have also been linked to lung damage in TB, with higher neutrophil counts correlating with more radiographic lung damage [51, 52]. Moreover, studies have assessed the association between cytokines (including IFNγ and TNFα, and several pro- and anti-inflammatory interleukins) and lung damage in TB. Several factors make it difficult to interpret and compare these results. Firstly, different studies used different measuring methods: for example, *M. tuberculosis*–induced cytokine production by peripheral blood mononuclear cells (PBMCs) measured in vitro may not parallel serum cytokine levels. Secondly, several cytokines are not limited to a single effector function. Only TNF-α and IL-1β in serum or bronchoalveolar fluid seem to unambiguously correlate with lung damage [53].

Patients with HIV-associated TB and low CD4 counts (CD4 < 200/μL) often present with atypical chest radiograph (CXR) findings, or even normal CXRs, and cavitation is rare [54]. This suggest that TB-related pulmonary damage might be reduced in HIV co-infected patients; the host immune response, necessary for protection against TB, is required for the development of cavities. Indeed, HIV co-infection affects several of the factors implicated in pulmonary damage. Although HIV's effect on the levels of several cytokines is variable across studies, HIV co-infection has been found to reduce sputum levels of several MMPs in conjunction with reduced cavitation on CXRs [50, 55]. Moreover, HIV-co-infection reduced the activity and lifespan of neutrophils [56]. Clinical studies assessing the effect of HIV co-infection on lung function impairment during and after tuberculosis are scarce as patients with HIV co-infection are often excluded or underrepresented in studies assessing lung function in TB. In a cross sectional study in Tanzania, assessing lung function in 501 patients with TB, of which 30% were HIV co-infected, abnormal lung function was indeed less common in those with HIV co-infection [57]. Other studies, done in Cameroon, Indonesia, and Kenya, including 48/269, 19/200, and

60/183 patients infected with HIV, found no effect of HIV co-infection on functional lung impairment in TB [58–60].

Restoration of *M. tuberculosis*-Specific Immune Response with ART

Antiretroviral therapy (ART) dramatically decreases morbidity and mortality in HIV-infected individuals. ART induces rapid reduction of plasma viral load, decreases of systemic immune activation and results in progressive repletion of CD4+ T cells. Multiple mechanisms contribute to the increase in CD4+ T cells in blood in response to ART. Over the first few months of ART, there is a redistribution of CD4+ T cells from the lymph nodes to the blood, leading to a rapid initial rise in CD4 counts. Moreover, homeostatic cell proliferation, decrease in cell death and increased thymic output also play a key role in the replenishment of CD4+ T cells [61, 62]. Although the CD4 absolute cell count at the time of treatment initiation is one of the main factors dictating the level to which CD4+ T cells are restored, other parameters such as the activation level of T cells at the time of treatment, age, and active co-infections also impact on the degree of reconstitution of the CD4 compartment. Nevertheless, ART-induced restoration of the immune system is often partial and persistent aberrant activation of both the adaptive and innate cells is observed even during fully suppressive ART [63, 64]. The dynamics of the reconstitution of *M. tuberculosis*-specific CD4+ cell responses during ART is not yet fully understood. While the absolute number of *M. tuberculosis*-specific CD4+ T cells in the peripheral blood of individuals with a pre-existing *M. tuberculosis* response increases with the restoration of the CD4 compartment, conflicting results have been reported regarding the ability of ART to restore the functional capacity of *M. tuberculosis*-specific responses. Suboptimal reconstitution of *M. tuberculosis*-specific immune responses could in part explain why ART-treated patients continue to have excess risk of TB disease exceeding that of HIV-uninfected people living in the same community despite achieving CD4 count levels above 500 cells/μL [30, 65].

Immune reconstitution on ART in patients with active TB, may result in tuberculosis-associated immune reconstitution inflammatory syndrome (TB-IRIS), an immunopathological reaction driven by recovering immune responses. TB-IRIS is covered in detail in chapter "The Tuberculosis-Associated Immune Reconstitution Inflammatory Syndrome (TB-IRIS)".

Summary

Immune responses mounted against *M. tuberculosis* are diverse and multifaceted involving all arms of the host immune system. Many different immune cell subsets have been associated with protection in experimental models of TB. However, to date, it is unclear exactly what type of immune responses are required to confer protection in humans. People living with HIV are at a much greater risk of developing active TB once infected, which increases as the degree of HIV-induced immune suppression increases. The hallmark of HIV infection is a progressive reduction of CD4+ T cell count. But HIV-induced aberrant chronic immune inflammation also impairs immune responses, altering the functionality of the remaining CD4+ T cells and impairing innate cell subsets. This leads to a generalized immune dysfunction favouring TB progression to disease. This highlights the importance of early antiretroviral treatment initiation for HIV-infected persons in order to protect against immune system damage and thereby reduce TB morbidity and mortality.

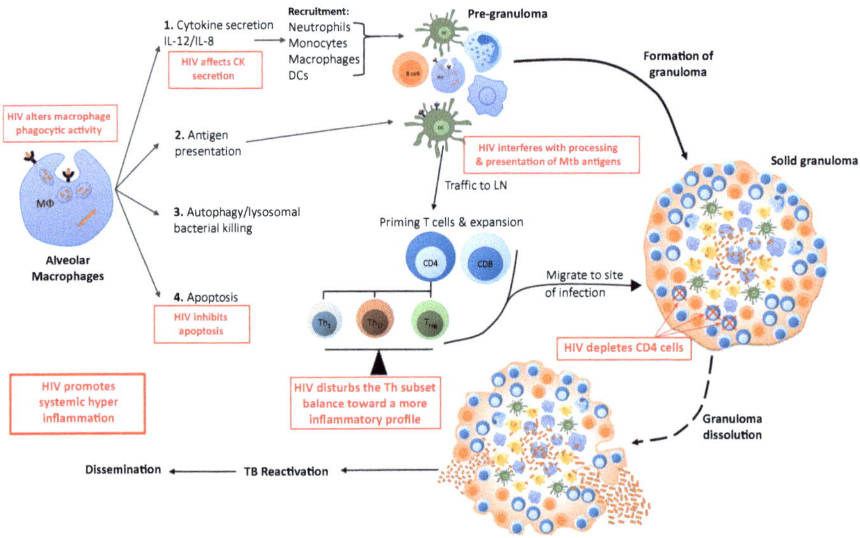

Figure: Overview of the immune response to *Mycobacterium tuberculosis* infection and impact of HIV. Tuberculosis infection is initiated when *M. tuberculosis* bacilli, present in exhaled droplets from another individuals are inhaled, recognized by Pattern Recognition Receptors (PRRs) and phagocytosed by resident alveolar macrophages in the respiratory tract. Once infected, cells mature and employ a number of processes to eliminate *M. tuberculosis* (e.g. autophagy, apoptosis) and recruit additional innate cells to the site of infection (via secretion of pro-inflammatory cytokines). During this process, loosely aggregated "pre-granulomas" are already formed. Infection of dendritic cells with *M. tuberculosis* induces their maturation and migration to the secondary draining lymph nodes, where the adaptive immune response is initiated by priming naïve CD4 or CD8 T lymphocytes. *M. tuberculosis*-specific CD4+ T cell responses include a wide array of T helper (Th) subsets endowed with distinct effector (Th1/Th17) or regulatory functions (Treg). These activated T cells migrate back to the lungs via blood, participate in granuloma formation and function (enhancing activation of macrophages). Solid granulomas are constituted by various immune cell types (primarily *M. tuberculosis*-infected macrophages, and highly differentiated cells such as multinucleated giant cells, foamy cells and epithelioid cells), surrounded by a rim of lymphocytes. It is likely that the clinical outcome of *M. tuberculosis* infection relies on the capacity of the immune system to reach a balanced response between effector and regulatory subsets; where pro-inflammatory responses enhance bacterial killing required to control *M. tuberculosis*, while anti-inflammatory responses limit pathology and inflammation during initial infection and latency. HIV-induced immune dysregulation most likely disrupt the granuloma structure. HIV-induced killing of resident CD4+ T cells probably results in a direct disruption of granuloma structure; and combined with alterations of T cell and macrophage function observed during HIV infection, this impairs the capacity of granulomas to contain *M. tuberculosis*, enhancing susceptibility to active disease and promoting *M. tuberculosis* dissemination.

Acknowledgement This figure was developed by Avuyonke Balfour.

References

1. Ernst JD (2012) The immunological life cycle of tuberculosis. Nat Rev Immunol 12(8):581–591
2. Deeks SG, Tracy R, Douek DC (2013) Systemic effects of inflammation on health during chronic HIV infection. Immunity 39(4):633–645
3. Kawai T, Akira S (2011) Toll-like receptors and their crosstalk with other innate receptors in infection and immunity. Immunity 34(5):637–650
4. Pahari S, Kaur G, Aqdas M, Negi S, Chatterjee D, Bashir H et al (2017) Bolstering immunity through pattern recognition receptors: a unique approach to control tuberculosis. Front Immunol 8:906
5. Lowe DM, Bandara AK, Packe GE, Barker RD, Wilkinson RJ, Griffiths CJ et al (2013) Neutrophilia independently predicts death in tuberculosis. Eur Respir J 42(6):1752–1757
6. Godfrey DI, Uldrich AP, McCluskey J, Rossjohn J, Moody DB (2015) The burgeoning family of unconventional T cells. Nat Immunol 16(11):1114–1123
7. Vorkas CK, Wipperman MF, Li K, Bean J, Bhattarai SK, Adamow M et al (2018) Mucosal-associated invariant and gammadelta T cell subsets respond to initial *Mycobacterium tuberculosis* infection. JCI Insight 3(19). https://doi.org/10.1172/jci.insight.121899
8. Kee SJ, Kwon YS, Park YW, Cho YN, Lee SJ, Kim TJ et al (2012) Dysfunction of natural killer T cells in patients with active *Mycobacterium tuberculosis* infection. Infect Immun 80(6):2100–2108
9. Behar SM, Divangahi M, Remold HG (2010) Evasion of innate immunity by *Mycobacterium tuberculosis*: is death an exit strategy? Nat Rev Microbiol 8(9):668–674
10. Queval CJ, Brosch R, Simeone R (2017) The macrophage: a disputed fortress in the battle against *Mycobacterium tuberculosis*. Front Microbiol 8:2284
11. Khan N, Vidyarthi A, Javed S, Agrewala JN (2016) Innate immunity holding the flanks until reinforced by adaptive immunity against *Mycobacterium tuberculosis* infection. Front Microbiol 7:328
12. Coleman CM, Wu L (2009) HIV interactions with monocytes and dendritic cells: viral latency and reservoirs. Retrovirology 6:51
13. Sattentau QJ, Stevenson M (2016) Macrophages and HIV-1: an unhealthy constellation. Cell Host Microbe 19(3):304–310
14. Cosgrove C, Ussher JE, Rauch A, Gartner K, Kurioka A, Huhn MH et al (2013) Early and nonreversible decrease of CD161++/MAIT cells in HIV infection. Blood 121(6):951–961

15. Kasprowicz VO, Cheng TY, Ndung'u T, Sunpath H, Moody DB, Kasmar AG (2016) HIV disrupts human T cells that target mycobacterial glycolipids. J Infect Dis 213(4):628–633
16. Pauza CD, Poonia B, Li H, Cairo C, Chaudhry S (2014) Gammadelta T Cells in HIV disease: past, present, and future. Front Immunol 5:687
17. Juno JA, Phetsouphanh C, Klenerman P, Kent SJ (2019) Perturbation of mucosal-associated invariant T cells and iNKT cells in HIV infection. Curr Opin HIV AIDS 14(2):77–84
18. Keane J, Gershon S, Wise RP, Mirabile-Levens E, Kasznica J, Schwieterman WD et al (2001) Tuberculosis associated with infliximab, a tumor necrosis factor alpha-neutralizing agent. N Engl J Med 345(15):1098–1104
19. Newport MJ, Huxley CM, Huston S, Hawrylowicz CM, Oostra BA, Williamson R et al (1996) A mutation in the interferon-gamma-receptor gene and susceptibility to mycobacterial infection. N Engl J Med 335(26):1941–1949
20. Sakai S, Kauffman KD, Sallin MA, Sharpe AH, Young HA, Ganusov VV et al (2016) CD4 T cell-derived IFN-gamma plays a minimal role in control of pulmonary *Mycobacterium tuberculosis* infection and must be actively repressed by PD-1 to prevent lethal disease. PLoS Pathog 12(5):e1005667
21. Barber DL, Mayer-Barber KD, Feng CG, Sharpe AH, Sher A (2011) CD4 T cells promote rather than control tuberculosis in the absence of PD-1-mediated inhibition. J Immunol 186(3):1598–1607
22. Barber DL, Sakai S, Kudchadkar RR, Fling SP, Day TA, Vergara JA et al (2019) Tuberculosis following PD-1 blockade for cancer immunotherapy. Sci Transl Med 11(475):eaat2702
23. Kozakiewicz L, Phuah J, Flynn J, Chan J (2013) The role of B cells and humoral immunity in *Mycobacterium tuberculosis* infection. Adv Exp Med Biol 783:225–250
24. Costello AM, Kumar A, Narayan V, Akbar MS, Ahmed S, Abou-Zeid C et al (1992) Does antibody to mycobacterial antigens, including lipoarabinomannan, limit dissemination in childhood tuberculosis? Trans R Soc Trop Med Hyg 86(6):686–692
25. Li H, Wang XX, Wang B, Fu L, Liu G, Lu Y et al (2017) Latently and uninfected healthcare workers exposed to TB make protective antibodies against *Mycobacterium tuberculosis*. Proc Natl Acad Sci U S A 114(19):5023–5028
26. Guadalupe M, Reay E, Sankaran S, Prindiville T, Flamm J, McNeil A et al (2003) Severe CD4+ T-cell depletion in gut lymphoid tissue during primary human immunodeficiency virus type 1 infection and substantial delay in restoration following highly active antiretroviral therapy. J Virol 77(21):11708–11717
27. Monteiro P, Gosselin A, Wacleche VS, El-Far M, Said EA, Kared H et al (2011) Memory CCR6+CD4+ T cells are preferential targets for productive HIV type 1 infection regardless of their expression of integrin beta7. J Immunol 186(8):4618–4630
28. Brenchley JM, Price DA, Douek DC (2006) HIV disease: fallout from a mucosal catastrophe? Nat Immunol 7(3):235–239
29. Corleis B, Bucsan AN, Deruaz M, Vrbanac VD, Lisanti-Park AC, Gates SJ et al (2019) HIV-1 and SIV infection are associated with early loss of lung interstitial CD4+ T cells and dissemination of pulmonary tuberculosis. Cell Rep 26(6):1409–1418. e5
30. Sonnenberg P, Glynn JR, Fielding K, Murray J, Godfrey-Faussett P, Shearer S (2005) How soon after infection with HIV does the risk of tuberculosis start to increase? A retrospective cohort study in South African gold miners. J Infect Dis 191(2):150–158
31. Geldmacher C, Ngwenyama N, Schuetz A, Petrovas C, Reither K, Heeregrave EJ et al (2010) Preferential infection and depletion of Mycobacterium tuberculosis-specific CD4 T cells after HIV-1 infection. J Exp Med 207(13):2869–2881
32. Cattamanchi A, Smith R, Steingart KR, Metcalfe JZ, Date A, Coleman C et al (2011) Interferon-gamma release assays for the diagnosis of latent tuberculosis infection in HIV-infected individuals: a systematic review and meta-analysis. J Acquir Immune Defic Syndr 56(3):230–238
33. Bunjun R, Riou C, Soares AP, Thawer N, Muller TL, Kiravu A et al (2017) Effect of HIV on the frequency and number of Mycobacterium tuberculosis-specific CD4+ T cells in blood and airways during latent M. tuberculosis infection. J Infect Dis 216(12):1550–1560

34. Neff CP, Chain JL, MaWhinney S, Martin AK, Linderman DJ, Flores SC et al (2015) Lymphocytic alveolitis is associated with the accumulation of functionally impaired HIV-specific T cells in the lung of antiretroviral therapy-naive subjects. Am J Respir Crit Care Med 191(4):464–473
35. Day CL, Mkhwanazi N, Reddy S, Mncube Z, van der Stok M, Klenerman P et al (2008) Detection of polyfunctional Mycobacterium tuberculosis-specific T cells and association with viral load in HIV-1-infected persons. J Infect Dis 197(7):990–999
36. Bell LC, Pollara G, Pascoe M, Tomlinson GS, Lehloenya RJ, Roe J et al (2016) In vivo molecular dissection of the effects of HIV-1 in active tuberculosis. PLoS Pathog 12(3):e1005469
37. Appay V, Kelleher AD (2016) Immune activation and immune aging in HIV infection. Curr Opin HIV AIDS 11(2):242–249
38. Flynn JL, Chan J, Lin PL (2011) Macrophages and control of granulomatous inflammation in tuberculosis. Mucosal Immunol 4(3):271–278
39. Sasindran SJ, Torrelles JB (2011) *Mycobacterium tuberculosis* infection and inflammation: what is beneficial for the host and for the bacterium? Front Microbiol 2:2
40. Crump JA, Ramadhani HO, Morrissey AB, Saganda W, Mwako MS, Yang LY et al (2012) Bacteremic disseminated tuberculosis in sub-saharan Africa: a prospective cohort study. Clin Infect Dis 55(2):242–250
41. Lieberman TD, Wilson D, Misra R, Xiong LL, Moodley P, Cohen T et al (2016) Genomic diversity in autopsy samples reveals within-host dissemination of HIV-associated *Mycobacterium tuberculosis*. Nat Med 22(12):1470–1474
42. Ryndak MB, Singh KK, Peng Z, Zolla-Pazner S, Li H, Meng L et al (2014) Transcriptional profiling of *Mycobacterium tuberculosis* replicating ex vivo in blood from HIV- and HIV+ subjects. PLoS One 9(4):e94939
43. Harding JS, Rayasam A, Schreiber HA, Fabry Z, Sandor M (2015) Mycobacterium-infected dendritic cells disseminate granulomatous inflammation. Sci Rep 5:15248
44. Oehlers SH, Cronan MR, Scott NR, Thomas MI, Okuda KS, Walton EM et al (2015) Interception of host angiogenic signalling limits mycobacterial growth. Nature 517(7536):612–615
45. Nusbaum RJ, Calderon VE, Huante MB, Sutjita P, Vijayakumar S, Lancaster KL et al (2016) Pulmonary tuberculosis in humanized mice infected with HIV-1. Sci Rep 6:21522
46. Diedrich CR, O'Hern J, Wilkinson RJ (2016) HIV-1 and the *Mycobacterium tuberculosis* granuloma: a systematic review and meta-analysis. Tuberculosis (Edinb) 98:62–76
47. Dannenberg AM Jr (2006) Pathogenesis of human pulmonary tuberculosis: insights from the rabbit model. ASM Press, Washington, DC
48. Hunter RL (2016) Tuberculosis as a three-act play: a new paradigm for the pathogenesis of pulmonary tuberculosis. Tuberculosis (Edinb) 97:8–17
49. Elkington PT, Ugarte-Gil CA, Friedland JS (2011) Matrix metalloproteinases in tuberculosis. Eur Respir J 38(2):456–464
50. Walker NF, Clark SO, Oni T, Andreu N, Tezera L, Singh S et al (2012) Doxycycline and HIV infection suppress tuberculosis-induced matrix metalloproteinases. Am J Respir Crit Care Med 185(9):989–997
51. Berry MP, Graham CM, McNab FW, Xu Z, Bloch SA, Oni T et al (2010) An interferon-inducible neutrophil-driven blood transcriptional signature in human tuberculosis. Nature 466(7309):973–977
52. Panteleev AV, Nikitina IY, Burmistrova IA, Kosmiadi GA, Radaeva TV, Amansahedov RB et al (2017) Severe tuberculosis in humans correlates best with neutrophil abundance and lymphocyte deficiency and does not correlate with antigen-specific CD4 T-cell response. Front Immunol 8:963
53. Stek C, Allwood B, Walker NF, Wilkinson RJ, Lynen L, Meintjes G (2018) The immune mechanisms of lung parenchymal damage in tuberculosis and the role of host-directed therapy. Front Microbiol 9:2603
54. Kwan CK, Ernst JD (2011) HIV and tuberculosis: a deadly human syndemic. Clin Microbiol Rev 24(2):351–376

55. Walker NF, Wilkinson KA, Meintjes G, Tezera LB, Goliath R, Peyper JM et al (2017) Matrix degradation in human immunodeficiency virus type 1—associated tuberculosis and tuberculosis immune reconstitution inflammatory syndrome: a prospective observational study. Clin Infect Dis 65(1):121–132
56. Lowe DM, Bangani N, Goliath R, Kampmann B, Wilkinson KA, Wilkinson RJ et al (2015) Effect of antiretroviral therapy on HIV-mediated impairment of the neutrophil antimycobacterial response. Ann Am Thorac Soc 12(11):1627–1637
57. Manji M, Shayo G, Mamuya S, Mpembeni R, Jusabani A, Mugusi F (2016) Lung functions among patients with pulmonary tuberculosis in Dar es Salaam—a cross-sectional study. BMC Pulm Med 16(1):58
58. Mbatchou Ngahane BH, Nouyep J, Nganda Motto M, Mapoure Njankouo Y, Wandji A, Endale M et al (2016) Post-tuberculous lung function impairment in a tuberculosis reference clinic in Cameroon. Respir Med 114:67–71
59. Ralph AP, Kenangalem E, Waramori G, Pontororing GJ, Sandjaja TE et al (2013) High morbidity during treatment and residual pulmonary disability in pulmonary tuberculosis: under-recognised phenomena. PLoS One 8(11):e80302
60. Mugo PN, Mecha J, Muhwa C (2018) Pulmonary function and quality of life in patients with treated smear positive pulmonary tuberculosis at three tuberculosis clinics in Nairobi, Kenya. Eur Respir J. 52:Suppl. 62, PA2749.
61. Corbeau P, Reynes J (2011) Immune reconstitution under antiretroviral therapy: the new challenge in HIV-1 infection. Blood 117(21):5582–5590
62. Wilson EM, Sereti I (2013) Immune restoration after antiretroviral therapy: the pitfalls of hasty or incomplete repairs. Immunol Rev 254(1):343–354
63. Hunt PW, Lee SA, Siedner MJ (2016) Immunologic biomarkers, morbidity, and mortality in treated HIV infection. J Infect Dis 214(Suppl 2):S44–S50
64. Nabatanzi R, Cose S, Joloba M, Jones SR, Nakanjako D (2018) Effects of HIV infection and ART on phenotype and function of circulating monocytes, natural killer, and innate lymphoid cells. AIDS Res Ther 15(1):7
65. Gupta A, Wood R, Kaplan R, Bekker LG, Lawn SD (2012) Tuberculosis incidence rates during 8 years of follow-up of an antiretroviral treatment cohort in South Africa: comparison with rates in the community. PLoS One 7(3):e34156

Clinical Manifestations of HIV-Associated Tuberculosis in Adults

Sean Wasserman, David Barr, and Graeme Meintjes

Abstract HIV-associated tuberculosis is a heterogenous disease that confronts clinicians with substantial diagnostic challenge. Clinical syndromes are frequently non-specific in terms of symptoms, physical examination, routine laboratory testing, and chest radiography. Further complicating management is the possibility of co-infection with other severe opportunistic infections, all of which may have clinical presentations that mimic tuberculosis. Early recognition and treatment is urgent because of more severe manifestations and rapid progression, particularly at low CD4 counts. This chapter describes clinical manifestations and diagnostic approaches for HIV-associated tuberculosis in adults, with an emphasis on practice in resource-limited, high-burden settings. Advanced immunosuppression and disseminated disease are considered separately from ambulant patients with preserved CD4 cell counts in order to highlight differences in clinical phenotype, differential diagnosis, and management strategies. Clinical features and evaluation of common extra-pulmonary manifestations are also covered.

Keywords HIV-associated tuberculosis · Mycobacteraemia · Disseminated tuberculosis · Extra-pulmonary tuberculosis · Lipoarabinomannan · GeneXpert · Rapid diagnostics

---

S. Wasserman (✉) · G. Meintjes
Wellcome Centre for Infectious Diseases Research in Africa, Institute of Infectious Disease and Molecular Medicine, Department of Medicine, University of Cape Town, Cape Town, South Africa

Division of Infectious Diseases and HIV Medicine, Department of Medicine, University of Cape Town, Cape Town, South Africa
e-mail: sean.wasserman@uct.ac.za; graemein@mweb.co.za

D. Barr
Institute of Infection and Global Health, University of Liverpool, Liverpool, UK
e-mail: David.Barr@liverpool.ac.uk

I. Sereti et al. (eds.), *HIV and Tuberculosis*,
https://doi.org/10.1007/978-3-030-29108-2_5

Introduction

Infection with *Mycobacterium tuberculosis* causes a broad spectrum of pathology, the manifestations of which are largely influenced by host immune response. Although tuberculosis has traditionally been understood as taking dichotomous form—with patients having either latent or active infection depending on the absence or presence of clinical symptoms—it is now recognised that latent tuberculosis is part of a spectrum that spans from the possible elimination of *M. tuberculosis* by the innate immune system, through immune containment (associated with tuberculin skin test or interferon-gamma release assay positivity), to asymptomatic yet culture positive states (so-called 'sub-clinical tuberculosis') through to active and symptomatic disease [1]. The term 'HIV-associated tuberculosis' describes the clinical manifestations of active disease caused by *M. tuberculosis* in HIV-infected patients, and will be the focus of this chapter.

Immune dysfunction associated with HIV infection (detailed in chapter "Immune Responses to *Mycobacterium tuberculosis* and the Impact of HIV Infection"—Immunology) has profound effects on the course and clinical phenotype of tuberculosis. Granuloma formation, the hallmark of *M. tuberculosis* infection, requires well-orchestrated immune responses and is central to local control of *M. tuberculosis*. Alteration of adaptive immunity in HIV infection, particularly through absolute and functional depletion of CD4+ cells, leads to granuloma disruption that underlies atypical clinical presentations observed in HIV-associated tuberculosis. HIV-infected individuals with advanced immunodeficiency (CD4 count <200 cells/μL) are generally unable to form organised granulomas leading to ineffective containment and unrestrained bacillary replication. Rather than causing localised apical cavitary lung disease, as is typically observed in HIV-uninfected patients or in HIV-infected patients with well-preserved CD4 counts, *M. tuberculosis* dissemination and aberrant inflammatory responses often results in non-cavitary pulmonary parenchymal disease, extra-pulmonary organ involvement, intra-thoracic adenopathy and haematogenous ('miliary') infiltration of the lung parenchyma and other organs. These clinical syndromes are frequently non-specific in terms of symptoms, are more challenging to diagnose, and can rapidly progress to severe and life-threatening illness. This is illustrated by post-mortem studies in resource-limited settings where almost half of deaths from tuberculosis are undiagnosed at the time of death [2]. Further complicating clinical management of suspected tuberculosis in patients with advanced HIV is the possibility of co-infection with other severe opportunistic infections such as *Pneumocystis jirovecii*, dimorphic fungi, cytomegalovirus, and bacterial sepsis, as well as HIV-associated malignancies like Kaposi sarcoma and lymphoma, all of which may have clinical presentations that mimic tuberculosis.

The aim of this chapter is to describe important clinical manifestations and diagnostic approaches in the clinical setting for HIV-associated tuberculosis in adults. The emphasis is on patients in resource-limited, high-burden settings with advanced HIV infection in whom early recognition and treatment of tuberculosis is critical, but may be hindered by the particular challenges encountered in this population.

Because of specific issues relating to diagnosis and management, neurological disease and tuberculosis-IRIS are addressed separately in chapters "Neurological Tuberculosis in HIV" and "The Tuberculosis-Associated Immune Reconstitution Inflammatory Syndrome (TB-IRIS)", respectively. Paediatric tuberculosis is also covered in a dedicated chapter.

Clinical and Radiological Features at Higher CD4 Counts

HIV-infected individuals with preserved CD4 cell counts, particularly >350 cells/μL, are more likely to manifest similar symptoms and signs as those who are HIV-negative. The predominant disease site is pulmonary; pleural effusion and lymphadenitis are also frequently encountered. Although less common than in severe immunosuppression, extra-pulmonary involvement of any site can occur at high CD4 counts—these are discussed in section "Specific Clinical Syndromes Associated with EPTB" on extra-pulmonary tuberculosis (EPTB) below. Even with relatively intact immune responses, pulmonary disease in HIV-associated tuberculosis has a heterogenous clinical presentation and considerable overlap with other diagnoses in radiological appearances.

Symptoms

Early pulmonary tuberculosis may be asymptomatic, but eventually leads to chronic productive cough accompanied by non-specific constitutional symptoms including weight loss, night sweats (often drenching), fever, chills, fatigue and anorexia, but not rigors. Persistent, slowly progressive respiratory symptoms are an important feature in HIV-infected patients with relatively well-preserved immunity, and should prompt intensive efforts to exclude tuberculosis, even in the absence of typical CXR changes and negative smear microscopy and Xpert. Pleuritic chest pain occurs if parenchymal inflammation abuts the parietal pleura, and is common with tuberculous pleural effusions. Cough may uncommonly be associated with haemoptysis related to endobronchial erosion or sloughing of caseous material from cavities. Haemoptysis is usually mild, but can present as a sudden and massive event as a result of erosion of an artery in the pulmonary circulation. Importantly, many patients with HIV co-infection may lack the typical symptoms associated with pulmonary tuberculosis, and are at risk of numerous diseases whose symptoms overlap with tuberculosis.

A meta-analysis including individual data from 8148 HIV-infected patients identified current cough of any duration, fever, night sweats, or weight loss as having the best predictive value for active tuberculosis [3]. These are recommended by WHO as a screening tool in national tuberculosis programs, to be used repeatedly in follow-up visits. However, the specificity of these symptoms is poor (50%) and their negative predictive value falls substantially when applied to patient populations with greater than 20% prevalence of TB, which is common in high burden clinical settings [3, 4]. The prevalence of completely asymptomatic sputum culture-positive tuberculosis amongst ambulant people living with HIV in a community setting may be as high as 15% [5]. Absence of typical symptoms therefore does not exclude pulmonary tuberculosis and this diagnosis needs to be aggressively pursued amongst symptomatic patients with HIV in high burden settings.

Clinical and Laboratory Examination

Clinical examination findings relate to the extent of lung involvement and chronicity of illness, and are often unremarkable. Fever is not universal. There may be auscultatory signs of consolidation and cavitation in advanced disease, but these are characteristically absent at initial presentation. Tuberculous pleural effusions are almost always unilateral, and strongly suggested by dullness on percussion and reduced breath sounds on the side of the effusion. Pallor and cachexia may be present. Digital clubbing is absent among patients with recent-onset symptoms, but may manifest after prolonged infection or repeated episodes in association with bronchiectasis and fibrosis. Routine laboratory tests are usually non-contributory: total white blood cell count is typically in the normal range, and there may be a normocytic or mildly microcytic anaemia of inflammation. Mild eosinophilia is occasionally seen.

Non-specific markers of systemic inflammation such as C-reactive protein (CRP) [6–8] and procalcitonin (PCT) [9] are frequently elevated, and may perform better than symptom screening for identifying patients with active tuberculosis [10]. In a cohort of Ugandan adults attending outpatient ART clinics (median CD4 count 137 cells/μL), point-of-care (POC) CRP had a higher specificity than symptom screening (87% vs. 21%; $P < 0.001$) for tuberculosis, and correctly identified more patients without active tuberculosis [10]. However, CRP and PCT do not reliably distinguish tuberculosis from bacterial infection or pneumocystis pneumonia [9, 11], and when negative do not have sufficient predictive value to confidently exclude tuberculosis in symptomatic HIV-infected patients in high burden settings [6–8], where the consequences of misdiagnosis are severe. For example, the negative predictive value (NPV) of CRP was only 72% in tuberculosis suspects in Kwazulu-Natal, South Africa, where the background prevalence of confirmed or possible tuberculosis was ~30%. HIV status did not influence the performance of CRP in this cohort [8]. Although the NPV of a very low CRP (<1.5 mg/L) is excellent (100% in an active case finding study in Cape Town, South Africa), this may not be useful for most patients because such very low CRP values may be infrequent in the population of interest [6].

Table 1 Chest X-ray features of HIV-associated tuberculosis in relation to CD4 cell count

CD4 > 350 cells/μL	CD4 < 200 cells/μL
Upper lobe parenchymal infiltrates and cavitation	Nodular infiltrates, may be localised or widespread
Pleural effusion	Consolidation
Fibrosis	Miliary infiltration
Bronchiectasis	Mediastinal and hilar adenopathy
	Pleural and/or pericardial effusion

Imaging

Chest radiography is an important diagnostic tool for assessing HIV-associated tuberculosis in patients with higher CD4 counts. The typical features, as seen in HIV-negative pulmonary tuberculosis, result from liquefaction of caseous necrosis with subsequent cavity formation, areas of parenchymal infiltrates, and progressive fibrosis and lung destruction, predominantly affecting upper lobes (Table 1) [12]. A pooled analysis of five cohort studies showed a significant association of cavitation and parenchymal infiltrates with CD4 cell counts ≥200 cells/μL in patients with HIV-associated tuberculosis [13]. Bronchogenic spread may result in ill-defined micronodules with segmental or lobar distribution, seen best on high resolution computed tomography (CT) imaging as a so-called "tree-in-bud" sign. Post-tuberculosis fibrocavitatory disease and bronchiectatic changes may lead to chronic respiratory symptoms with negative *M. tuberculosis* cultures and predispose to chronic lung infections, including *Aspergillus* sp. and non-tuberculous mycobacteria.

Approach to Diagnosis of Pulmonary Tuberculosis in Ambulant Patients with Higher CD4 Counts

The diagnostic approach for suspected tuberculosis in HIV-infected patients with CD4 counts ≥350 cells/μL is similar to HIV-negative patients, and is presented in Fig. 1. Chest X-ray is central to diagnosis but interpretation is reader-dependent, and there may non-specific changes and the absence of abnormalities even in the context of positive smear microscopy and culture [13, 14]. There is generally a less rapidly-progressive course compared to patients with severe immune compromise, and therefore opportunity to confirm the diagnosis with molecular or microbiological testing.

Spontaneously expectorated sputum is often available for testing, particularly when cavitary disease is present, as cavitation is associated with high bacillary load in sputum and greater chance of diagnosis by smear microscopy. If this is not possible sputum induction with hypertonic saline should be attempted. Although sputum induction with nebulized hypertonic saline potentially identifies approximately

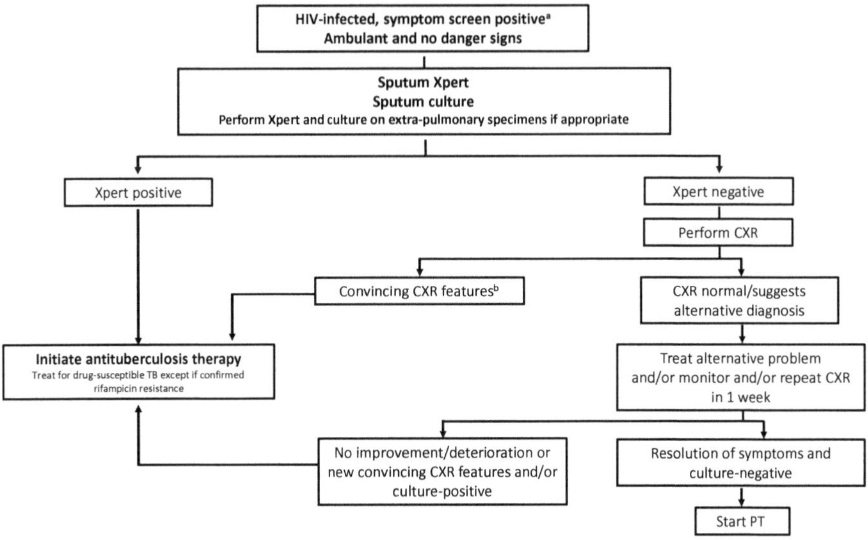

Fig. 1 Diagnostic approach for suspected HIV-associated tuberculosis at CD4 ≥ 350 cells/μL (**a**) Current cough, night sweats, fever, weight loss. (**b**) Cavitations, nodular infiltrate, pleural effusion (lymphocytic). *CXR* chest X-ray, *PT* preventive therapy

25% additional cases in symptomatic patients who are smear-negative or unable to spontaneously produce sputum [15], a negative smear never excludes the diagnosis of tuberculosis in HIV co-infected patients. Sputum Xpert-MTB/RIF (or the more sensitive Xpert-MTB/RIF Ultra) is the preferred initial diagnostic test for ambulatory patients, particularly those with higher CD4 counts. This is due to superior sensitivity to smear microscopy, rapid turnaround time, and ability to detect the presence of rifampicin resistance (discussed in chapter "Diagnosis of HIV-Associated Tuberculosis"). On a population level, screening symptomatic ambulatory HIV-infected patients with Xpert detects more cases of tuberculosis compared to fluorescence microscopy, and results in lower mortality amongst those with clinical stage 3 or 4 disease [16]. However, Xpert's imperfect sensitivity and negative predictive value is insufficient to exclude HIV-associated tuberculosis in high burden settings and may not influence empiric management decisions [17]. It is therefore important to also perform sputum culture, for both diagnostic and monitoring reasons, as well as Xpert and culture testing on extra-pulmonary samples when appropriate (and if available). With HIV-infected patients now routinely being admitted to intensive care units, tuberculosis is an important consideration in this patient population in high burden settings. A study in Cape Town, South Africa, diagnosed tuberculosis in 15% (46/317 screened) of mechanically ventilated patients, a quarter of whom were HIV-infected. Xpert performed well on tracheal aspirates, identifying all patients with positive *M. tuberculosis* cultures [18].

Occasionally empiric antituberculosis treatment is started without confirmation while awaiting results of microbiological testing. This may be considered under the following circumstances: strongly suggestive chest X-ray abnormalities, poor response to empiric therapy for an alternative condition (e.g. bacterial pneumonia), severe illness, or rapid deterioration. A subset of patients with typical symptoms and radiological features of pulmonary tuberculosis have persistently negative *M. tuberculosis* cultures with no alternative aetiology identified. These patients are often diagnosed with 'culture-negative' pulmonary tuberculosis and started on empiric antituberculosis therapy. Potential pitfalls of empiric treatment include missing another serious alternative diagnosis or drug-resistant tuberculosis, and exposing patients unnecessarily to prolonged courses of antituberculosis therapy. It is therefore critical to monitor clinical response to therapy and review sputum culture results when empiric therapy is initiated.

The differential diagnosis for HIV-associated pulmonary tuberculosis at higher CD4 counts includes lung malignancy (which occurs more frequently amongst HIV-infected vs. HIV-uninfected smokers [19]) and chronic infections such nontuberculous mycobacterial lung disease and aspergillosis. Other important causes of symptoms suggestive of tuberculosis include weight loss due to food insecurity, viral upper respiratory tract infection, and post-tuberculosis chronic lung disease [20]. Although seen more commonly amongst patients with CD4 <200 cells/μL, pneumocystis pneumonia remains a possibility in those with higher CD4 counts. Important differential diagnoses for tuberculosis in HIV-infected patients with higher CD4 counts are listed in Table 2.

Clinical Manifestations of HIV-Associated Tuberculosis with Advanced Immunosuppression

By contrast, the clinical presentation of tuberculosis in HIV co-infected patients with advanced immunosuppression (CD4 count < 200 cells/μL) is associated with atypical pulmonary manifestations, frequent extra-pulmonary involvement, and a higher prevalence of disseminated disease with rapid progression. With improved access to antiretroviral therapy (ART) and higher CD4 count thresholds for treatment initiation the clinical presentation of HIV-associated tuberculosis may begin to shift towards the typical syndrome seen in immune competent individuals [21]. However, the majority of cases of HIV-associated tuberculosis still occur at low CD4 counts in high burden settings. Analysis of a large dataset from the tuberculosis registry in Cape Town, South Africa, found that 25% of patients on ART had CD4 $\leq$ 90 cells/μL at the time of initiation of antituberculosis therapy; amongst ART-naïve patients, the lower CD4 count quartile was 67 cells/μL [21]. Another study in the Western Cape Province in South Africa showed that over half of patients with CD4 < 50 cells/μL were ART-experienced, many having disengaged from ART care [22]. Thus, even with substantial increases in ART coverage, a large proportion of patients with HIV-associated tuberculosis are expected to manifest atypical, non-specific, and severe presentations, which present important clinical and diagnostic challenges.

Table 2 Differential diagnosis for HIV-associated tuberculosis at higher CD4 counts

Disease	Characteristics
Bacterial pneumonia	Short duration symptoms (≤7 days) with focal or multi-focal consolidation on CXR. Prominent inflammatory response (rigors, chest signs) seen in lobar/pneumococcal pneumonia. Early response to antibiotics
Pulmonary non-tuberculous mycobacterial infection	Predisposing structural lung disease, older age patient Indolent onset, chronic respiratory symptoms, weight loss and fevers leading to extensive pathology on chest X-ray at time of presentation. Requires confirmation with >1 sputum culture of same species. Suspect if no improvement on antituberculosis therapy with persistent smear positivity without culture confirmation of tuberculosis
Cryptococcus[a]	Can cause primary respiratory disease (without or with meningitis) with non-specific sub-acute symptoms and diverse chest X-ray features
Nocardia[a]	Subacute respiratory symptoms with fever; causes nodules and lung abscess on CXR. May be associated with skin and CNS lesions. Culture of sputum or aspirate shows branching Gram-positive bacilli that are weakly acid fast on modified ZN staining
Pneumocystis pneumonia[a]	Subacute to acute respiratory symptoms with prominent dyspnoea and tachypnoea. Typically normal auscultation but desaturation on mild exertion. Chest X-ray shows bilateral infiltrates, without adenopathy or pleural effusions
Chronic pulmonary aspergillosis	Fever not prominent unless invasive disease. Haemoptysis common. Fungal ball can sometimes be seen in cavity on CXR
Chronic obstructive airways disease	History of smoking or biomass smoke exposure. Long standing exertional dyspnoea often with wheeze. Prolonged expiration, barrel chest, hyper-resonance, hyper-expansion on CXR
Lung cancer	Older patient with risk factors. Often minimal symptoms until advanced disease, fever not typical unless secondary pneumonia. May be evidence of metastases (bone pain, irregular liver, brain space occupying lesions), local mass effects or paraneoplastic phenomena
Kaposi sarcoma[a]	Majority have associated muco-cutaneous disease. Diverse presentations, including haemoptysis, chest pain, and severe respiratory distress. CXR typically shows 'flame-shaped' infiltrate extending from hilum adjacent to vascular and bronchial trees; septal and nodular patterns; associated effusions which are often blood stained when aspirated

[a]More common at lower CD4 counts

Disseminated Tuberculosis and Mycobacteraemia

Post-mortem series of hospitalised patients dying of HIV-associated tuberculosis show that in ~90% of cases tuberculosis pathology is disseminated, i.e. involving several organ systems, in particular the reticulo-endothelial system and tissues with high blood flow such as the kidneys [2]. There is a broad spectrum of illness severity, but disseminated tuberculosis is often associated with organ dysfunction and decline in patient global functional status (e.g. the WHO danger signs 'inability to walk unaided' and respiratory rate >30 breaths/min).

Many studies performing routine mycobacterial blood cultures in hospitalised HIV-infected patients suggest that a large proportion of those with critical illness have active mycobacteraemic dissemination at time of admission [David Barr, personal communication]. In the spectrum of tuberculosis infection, these are patients who are likely to have the highest total body bacilli-burden, and *M. tuberculosis* blood stream infection (BSI) in the range of ten to ten-thousand bacilli per ml have been recorded [David Barr, personal communication]. Although localised pauci-bacillary tuberculosis infection can cause acute life-threatening illness (for example, tuberculous meningitis, bowel perforation, or rupture of a great vessel), high bacilli-burden disseminated tuberculosis, with or without current BSI, is associated with high mortality and can be considered the major mode of disease in critically unwell patients with HIV-associated tuberculosis.

Due to profound immunosuppression, granuloma formation may be atypical or absent, and serositis may be blunted, resulting in non-specific clinical presentation. Classic miliary pathology—which results from a sudden, large bacteraemic event, with granuloma formation at sites of seeding—is an unreliable indicator of HIV-associated tuberculosis BSI. Indeed, in one study of disseminated tuberculosis, a positive tuberculosis blood culture was associated with *absence* of miliary shadowing, but a higher probability of death [23]. The absence of organised granulomas does not denote an absence of inflammation; rather hospitalised patients with severe HIV-associated tuberculosis and BSI have marked innate immune cell activation and dysfunction [24], and are at high-risk of secondary bacterial infections.

Clinical and Radiological Features

As outlined above, classical tuberculosis symptoms may be absent in HIV-associated tuberculosis and overlap with other conditions affecting these patients. The WHO standardised screening rule for tuberculosis in HIV-infected individuals (any of cough, fever, night sweats, or weight loss, developed and validated through meta-analysis of predominantly outpatient data [3]) has no predictive value for critically unwell inpatients [25, 26]. The presentations of disseminated tuberculosis are diverse, and depend on organ involvement and intensity of systemic inflammatory response. Non-classical symptoms such as vomiting, diarrhoea, generalised pain and weakness may have higher associated likelihood ratios for tuberculosis BSI than cough or night sweats. A typical history elicited in patients with HIV-associated tuberculosis BSI includes a sub-acute onset—often with multiple prior visits to health facilities—followed by a more rapid decline in functional status associated with anorexia, prolonged fevers, night sweats, and generalised weakness. Despite patients typically being young adults, inability to walk unaided and/or self-feed are common, and strongly associated with early mortality. Fever is almost universal [27], but unlike other, non-mycobacterial BSIs, history of pronounced rigors is unusual.

Patients with tuberculosis BSI can present with a syndrome of septic shock; indeed, tuberculosis is the most common microbiological diagnosis in studies of severe sepsis in HIV-infected inpatients in sub-Saharan Africa [28, 29]. However, only a minority of critically unwell patients with tuberculosis BSI have frank hypotension, and the "hot & flushed" appearance of acute, systemic vasodilation is unusual. High fever (e.g. temperature >39.0 °C) is not predictive of mortality and many patients are hypothermic, with cool peripheries but relatively normal capillary refill. Raised lactate is common and strongly associated with mortality, but not closely related to hypotension, and may instead reflect accelerated *aerobic* metabolism in activated innate immune cells [30], rather than anaerobic respiration from tissue hypoperfusion. In contrast to frank hypotension, postural-drop in blood pressure can frequently be elicited, and moderate tachycardia is the norm. Patients with tuberculosis BSI are often profoundly fluid depleted, with near universal hyponatraemia suggesting a sub-acute loss of total body sodium and water. Lethargy and limited ability to mobilise are very common in tuberculosis BSI, but wasting is not universal. Reduced levels of consciousness may reflect metabolic effects of systemic infection, but meningitis or raised intra-cranial pressure should be ruled-out with lumbar puncture and/or brain imaging, and hypoglycaemia excluded.

Despite respiratory symptoms often being absent, the majority of HIV-infected patients with disseminated disease will have pulmonary involvement as evidenced by positive sputum cultures or radiological abnormalities. When present, respiratory symptoms may include dry cough, dyspnoea, reduced effort tolerance, and pleuritic chest pain. Raised respiratory rate is common, and often associated with nasal flare, but not with other signs of respiratory failure such as inability to complete sentences. The raised respiratory rate can often be explained by metabolic acidosis, in turn associated with reduced glomerular filtration rate due to fluid and sodium depletion, and/or raised lactate. Acute renal impairment is the most common organ dysfunction in tuberculosis associated sepsis. Tachypnoea may also reflect impaired oxygenation due to extensive tuberculous infiltration or massive pleural or pericardial effusion. Crepitations and evidence of pleural effusion may be present, but auscultation is often normal. Acute respiratory distress syndrome is rare in tuberculosis BSI and marked hypoxia is more suggestive of pneumocystis pneumonia, which is an important competing or concurrent diagnosis.

As discussed above, chest radiology becomes atypical and non-specific at lower CD4 counts [31], which is the case in critically unwell tuberculosis BSI patients. Chest X-rays may be completely normal in up to 30% despite a positive sputum *M. tuberculosis* culture [32]. Abnormalities may include diffuse and lower lobe opacities, subtle nodular infiltrates, a miliary pattern, or consolidation, with or without intrathoracic lymphadenopathy or pleural effusions (Figs. 2 and 3) [13, 31]. Classic miliary shadowing is not generally seen in tuberculosis BSI in patients with advanced HIV; less symmetrical, but still diffuse, interstitial nodule infiltrates without uniformity of nodule size and often sparing some lung zones is more typical (and may incorrectly be described as miliary by some clinicians). Non-specific airspace opacification is also frequent, and distinguishing from bacterial pneumonia is difficult in many cases. The chest X-ray features of HIV associated tuberculosis are

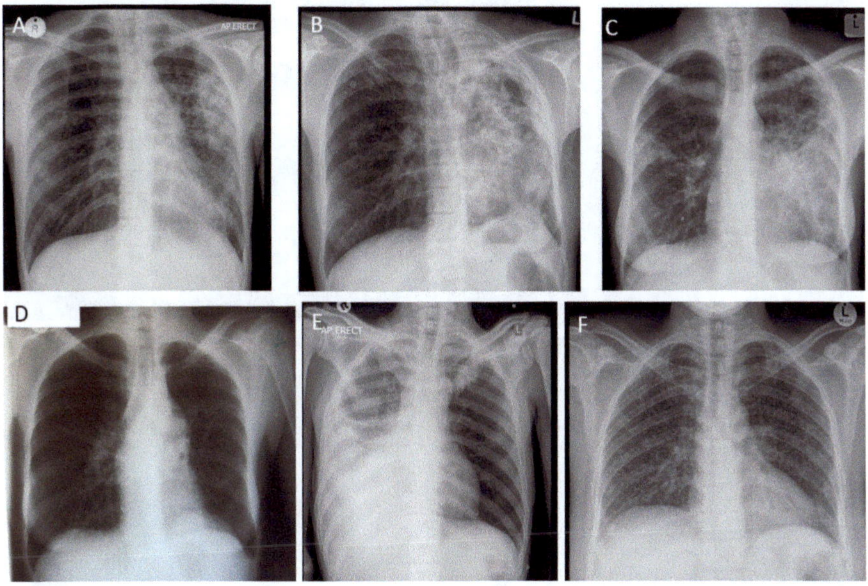

Fig. 2 Illustrative examples of chest X-ray features in HIV-associated tuberculosis. Images A and B: 32-year-old woman, who had an initial episode of culture-positive tuberculosis in 2015 with a CD4 count of 390 cells/μL. Chest X-ray from that time shows bilateral nodular opacification, worse on left with areas of confluence, and hilar adenopathy. (**a**) She re-presented with progressive respiratory symptoms and constitutional symptoms in 2019, after being on ART for 3 years and with a CD4 count of 1274 cells/μL. CXR shows volume loss and pleural thickening with widespread fibrosis and cystic changes. (**b**) Culture was again positive for *M. tuberculosis*. (**c**) 35 year old woman, CD4 count 479 cells/μL and virally suppressed on ART. Presented with new constitutional symptoms and sputum culture was positive for TB. CXR shows confluent opacification with areas of breakdown on the left and bilateral nodular infiltrates. (**d**) 29 year old woman with constitutional symptoms, newly diagnosed with HIV, CD4 109. Induced sputum culture positive *M. tuberculosis*. CXR shows extensive mediastinal and hilar lymphadenopathy, and a small left sided pleural effusion. (**e**) 54 year old man, newly diagnosed HIV with CD4 68 cells/μL. Presented with chest pain and shortness of breath. CXR shows pleural effusion with surrounding nodules on the right. (**f**) 39 year old woman, CD4 unknown on second line ART, with radiological abnormalities noticed incidentally after presenting with upper gastrointestinal bleeding. CXR shows classical miliary pattern

listed in Table 1. As a consequence of non-distinctive clinical and radiological features, the differential diagnosis for pulmonary involvement in patients with advanced immunosuppression is much broader compared to those with higher CD4 counts. Besides bacterial and *Pneumocystis jiroveci* pneumonia, which are common causes of respiratory disease in advanced HIV and can be co-pathogens with tuberculosis [33], dimorphic fungal infection, cryptococcosis, nocardiosis, pulmonary cytomegalovirus, and Kaposi sarcoma should also be considered.

Although often accompanied by organ dysfunction, disseminated tuberculosis may have no localising features, and in high burden settings tuberculosis is a common cause of pyrexia of unknown origin in hospitalised patients with HIV [34].

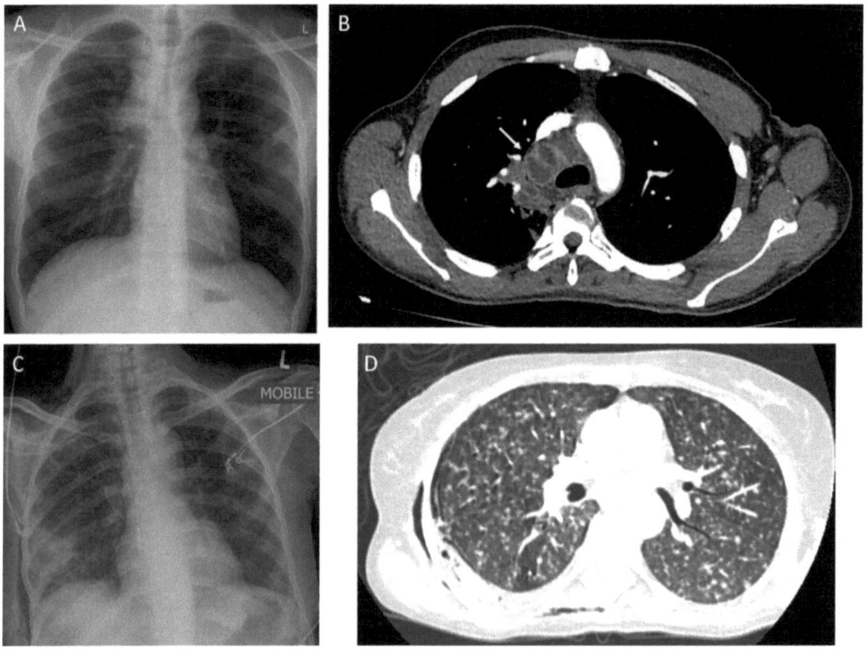

Fig. 3 Illustrative examples of chest X-ray and corresponding CT images in HIV-associated tuberculosis. (**a**, **b**) 38 year old man, CD4 297 on second line ART. Previous pulmonary tuberculosis. Presented with constitutional symptoms and persistent fever. Extensive right mediastinal adenopathy on CXR and CT chest which is ring enhancing with necrotic centres (arrow). (**c**, **d**) 37 year old woman, CD4 count 56 cells/μL, ART-naïve. Admitted to hospital after sustaining polytrauma. Found to have persistent fever with extensive bilateral nodular infiltrates on CXR (**c**) and CT chest (**d**). Sputum Xpert was positive

Anaemia is nearly universal [35], often profound enough to be seen on examination. All EPTB presentations (discussed below in section "Specific Clinical Syndromes Associated with EPTB") may be found, but findings are generally more subtle: for example, pleural effusions are common but may be small and bilateral; traces of ascitic fluid below the threshold for shifting dullness, and small pericardial effusions not detectable on examination. These effusions can be difficult to distinguish from the effects of hypoalbuminaemia. Adenopathy—particularly nodes >2 cm at multiple sites – are predictive of tuberculosis BSI in high HIV-tuberculosis burden settings. Significant hepato- or splenomegaly may be detected, but tender right upper quadrant or generalised abdominal tenderness without detectable enlargement is more common. Ultrasound imaging is frequently available in low resource settings and can give important clues to tuberculosis BSI, including splenic microabscesses and abdominal adenopathy (Fig. 4) [Niel van Hoving, unpublished]. Cold abscesses and musculoskeletal disease are rare in tuberculosis BSI (but may develop later in the context of IRIS). Unlike in classic miliary tuberculosis, retinal involvement is very uncommon in HIV-associated tuberculosis BSI.

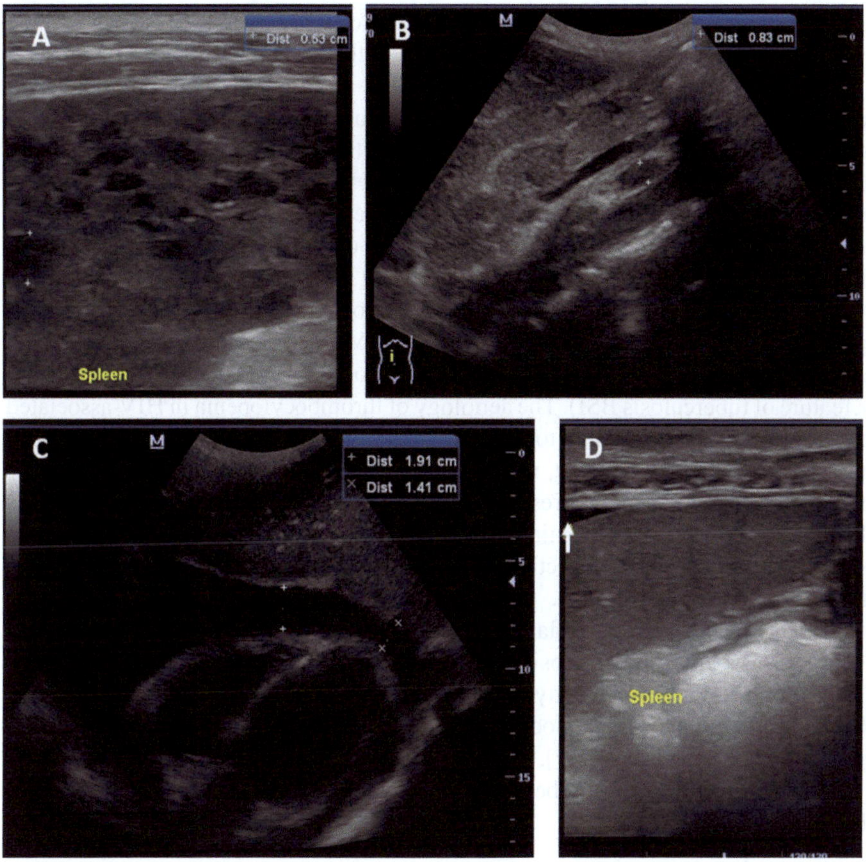

Fig. 4 Illustrative examples of ultrasound images in HIV-associated tuberculosis. (**a**, **b**) 36 year old man, presenting with cough, fever, night sweats and weight loss. Interrupted ART, CD4 count = 1. Ultrasound findings: Splenic micro abscesses (max diameter = 0.53 cm) (**a**); para-aortic lymph node below take-off of superior mesenteric artery (max diameter = 0.83 cm) (**b**). (**c**) 42 year old man, presenting with cough, night sweats and weight loss. ART-naïve, CD4 count = 164. Ultrasound findings: Pericardial effusion (1.41 cm at apex). (**d**) 52 year old man, presenting with cough, fever, night sweats and weight loss. On ART, CD4 count = 68. Ultrasound findings: Hyperechoic splenic lesions and ascites (arrow)

Routine Laboratory Investigations

As mentioned previously, hyponatraemia occurs frequently in severe HIV-associated tuberculosis. Tuberculosis is a well-known cause of inappropriate anti-diuretic hormone excretion, however, a diagnosis of SIADH is unfounded in a patient who is clinically fluid-depleted, and most tuberculosis BSI patients' low sodium levels respond to normal saline infusion. Acute kidney injury is usually reversible with extracellular fluid replenishment. Aggressive fluid resuscitation (fluid boluses totalling more than 2L in the 6 h following presentation) should be avoided as this led to worse outcomes in a randomised controlled trial of sepsis bundle of care in Zambia which recruited a large proportion of patients with tuberculosis BSI [36].

Anaemia associated with tuberculosis BSI is multifactorial, but predominantly caused by an inflammatory response to tuberculosis [37]. This is mediated by hepcidin, which sequesters free iron in an attempt to limit extracellular bacterial growth [35]. Nutritional deficiency, bone marrow suppression from HIV itself and infiltration by tuberculosis are other contributors to anaemia, and may lead to pancytopaenia. Blood transfusions, which anecdotally have been linked to deterioration in HIV-associated tuberculosis, appear to be safe [38], although optimal transfusion thresholds have not been defined. Thrombocytopenia, and a downward trend in platelet count, is a regular feature of severe HIV-associated tuberculosis (and, by contrast, the reactive thrombocytosis of tuberculosis in HIV-negative patients is not a feature of tuberculosis BSI). The aetiology of thrombocytopenia in HIV-associated tuberculosis is likely multifactorial, but increased consumption and destruction of platelets is the leading culprit, with bone-marrow infiltration being another major factor when pancytopenia is present. A high proportion of patients with tuberculosis BSI have (sub-clinical) disseminated intravascular coagulopathy if full coagulation screens are checked [39]. Active bleeding is uncommon; symptomatic venous thromboembolism is common.

Biochemical markers of inflammation (e.g. CRP) are reliably raised in inpatients with HIV-associated tuberculosis; those with BSI generally have CRP greater than 100 mg/L. However, specificity is poor, and cannot distinguish from other severe bacterial infections or pneumocystis pneumonia, even in settings with high prior probability of tuberculosis [11]. Neutrophilia (or total white cell count) greater than 12×10^9/L is not typical of tuberculosis BSI and suggests (possibly superimposed) bacterial infection.

Approach to Diagnosis of Disseminated Tuberculosis in Inpatients with Advanced HIV

In contrast to the chronic indolent course of pulmonary tuberculosis in immunocompetent people, disseminated tuberculosis can progress rapidly in patients with advanced HIV, necessitating more urgent diagnosis and treatment initiation. Risk of tuberculosis increases exponentially as CD4 counts fall and degree of immunosuppression is a factor in empiric management decisions. However, other severe infections that occur more commonly at lower CD4 counts (such as invasive pneumococcal disease, salmonellosis, pneumocystis) may have similar presentations and are important considerations (Table 3).

WHO has proposed an algorithm to diagnose smear-negative tuberculosis in seriously ill patients with HIV and suspected tuberculosis [40]. “Suspected tuberculosis” is defined by presence of TB symptoms (cough, fever, weight loss, night sweats), while “seriously-ill” is defined by the presence of at least one “danger

Table 3 Differential diagnosis for disseminated HIV-associated tuberculosis at low CD4 counts

Disease	Characteristics
Bacterial sepsis	Invasive pneumococcal disease and non-typhi salmonella are important causes, and may present with a non-specific 'sepsis syndrome' indistinguishable from tuberculosis BSI. Hypotension/shock unresponsive to fluid replacement may be present which is unusual with tuberculosis
Disseminated non-tuberculous mycobacterial infection	Distinguished from disseminated tuberculosis by prominent weight loss, hepatosplenomegaly, and cytopenias. May cause chronic diarrhoea due to small bowel involvement. Diagnosed by culture of NTM organism from sterile site. Usually seen only in patients with CD4 count <50 cells/μL
Lymphoma, multicentric Castleman's disease	Non-specific systemic 'B' symptoms and fever with peripheral and/or intra-thoracic or intra-abdominal adenopathy, cytopenias
Disseminated dimorphic fungal infection	Prominent systemic symptoms (weight loss and fever), skin lesions, hepatosplenomegaly, cytopenias. Nodular infiltration (can appear miliary) on chest X-ray
Pneumocystis pneumonia	Subacute to acute respiratory symptoms with prominent dyspnoea and tachypnoea. Typically normal auscultation but desaturation on mild exertion. Chest X-ray shows bilateral 'ground glass' infiltrates, without adenopathy or pleural effusions, and may cause pneumatoceles, pneumothorax or pneumomediastinum
Kaposi sarcoma	In addition to pulmonary involvement (described above), may manifest as nodal disease, organomegaly, serositis, and/or gastrointestinal lesions with bleeding. Majority have visible mucocutaneous involvement
Cytomegalovirus	Relatively uncommon. Manifests in a specific organ site, mainly as colitis and/or retinitis, and very rarely pneumonitis. Presents with fever and symptoms relating to involved organ, generally in patients with CD4 count <50 cells/μL

sign": respiratory rate >30, heart rate >120, temperature >39 °C, and inability to walk unaided. As discussed above the TB symptom screening rule has limited or no value in critically-ill inpatients. Two of the WHO danger signs, inability to walk and temperature >39 °C, were significant predictors of tuberculosis in a large cohort of HIV-infected inpatients in South Africa [41]. In this study a likely radiological diagnosis of tuberculosis (assessed by an experienced radiologist) was also independently associated with confirmed infection. However, classic symptoms of weight loss and subjective fever had poor specificity, suggesting that screening should be applied more broadly in severely ill patients with HIV and a high pre-test probability of tuberculosis; in some settings this may be up to 30% of hospitalised HIV-infected patients regardless of presenting complaint [42]. Only one of the WHO danger-signs—inability to walk unaided—has empirical support as a predictor of mortality in HIV-associated TB [43].

Sputum Xpert testing is recommended as the initial diagnostic procedure in the WHO algorithm, but samples are difficult to obtain from critically unwell, non-ambulant patients even with availability of sputum induction. Consequently, the real-world diagnostic yield of sputum Xpert testing in this patient population is low, identifying only 28% of confirmed tuberculosis cases, and fewer than 20% of those with confirmed tuberculosis BSI, in one cohort from a high burden setting [42, 44]. Because of this difficulty in obtaining samples from extrapulmonary sites, accurate clinical prediction rules for tuberculosis that use easily obtainable clinical parameters and incorporate point-of-care or rapid laboratory diagnostics are required, particularly in resource-limited settings where most cases of severe HIV-associated tuberculosis occur. Application of rapid diagnostic tests for tuberculosis to non-sputum samples has thus emerged as an important strategy in the assessment of acutely unwell patients with advanced HIV.

POC urine-lipoarabinomannan (u-LAM) is highly specific for tuberculosis [45], can provide a rapid diagnosis, and greatly increases diagnostic yield of tuberculosis in hospitalised patients with HIV [44]. Amongst HIV-infected inpatients with suspected tuberculosis in a high burden setting, u-LAM detected two thirds of confirmed cases overall and half of cases with negative sputum smear microscopy [46]. In another study involving unselected HIV-infected medical admissions with a high prevalence of EPTB, u-LAM detected 40% of confirmed tuberculosis cases [42]. When combined with Xpert testing of concentrated urine (spun in a centrifuge with testing of the residual pellet) the overall sensitivity may rise to 70%. For those with CD4 counts <100 cells/μL the combined urine tests plus sputum Xpert had the ability to rapidly diagnose tuberculosis in ~80% of unselected HIV-infected admissions [42, 47], and their use was associated with substantial absolute reduction in mortality for critically ill subgroups [26].

Because of non-specific clinical and laboratory features of tuberculosis, low diagnostic yield of sputum Xpert, and need for urgent treatment initiation, POC u-LAM is indicated in all non-ambulant or hospitalised HIV-infected patients, irrespective of CD4 count. A positive u-LAM is an excellent rule-in test in high burden settings [47], and should prompt rapid initiation of antituberculosis therapy [48]. Although not yet recommended by WHO, urine Xpert has good specificity for tuberculosis in HIV-infected patients [42, 47], and is increasingly recognised as a useful test to rapidly identify inpatients with tuberculosis, particularly for patients who are u-LAM negative. Xpert testing also performs well on needle aspirates of lymph nodes (the sensitivity is lower on pleural and pericardial fluid), and therefore can also be useful when lymphadenopathy is present in critically unwell patients.

POC ultrasound is an inexpensive and non-invasive method to detect indirect evidence of disseminated tuberculosis. Features such as intra-abdominal adenopathy, ascites, and splenic micro-abscesses (Fig. 4) have good specificity for disseminated tuberculosis in high prevalence settings [Niel van Hoving, unpublished], and can be used as a basis to initiate empiric therapy. It is, however, important to recognise that none of these rapid diagnostics have 100% sensitivity in (culture confirmed) tuberculosis BSI patients: negative tests do not rule out tuberculosis and their use in combination is advised to maximise yield. Despite this limitation, negative u-LAM (particularly in combination with a negative urine Xpert and normal abdominal ultrasound examination) identifies patients at lower risk of early mortality [44, 45, 47, 49] in whom a definitive diagnosis of tuberculosis can be sought. In these situations,

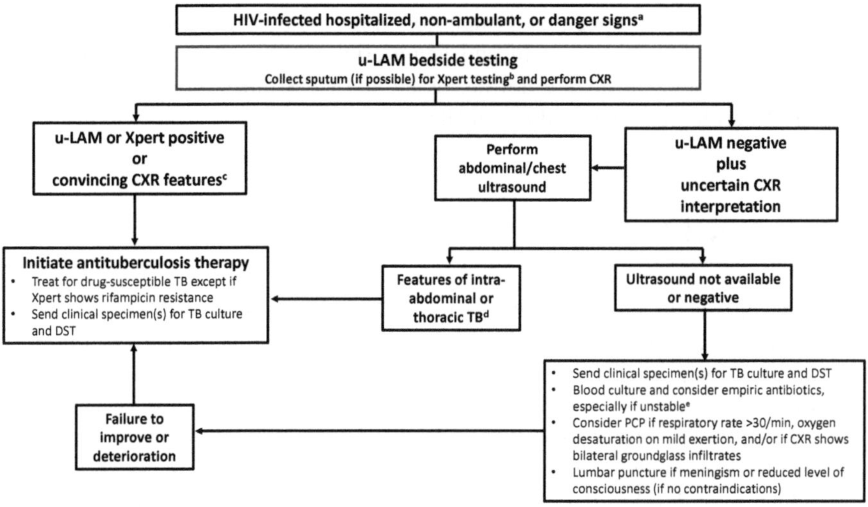

Fig. 5 Diagnostic approach for non-ambulant HIV-infected patients (**a**) Danger signs: inability to walk, respiratory rate > 30 breaths/min, heart rate > 120 beats/min, and temperature > 39°C. (**b**) Perform urine Xpert testing if available. (**c**) Miliary pattern, nodular infiltrate, mediastinal adenopathy, cavitations, nodular infiltrate. (**d**) Intra-abdominal adenopathy, ascites, splenic lesions, pleural or pericardial effusion. (**e**) Hypotension not responding to fluids, hypoxia, hyperlactaemia, decreased level of consciousness. *CXR* chest x-ray, *DST* drug susceptibility testing, *u-LAM* urine lipoarabinomannan, *PCP* pneumocystis pneumonia

assessment of clinical response to a trial of empiric antibiotic therapy is a helpful diagnostic manoeuvre: failure to respond or deterioration should prompt treatment for presumptive tuberculosis. Other opportunistic infections should be simultaneously investigated in patients with CD4 <200 cells/μL. Regardless of whether empiric antituberculosis therapy is initiated, an attempt should always be made to confirm the tuberculosis diagnosis and confirm rifampicin susceptibility with molecular and/or microbiological tests. These include mycobacterial cultures of induced sputum, blood, and urine if no localised disease is apparent. Culture and histological examination of lymph nodes, pleura, and bone marrow aspirate may be indicated under appropriate clinical circumstances. An algorithm for the evaluation of tuberculosis in non-ambulatory patients is shown in Fig. 5. Note, this approach applies to patients who reside in or originate from high-burden settings; the positive predictive value of rapid diagnostics in countries with low tuberculosis prevalence is limited and there is a higher threshold for empiric antituberculosis treatment initiation in those settings.

Specific Clinical Syndromes Associated with EPTB

Extra-pulmonary tuberculosis (EPTB) is common with advanced HIV, with or without concomitant pulmonary involvement [50]. Virtually any organ system can be affected, but the most frequent discreet sites are lymph nodes, pleura, intra-abdominal (peritoneal, intestinal, lymph node), the central nervous system (CNS), and pericardium. Skeletal system and genitourinary tuberculosis occur less frequently. These latter forms are not discussed and CNS tuberculosis is covered in chapter "Neurological Tuberculosis in HIV".

Tuberculous Adenitis

Lymphadenitis is the most frequent form of EPTB. In advanced HIV infection this is often multifocal, involving mediastinal and intra-abdominal nodes, with prominent constitutional symptoms. Extrathoracic lymph nodes are affected in around a fifth of patients with HIV-associated tuberculosis, commonly in the cervical and axillary regions. Involved nodes are typically soft and fixed to underlying tissue. Groups of nodes can become matted together and coalesce to form massive swellings which may develop fluctuant areas and open onto overlying skin as chronic discharging sinus tracts. Empiric therapy is often considered in high prevalence settings and with compatible clinical presentations, especially when other features of tuberculosis are present (such as constitutional symptoms and chest X-ray abnormalities).

A rapid diagnosis can be made by Xpert testing of needle aspirates [51, 52]; *M. tuberculosis* culture should be performed on a node biopsy specimen if Xpert testing is negative or unavailable. Lymphoma is the most important alternative diagnosis to consider in HIV-infected patients with adenopathy and constitutional symptoms, and should be excluded with an excisional lymph node biopsy particularly in those not improving on empiric antituberculous therapy. Other causes of significant adenopathy in HIV include Kaposi sarcoma, multicentric Castleman's disease and infection due to cryptococcus or dimorphic fungi. Malignancy is an important consideration, particularly in older patients who fail to respond to empiric therapy for presumptive tuberculous adenitis.

Pleural Tuberculosis

Pleural effusions are more common in HIV-associated tuberculosis than in HIV-uninfected patients and occur more frequently in those with CD4 cell counts <200 cells/μL [53, 54]. Symptoms include pleuritic chest pain and dry cough, which may progress to cause dyspnoea as the effusion enlarges; the indolent nature of the disease can allow effusions to become massive, occasionally leading to respiratory compromise and requiring urgent intercostal needle drainage. Simple pleural effusions may become complicated to form empyema, either due to superadded bacterial infection or from tuberculosis itself. HIV-infected patients with pleural tuberculosis have an increased frequency of systemic symptoms and more severe disease than HIV-uninfected patients [54]. Diagnosis usually involves a combination of clinical, biochemical, and microbiological testing. Chest radiology most commonly reveals a unilateral effusion (although bilateral effusions are more common in HIV), that varies in size from obliteration of the costophrenic angle to complete whiteout of a hemithorax; HIV-infection is associated with a higher frequency of concomitant parenchymal disease [54, 55].

The pleural fluid profile is not different in HIV: exudative with an elevated white cell count dominated by lymphocytes [54]. Adenosine deaminase (ADA), a cheap and accessible T-lymphocyte enzyme assay, has a specificity and sensitivity of ≥90% for tuberculosis at a cutoff value of ≥35 μm/L [56], and performs well even at low CD4 counts [57]. In high burden settings, the finding of a unilateral lymphocytic pleural effusion with elevated ADA in a young HIV-infected person has a very high positive predictive value for tuberculosis, and can be used as a basis for empiric antituberculosis therapy. Mycobacterial load in pleural fluid is low, and culture is usually negative; Xpert also performs poorly in pleural fluid [51, 52]. The definitive diagnostic procedure for suspected tuberculous pleural effusion is pleural biopsy, which has a high culture yield in HIV [58] and may also provide strongly supportive histological evidence (granulomatous inflammation and caseous necrosis). Other causes of pleural effusion in HIV include parapneumonic effusion (due to bacterial organisms), Kaposi sarcoma, cryptococcosis and primary effusion lymphoma; these (plus drug-resistant tuberculosis) should be excluded by pleural biopsy in patients not responding to empiric antituberculosis therapy for pleural effusion.

Abdominal Tuberculosis

Abdominal involvement is a common feature of HIV-associated tuberculosis, occurring either in association with disseminated disease or as a distinct syndrome. Any structure of the gastrointestinal tract may be affected, and the clinical presentation is strongly influenced by predominant disease site [59]. Abdominal symptoms and signs are non-specific in the context of disseminated tuberculosis, and include generalised pain, tenderness, and fever, reflecting peritoneal disease, intra-abdominal adenopathy, and hepato-splenic microabscesses. If present, organomegaly and ascites are mild. Massive hepatosplenomegaly is not expected in abdominal tuberculosis, and should raise suspicions of alternative diagnoses such as disseminated non-tuberculous mycobacteria, deep fungal infection, or lymphoma in HIV-infected patients. Abdominal tuberculosis may present acutely with a syndrome mimicking bacterial peritonitis. Perforation from ileal tuberculosis, and invasive salmonellosis, should be considered. Urgent treatment should be started with third generation cephalosporin (after performance of blood cultures), along with imaging and surgical review if available.

Yields from culture and Xpert testing of ascitic fluid are low, and diagnosis depends on detection of tuberculosis from other sites or indirect evidence. If ascites is sufficiently large for drainage, the fluid is characteristically exudative and lymphocytic. ADA levels ≥40 IU/L have a high sensitivity (100%) and specificity

(97%) for tuberculous peritonitis [60]. Use of abdominal ultrasound is an important diagnostic strategy for non-ambulant patients with HIV-associated tuberculosis: the presence of three abnormalities on POC ultrasound (including intra-abdominal adenopathy, ascites, splenic lesions, pleural or pericardial effusion) has a positive likelihood ratio >4.5, increasing the post-test probability of tuberculosis to >70% in high burden settings [Niel van Hoving, unpublished].

Tuberculous Pericarditis

The incidence of tuberculous pericarditis has risen dramatically in HIV-endemic regions. In developing countries, tuberculosis is responsible for over 50% of cases of pericarditis, and up to 75% of patients with large pericardial effusions in sub-Saharan Africa are HIV-infected, with tuberculosis the cause in most [61]. Pericardial tuberculosis manifests as three clinical syndromes, namely pericardial effusion (the most frequent), constrictive pericarditis, or as effusive-constrictive disease [62]. In addition to non-specific symptoms there is broad overlap of examination findings, and these syndromes can be difficult to differentiate. Chest pain, cough, reduced effort tolerance and classical constitutional symptoms associated with tuberculosis are common. Large pericardial effusions resulting in cardiac tamponade may mimic heart failure: cardiac dullness, pericardial friction rub, and pulsus paradoxus may be elicited. These signs, plus diastolic lift, may also be present in constrictive disease. In HIV, pericardial tuberculosis is more frequently associated with disseminated disease, as well as more severe manifestations including myopericarditis, dyspnoea, and haemodynamic instability [62]. HIV-infected patients also have a higher 6-month mortality (40%) than those who are HIV-negative [63].

Diagnosis of tuberculosis pericarditis is challenging, requiring a combination of clinical features, electrocardiography, and cardiac imaging. Specific physical signs and electrocardiographic abnormalities may provide clues to the presence of pericardial effusion, but confirmation requires cardiac ultrasound (Fig. 4c) (or echocardiography when available). Chest radiography is a useful adjunct to distinguish tuberculous pericarditis from other causes, with evidence of active pulmonary tuberculosis in 30% [64]. Intrapericardial fibrin strands and pericardial thickening on cardiac ultrasound or echocardiography are highly specific for tuberculosis [65, 66]. Other possible causes of pericardial effusion in HIV include pyogenic (may be as co-infection with tuberculosis), cryptococcosis, Kaposi sarcoma and lymphoma, as well as malignancy (particularly in older patients with higher CD4 counts). Therefore, pericardiocentesis is recommended in all patients with suspected tuberculous pericarditis, even when the diagnosis is suggested by imaging, provided effusion is sufficiently large. As with tuberculous pleural effusions and ascites, pericardial fluid is typically exudative with lymphocyte-predominant pleocytosis. The sensitivity (96%) and negative likelihood ratio (0.05) of ADA at a cutoff ≥35 IU/L was excellent in a South African cohort with pericardial effusion and a high prevalence of HIV infection, making this a valuable indirect test for a tuberculous aetiol-

Table 4 Diagnostic criteria for HIV-associated tuberculous pericarditis in high burden settings (Adapted from Mayosi BM, Burgess LJ, Doubell AF. Tuberculous pericarditis. Circulation 2005; 112(23): 3608–16)

Definite tuberculous pericarditis	Probable tuberculous pericarditis
Positive microbiological (AFBs or positive *M. tuberculosis* culture) or molecular (Xpert) test on pericardial fluid or tissue.	Presence of pericardial effusion plus evidence (microbiological, molecular, or antigen) of tuberculosis elsewhere
	Lymphocyte-predominant pericardial fluid with ADA ≥ 35 IU/L
	Caseous granulomatous inflammation with or without visible AFBs on histological examination of pericardial tissue
	Presence of pericardial effusion without confirmed tuberculosis and good response to empiric antituberculosis therapy

ogy [67]. Yield of smear microscopy and culture on pericardial fluid is variable but generally low [68], and overall diagnostic accuracy of Xpert is disappointing (sensitivity 64%) precluding it as a rule-out test [67]. Pericardial biopsy should be considered in equivocal cases, older patients, and where there is poor clinical response to empiric antituberculosis therapy. Proposed diagnostic criteria for tuberculous pericarditis are shown in Table 4.

Conclusions and Future Directions

HIV-associated tuberculosis is a heterogenous disease that confronts clinicians with substantial diagnostic challenge. Early recognition and treatment is urgent because of more severe manifestations and rapid progression in this population, particularly at low CD4 counts. Urine-based rapid diagnostics are an important advance but are not currently able to reliably exclude tuberculosis. Accurate clinical decision rules that integrate POC diagnostic tests on accessible clinical samples plus imaging (chest X-ray and ultrasound) are needed in high-burden, resource-limited settings to assist clinicians in early identification of HIV-associated tuberculosis. These should be validated in prospective cohort studies to assess impact on clinical outcomes and understand their optimal use. Machine learning algorithms are already being applied to large clinical and radiological datasets and in the near future may inform treatment decisions through widely-available platforms such as smartphone applications. For now, clinicians should have an especially high index of suspicion for tuberculosis in any HIV-infected patient with rapid loss of weight and a reduced level of function. The decision to start empiric antituberculosis therapy depends mainly on the severity of the clinical presentation and access to diagnostic tests, with a lower threshold in ill patients or those with a rapidly deteriorating condition. When starting empiric treatment, clinicians should consider and exclude important alternative diagnoses and monitor patients carefully for response to treatment.

References

1. Barry CE 3rd, Boshoff HI, Dartois V et al (2009) The spectrum of latent tuberculosis: rethinking the biology and intervention strategies. Nat Rev Microbiol 7:845–855
2. Gupta RK, Lucas SB, Fielding KL et al (2015) Prevalence of tuberculosis in post-mortem studies of HIV-infected adults and children in resource-limited settings: a systematic review and meta-analysis. AIDS (London, England) 29:1987–2002
3. Getahun H, Kittikraisak W, Heilig CM et al (2011) Development of a standardized screening rule for tuberculosis in people living with HIV in resource-constrained settings: individual participant data meta-analysis of observational studies. PLoS Med 8:e1000391
4. Hamada Y, Lujan J, Schenkel K et al (2018) Sensitivity and specificity of WHO's recommended four-symptom screening rule for tuberculosis in people living with HIV: a systematic review and meta-analysis. The Lancet HIV 5:e515–ee23
5. Wood R, Middelkoop K, Myer L et al (2007) Undiagnosed tuberculosis in a community with high HIV prevalence: implications for tuberculosis control. Am J Respir Crit Care Med 175:87–93
6. Lawn SD, Kerkhoff AD, Vogt M et al (2013) Diagnostic and prognostic value of serum C-reactive protein for screening for HIV-associated tuberculosis. Int J Tuberc Lung Dis 17:636–643
7. Drain PK, Mayeza L, Bartman P et al (2014) Diagnostic accuracy and clinical role of rapid C-reactive protein testing in HIV-infected individuals with presumed tuberculosis in South Africa. Int J Tuberc Lung Dis 18:20–26
8. Wilson D, Badri M, Maartens G (2011) Performance of serum C-reactive protein as a screening test for smear-negative tuberculosis in an ambulatory high HIV prevalence population. PLoS One 6:e15248
9. Nyamande K, Lalloo UG (2006) Serum procalcitonin distinguishes CAP due to bacteria, Mycobacterium tuberculosis and PJP. Int J Tuberc Lung Dis 10:510–515
10. Yoon C, Davis JL, Huang L et al (2014) Point-of-care C-reactive protein testing to facilitate implementation of isoniazid preventive therapy for people living with HIV. J Acquir Immune Defic Syndr 1999 65(5):551–556
11. Mendelson F, Griesel R, Tiffin N et al (2018) C-reactive protein and procalcitonin to discriminate between tuberculosis, *Pneumocystis jirovecii* pneumonia, and bacterial pneumonia in HIV-infected inpatients meeting WHO criteria for seriously ill: a prospective cohort study. BMC Infect Dis 18:399
12. Van Dyck P, Vanhoenacker FM, Van den Brande P et al (2003) Imaging of pulmonary tuberculosis. Eur Radiol 13:1771–1785
13. Perlman DC, el-Sadr WM, Nelson ET et al (1997) Variation of chest radiographic patterns in pulmonary tuberculosis by degree of human immunodeficiency virus-related immunosuppression. Clin Infect Dis 25:242–246
14. Pedro-Botet J, Gutierrez J, Miralles R et al (1992) Pulmonary tuberculosis in HIV-infected patients with normal chest radiographs. AIDS (London, England) 6:91–93
15. Organization WH (2007) Improving the diagnosis and treatment of smear-negative pulmonary and extrapulmonary tuberculosis among adults and adolescents. WHO, Geneva
16. Ngwira LG, Corbett EL, Khundi M et al (2019) Screening for tuberculosis with Xpert MTB/RIF versus fluorescent microscopy among adults newly diagnosed with HIV in rural Malawi: a cluster randomized trial (Chepetsa). Clin Infect Dis 68(7):1176–1183
17. Hermans SM, Babirye JA, Mbabazi O et al (2017) Treatment decisions and mortality in HIV-positive presumptive smear-negative TB in the Xpert MTB/RIF era: a cohort study. BMC Infect Dis 17:433
18. Calligaro GL, Theron G, Khalfey H et al (2015) Burden of tuberculosis in intensive care units in Cape Town, South Africa, and assessment of the accuracy and effect on patient outcomes of the Xpert MTB/RIF test on tracheal aspirate samples for diagnosis of pulmonary tuberculosis:

a prospective burden of disease study with a nested randomised controlled trial. Lancet Respir Med 3:621–630

19. Sigel K, Makinson A, Thaler J (2017) Lung cancer in persons with HIV. Curr Opin HIV AIDS 12:31–38
20. Hanifa Y, Toro Silva S, Karstaedt A et al (2019) What causes symptoms suggestive of tuberculosis in HIV-positive people with negative initial investigations? Int J Tuberc Lung Dis 23(2):157–165
21. Kaplan R, Hermans S, Caldwell J et al (2018) HIV and TB co-infection in the ART era: CD4 count distributions and TB case fatality in Cape Town. BMC Infect Dis 18:356
22. Osler M, Hilderbrand K, Goemaere E et al (2018) The continuing burden of advanced HIV disease over 10 years of increasing antiretroviral therapy coverage in South Africa. Clin Infect Dis 66:S118–Ss25
23. Crump JA, Reller LB (2003) Two decades of disseminated tuberculosis at a university medical center: the expanding role of mycobacterial blood culture. Clin Infect Dis 37:1037–1043
24. Janssen S, Schutz C, Ward A et al (2017) Mortality in severe human immunodeficiency virus-tuberculosis associates with innate immune activation and dysfunction of monocytes. Clin Infect Dis 65:73–82
25. Griesel R, Stewart A, van der Plas H et al (2018) Optimizing tuberculosis diagnosis in HIV-infected inpatients meeting the criteria of seriously Ill in the WHO algorithm. Clin Infect Dis 66(9):1419–1426
26. Gupta-Wright A, Corbett EL, van Oosterhout JJ et al (2018) Rapid urine-based screening for tuberculosis in HIV-positive patients admitted to hospital in Africa (STAMP): a pragmatic, multicentre, parallel-group, double-blind, randomised controlled trial. Lancet 392:292–301
27. Maartens G, Willcox PA, Benatar SR (1990) Miliary tuberculosis: rapid diagnosis, hematologic abnormalities, and outcome in 109 treated adults. Am J Med 89:291–296
28. Jacob ST, Pavlinac PB, Nakiyingi L et al (2013) *Mycobacterium tuberculosis* bacteremia in a cohort of HIV-infected patients hospitalized with severe sepsis in Uganda-high frequency, low clinical suspicion [corrected] and derivation of a clinical prediction score. PLoS One 8:e70305
29. Muchemwa L, Shabir L, Andrews B et al (2017) High prevalence of *Mycobacterium tuberculosis* bacteraemia among a cohort of HIV-infected patients with severe sepsis in Lusaka, Zambia. Int J STD AIDS 28:584–593
30. Gleeson LE, Sheedy FJ, Palsson-McDermott EM et al (2016) Cutting edge: *Mycobacterium tuberculosis* induces aerobic glycolysis in human alveolar macrophages that is required for control of intracellular bacillary replication. J Immunol 196:2444–2449
31. Keiper MD, Beumont M, Elshami A et al (1995) CD4 T lymphocyte count and the radiographic presentation of pulmonary tuberculosis. A study of the relationship between these factors in patients with human immunodeficiency virus infection. Chest 107:74–80
32. Greenberg SD, Frager D, Suster B et al (1994) Active pulmonary tuberculosis in patients with AIDS: spectrum of radiographic findings (including a normal appearance). Radiology 193:115–119
33. Wasserman S, Engel ME, Griesel R, Mendelson M (2016) Burden of pneumocystis pneumonia in HIV-infected adults in sub-Saharan Africa: a systematic review and meta-analysis. BMC Infect Dis 16:482
34. Rupali P, Abraham OC, Zachariah A et al (2003) Aetiology of prolonged fever in antiretroviral-naive human immunodeficiency virus-infected adults. Natl Med J India 16:193–199
35. Kerkhoff AD, Meintjes G, Burton R et al (2016) Relationship between blood concentrations of hepcidin and anaemia severity, mycobacterial burden and mortality in patients with HIV-associated tuberculosis. J Infect Dis 213(1):61–70
36. Andrews B, Semler MW, Muchemwa L et al (2017) Effect of an early resuscitation protocol on in-hospital mortality among adults with sepsis and hypotension: a randomized clinical trial. JAMA 318:1233–1240
37. Kerkhoff AD, Meintjes G, Opie J et al (2016) Anaemia in patients with HIV-associated TB: relative contributions of anaemia of chronic disease and iron deficiency. Int J Tuberc Lung Dis 20:193–201

38. Kerkhoff AD, Lawn SD, Schutz C et al (2015) Anemia, blood transfusion requirements and mortality risk in human immunodeficiency virus-infected adults requiring acute medical admission to hospital in South Africa. Open Forum Infect Dis 2:ofv173
39. Janssen S, Schutz C, Ward AM et al (2017) Hemostatic changes associated with increased mortality rates in hospitalized patients with HIV-associated tuberculosis: a prospective cohort study. J Infect Dis 215:247–258
40. World Health Organization (2017) Guidelines for treatment of drug-susceptible tuberculosis and patient care, 2017 update. WHO, Geneva
41. Griesel R, Stewart A, van der Plas H et al (2018) Optimizing tuberculosis diagnosis in human immunodeficiency virus-infected inpatients meeting the criteria of seriously ill in the World Health Organization algorithm. Clin Infect Dis 66:1419–1426
42. Lawn SD, Kerkhoff AD, Burton R et al (2015) Rapid microbiological screening for tuberculosis in HIV-positive patients on the first day of acute hospital admission by systematic testing of urine samples using Xpert MTB/RIF: a prospective cohort in South Africa. BMC Med 13:192
43. Gupta-Wright A, Corbett EL, Wilson D et al (2019) Risk score for predicting mortality including urine lipoarabinomannan detection in hospital inpatients with HIV-associated tuberculosis in sub-Saharan Africa: derivation and external validation cohort study. PLoS Med 16:e1002776
44. Kerkhoff AD, Barr DA, Schutz C et al (2017) Disseminated tuberculosis among hospitalised HIV patients in South Africa: a common condition that can be rapidly diagnosed using urine-based assays. Sci Rep 7:10931
45. Gupta-Wright A, Peters JA, Flach C et al (2016) Detection of lipoarabinomannan (LAM) in urine is an independent predictor of mortality risk in patients receiving treatment for HIV-associated tuberculosis in sub-Saharan Africa: a systematic review and meta-analysis. BMC Med 14:53
46. Shah M, Variava E, Holmes CB et al (2009) Diagnostic accuracy of a urine lipoarabinomannan test for tuberculosis in hospitalized patients in a High HIV prevalence setting. J Acquir Immune Defic Syndr (1999) 52:145–151
47. Lawn SD, Kerkhoff AD, Burton R et al (2017) Diagnostic accuracy, incremental yield and prognostic value of Determine TB-LAM for routine diagnostic testing for tuberculosis in HIV-infected patients requiring acute hospital admission in South Africa: a prospective cohort. BMC Med 15:67
48. World Health Organization (2015) Guidelines on the management of latent tuberculosis infection. WHO, Geneva
49. Lawn SD, Kerkhoff AD, Vogt M et al (2013) HIV-associated tuberculosis: relationship between disease severity and the sensitivity of new sputum-based and urine-based diagnostic assays. BMC Med 11:231
50. Iseman MD (2000) A clinician's guide to tuberculosis. Lippincott Williams & Wilkins, Philadelphia
51. World Health Organization. Automated real-time nucleic acid amplification technology for rapid and simultaneous detection of tuberculosis and rifampicin resistance: Xpert MTB/RIF assay for the diagnosis of pulmonary and extrapulmonary TB in adults and children. Policy update. Geneva, 2013
52. Lawn SD, Mwaba P, Bates M et al (2013) Advances in tuberculosis diagnostics: the Xpert MTB/RIF assay and future prospects for a point-of-care test. Lancet Infect Dis 13:349–361
53. Batungwanayo J, Taelman H, Allen S et al (1993) Pleural effusion, tuberculosis and HIV-1 infection in Kigali, Rwanda. AIDS (London, England) 7:73–79
54. Aljohaney A, Amjadi K, Alvarez GG (2012) A systematic review of the epidemiology, immunopathogenesis, diagnosis, and treatment of pleural TB in HIV-infected patients. Clin Dev Immunol 2012:842045
55. Luzze H, Elliott AM, Joloba ML et al (2001) Evaluation of suspected tuberculous pleurisy: clinical and diagnostic findings in HIV-1-positive and HIV-negative adults in Uganda. Int J Tuberc Lung Dis 5:746–753

56. Aggarwal AN, Agarwal R, Sehgal IS et al (2019) Adenosine deaminase for diagnosis of tuberculous pleural effusion: a systematic review and meta-analysis. PLoS One 14:e0213728
57. Baba K, Hoosen AA, Langeland N et al (2008) Adenosine deaminase activity is a sensitive marker for the diagnosis of tuberculous pleuritis in patients with very low CD4 counts. PLoS One 3:e2788
58. Heyderman RS, Makunike R, Muza T et al (1998) Pleural tuberculosis in Harare, Zimbabwe: the relationship between human immunodeficiency virus, CD4 lymphocyte count, granuloma formation and disseminated disease. Tropical Med Int Health 3:14–20
59. Sharma MP, Bhatia V (2004) Abdominal tuberculosis. Indian J Med Res 120:305–315
60. Riquelme A, Calvo M, Salech F et al (2006) Value of adenosine deaminase (ADA) in ascitic fluid for the diagnosis of tuberculous peritonitis: a meta-analysis. J Clin Gastroenterol 40:705–710
61. Magula NP, Mayosi BM (2003) Cardiac involvement in HIV-infected people living in Africa: a review. Cardiovasc J South Afr 14:231–237
62. Mayosi BM, Wiysonge CS, Ntsekhe M et al (2006) Clinical characteristics and initial management of patients with tuberculous pericarditis in the HIV era: the Investigation of the Management of Pericarditis in Africa (IMPI Africa) registry. BMC Infect Dis 6:2
63. Mayosi BM, Wiysonge CS, Ntsekhe M et al (2008) Mortality in patients treated for tuberculous pericarditis in sub-Saharan Africa. S Afr Med J 98:36–40
64. Reuter H, Burgess LJ, Doubell AF (2005) Role of chest radiography in diagnosing patients with tuberculous pericarditis. Cardiovasc J South Afr 16:108–111
65. George S, Salama AL, Uthaman B et al (2004) Echocardiography in differentiating tuberculous from chronic idiopathic pericardial effusion. Heart 90:1338–1339
66. Liu PY, Li YH, Tsai WC et al (2001) Usefulness of echocardiographic intrapericardial abnormalities in the diagnosis of tuberculous pericardial effusion. Am J Cardiol 87:1133–1135. a10
67. Pandie S, Peter JG, Kerbelker ZS et al (2014) Diagnostic accuracy of quantitative PCR (Xpert MTB/RIF) for tuberculous pericarditis compared to adenosine deaminase and unstimulated interferon-γ in a high burden setting: a prospective study. BMC Med 12:101
68. Mayosi BM, Burgess LJ, Doubell AF (2005) Tuberculous pericarditis. Circulation 112:3608–3616

The Tuberculosis-Associated Immune Reconstitution Inflammatory Syndrome (TB-IRIS)

Irini Sereti, Gregory P. Bisson, and Graeme Meintjes

Abstract Initiation of antiretroviral therapy (ART) in HIV infected people with tuberculosis (TB) significantly improves their overall prognosis, but paradoxical worsening of the clinical or radiographic manifestations of TB can occur during the initial weeks of ART, a phenomenon that is called immune reconstitution inflammatory syndrome or IRIS. Paradoxical TB-IRIS occurs in patients initiating ART while established on TB treatment and presents with systemic and/or localized symptoms and signs. Patients with more severe CD4 lymphopenia, disseminated TB and more rapid initiation of ART after TB diagnosis are at higher risk. TB-IRIS is thought to represent an aberrant inflammatory response in patients with low CD4 counts and high mycobacterial burden when HIV plasma viremia is suppressed leading to hyperactivation of the innate immune system, especially myeloid cells with inflammasome activation. This triggers exuberant TB-specific CD4 T cell responses and release of proinflammatory cytokines likely resulting in activation of tissue macrophages with production of matrix metalloproteinases contributing to tissue pathology. Apart from drainage of suppurative lesions, treatment with prednisone can alleviate many of the symptoms. Prednisone was also shown to reduce the incidence of paradoxical TB-IRIS in a placebo-controlled clinical trial that enrolled high risk patients with CD4 counts less than 100 cells/μL when starting ART.

The work of IS was supported by the Intramural Research Program of NIAID/NIH.

I. Sereti (✉)
National Institutes of Health, National Institute of Allergy and Infectious Diseases, Bethesda, MD, USA
e-mail: isereti@niaid.nih.go

G. P. Bisson
Perelman School of Medicine at the University of Pennsylvania, Philadelphia, PA, USA
e-mail: bisson@mail.med.upenn.edu

G. Meintjes
Wellcome Centre for Infectious Diseases Research in Africa, Institute of Infectious, Disease and Molecular Medicine, University of Cape Town, Cape Town, South Africa

Division of Infectious Diseases and HIV Medicine, Department of Medicine, University of Cape Town, Cape Town, South Africa
e-mail: graemein@mweb.co.za

I. Sereti et al. (eds.), *HIV and Tuberculosis*,
https://doi.org/10.1007/978-3-030-29108-2_6

Keywords Immune reconstitution inflammatory syndrome · Cytokines · Tuberculosis · HIV · Prednisone · Inflammation

Introduction

When patients with advanced HIV infection initiate antiretroviral therapy (ART), provided adherence is optimal and their virus does not harbor primary drug resistance, there is predictably a rapid decline in HIV viral load and recovery of immune function in the vast majority. Since the mid-1990's, combination ART, usually with a combination of 3 drugs, has made it possible to achieve sustained suppression of HIV viral load in plasma to below 50 copies/mL. This allows for both quantitative and qualitative reversal of the immune suppression caused by HIV. After initiation of ART, the majority of patients achieve a CD4 count greater 200 cells/mm^3 [1]. Autran and colleagues [2] demonstrated that there are three phases of T cell reconstitution. There is an early rise of memory CD4 T cells. This is thought to reflect recirculation of cells previously recruited into productively infected tissues once viral replication is suppressed. There is little *de novo* production of immune cells in the first 2–3 months of ART [3]. The second phase is characterized by a reduction in T cell activation with improved CD4 T cell reactivity to recall antigens. There is a later rise of naïve CD4 T cells while CD8 T cells decline [2]. The regeneration of naïve CD4 T cells is accompanied by restoration in the diversity of the CD4 T cell receptor repertoire [4].

Therefore, while it may take years for patients with a very low CD4 count to recover within the normal range, even in the first few weeks of ART there is usually a rise in CD4 counts accompanied by other functional improvements in immune responses involving CD4 cells and other components of the immune system. During early ART when there is rapid decline in HIV viral load and early immune recovery, patients who have concurrent active TB may manifest inflammatory features associated with TB disease. Such immunopathological reactions are thought to result from the recovering immune system reacting to *Mycobacterium tuberculosis* antigen present in tissue at the site of disease, and this condition has been termed the TB-associated immune reconstitution inflammatory syndrome (TB-IRIS).

Two forms of TB-IRIS are recognized: paradoxical and unmasking [5, 6]. Paradoxical TB-IRIS occurs in patients diagnosed with active TB and started on TB treatment prior to starting ART. Typically they have had a favorable clinical response to TB treatment, then after starting ART present with clinical deterioration manifesting as recurrent, worsening or new TB symptoms and signs or radiographic deterioration. Unmasking TB-IRIS occurs in patients with active but undiagnosed TB prior to ART, who develop an exaggerated inflammatory presentation of TB during early ART (Fig. 1). Not all patients diagnosed with active TB during early ART are regarded as having unmasking TB-IRIS, but rather a subset with more inflammatory presentations. Consensus case definitions for both forms of TB-IRIS were developed by the International Network for the Study of HIV-associated IRIS

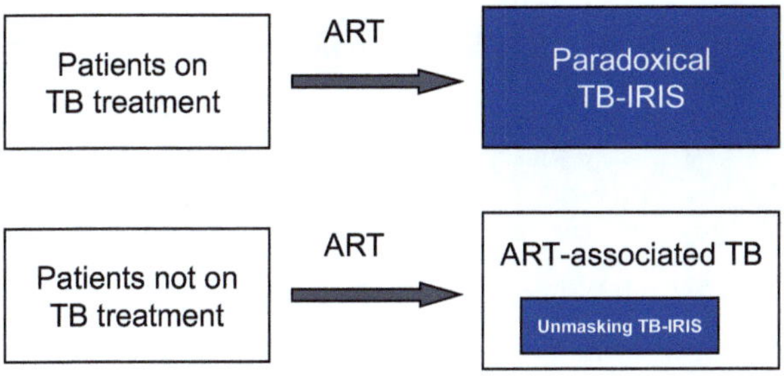

Fig. 1 Schema illustrating timing of initiation of TB treatment and antiretroviral therapy (ART) in relation to different forms of TB-IRIS

(INSHI) [5]. The major focus of this chapter is paradoxical TB-IRIS (Section 1), but in Section 2 we discuss unmasking TB-IRIS. In Section 3 the immune mechanisms underlying TB-IRIS are discussed.

IRIS is also described in association with a wide range of other infections, including fungal infections (e.g. cryptococcosis), viral infections (e.g. hepatitis B), protozoal infections (e.g. toxoplamosis) and malignancies (e.g. Kaposi's sarcoma) [7]. However, TB-IRIS is the most significant form of IRIS in ART programmes globally, given that TB is the most common co-infection affecting patients with advanced HIV and because of the severity of TB-IRIS complications.

Section 1: Clinical Aspects of Paradoxical TB-IRIS

Clinical Features

The most common manifestations observed in paradoxical TB-IRIS are pulmonary and nodal. Pulmonary features manifest with recurrent or worsening respiratory symptoms, such as cough, chest pain and dyspnea. The chest radiograph typically shows worsening pulmonary infiltration (example in Fig. 2). This may include non-confluent infiltrates, consolidation or miliary infiltrates. New cavitation due to TB-IRIS has also been described. Worsening of pulmonary inflammation may also be assessed using FDG PET-CT scanning (examples in Fig. 3). Patients may develop deterioration in pulmonary function assessed by spirometry [8]. Nodal TB-IRIS manifests with new or increasing enlargement of TB lymph nodes. This most frequently affects cervical nodes but may also involve thoracic, abdominal, axillary or inguinal nodes. Frequently, the nodes enlarge with accompanying features of acute inflammation such as red and tender overlying skin (example in Fig. 4). Often, nodes enlarging due to TB-IRIS suppurate and become fluctuant within a few weeks, and chronic draining sinuses may form. TB-IRIS may also manifest with the formation of tubercular abscesses such as psoas abscesses, that may be chronic (example in Fig. 5).

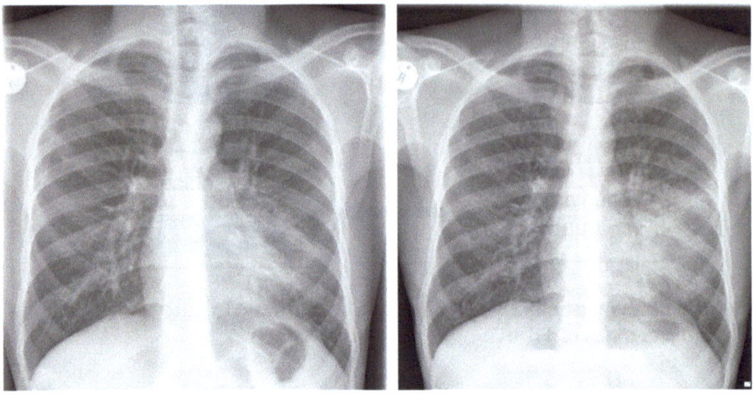

Fig. 2 This patient with HIV-associated pulmonary TB started ART after TB treatment. Three weeks after starting ART he developed recurrent fatigue, drenching night sweats, and left sided pleuritic chest pain and weight loss of 3 kg. Chest radiograph showed extension of the left sided infiltrate (**b**) compared to the baseline radiograph (**a**). The TB-IRIS was treated with prednisone with rapid symptom resolution over the next 2 weeks

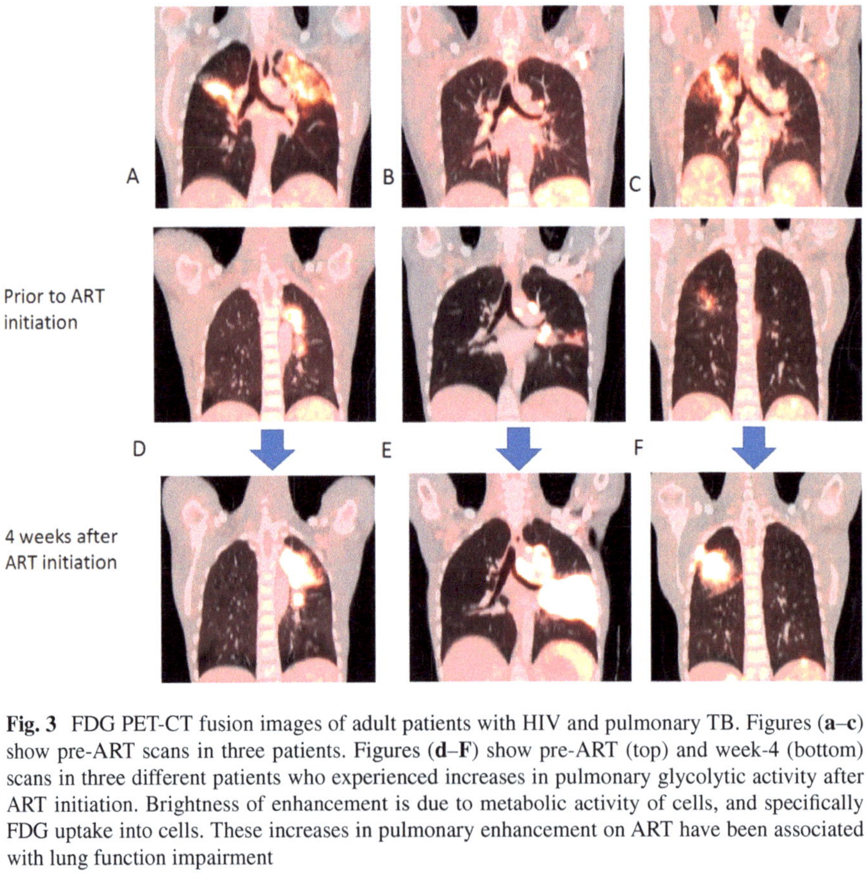

Fig. 3 FDG PET-CT fusion images of adult patients with HIV and pulmonary TB. Figures (**a–c**) show pre-ART scans in three patients. Figures (**d–F**) show pre-ART (top) and week-4 (bottom) scans in three different patients who experienced increases in pulmonary glycolytic activity after ART initiation. Brightness of enhancement is due to metabolic activity of cells, and specifically FDG uptake into cells. These increases in pulmonary enhancement on ART have been associated with lung function impairment

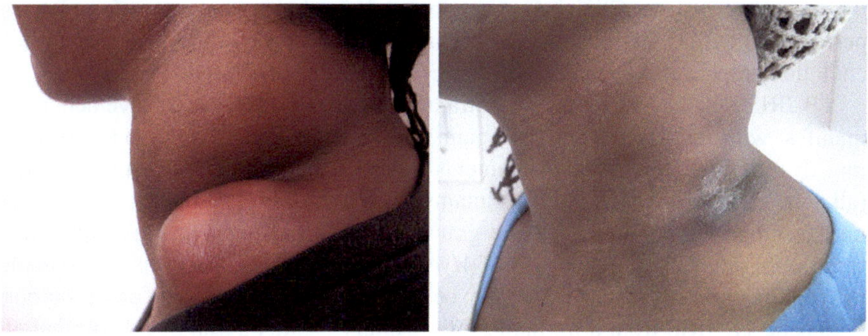

Fig. 4 This patient developed massive enlargement of cervical nodes due to TB-IRIS soon after starting ART while on TB treatment. The nodes had red overlying skin (panel A) and were tender to palpation. After needle aspiration of pus there was chronic drainage through a sinus for several months with eventual decrease in size and subsequent resolution

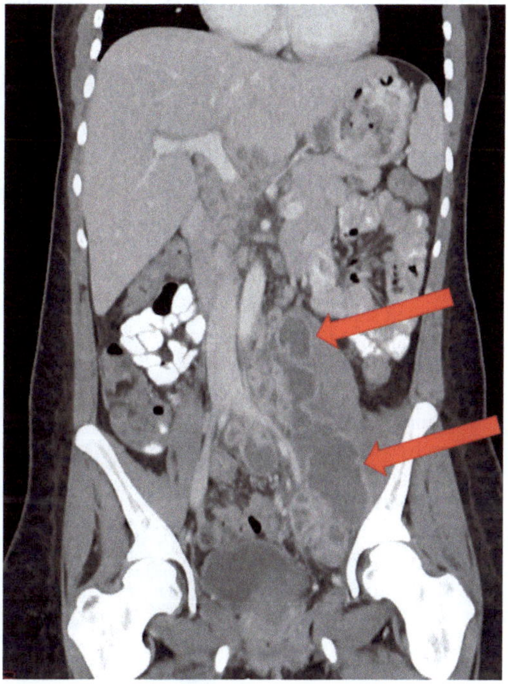

Fig. 5 30-year-old woman with paradoxical TB-IRIS manifesting as severe left thigh pain with inability to walk. Computerized tomography showing an extensive multiloculated psoas abscess that required drainage by Interventional Radiology. More details of the clinical history have been previously reported [54]

Serositis is another common feature of TB-IRIS. This may manifest with new or enlarging pleural effusions, pericardial effusions or ascites. Rapid enlargement of pleural effusions may cause significant dyspnea, while rapid enlargement of pericardial effusions may be life threatening due to development of cardiac tamponade requiring emergency intervention. TB-IRIS may also manifest with the clinical sign of perotinism presumably due to the development of IRIS-related inflammation on the peritoneal surface. Clinicians must obviously exclude other causes (e.g. bowel perforation) before ascribing symptoms and signs to TB-IRIS. Chylous ascites and chylothorax have also been described as complications of TB-IRIS [9].

TB-IRIS may affect a number of intra-abdominal organs. Liver involvement may manifest with right upper quadrant pain, nausea and vomiting and tender hepatomegaly [10]. Liver function derangements are predominantly of cholestatic enzymes (alkaline phosphatase and gamma-glutamyl transferase) [11]. Patients may develop mild jaundice. Liver histology shows granulomas. It is presumed that this inflammatory response is targeting antigens of *Mycobacterium tuberculosis* in liver tissue in the context of disseminated TB. It may be challenging to differentiate presentations of hepatic TB-IRIS from drug-induced liver injury—either a trial of interruption of hepatotoxic medications or a liver biopsy may be required to differentiate. Splenic TB-IRIS may manifest as splenomegaly or abscess formation in the spleen visualized on imaging (example in Fig. 6). There are rare case reports of splenic rupture due to TB-IRIS [12]. Intestinal involvement is not frequently apparent, but there is a case report of intestinal perforation due to TB-IRIS and some patients do develop diarrhea at the time of TB-IRIS onset perhaps reflecting intestinal involvement.

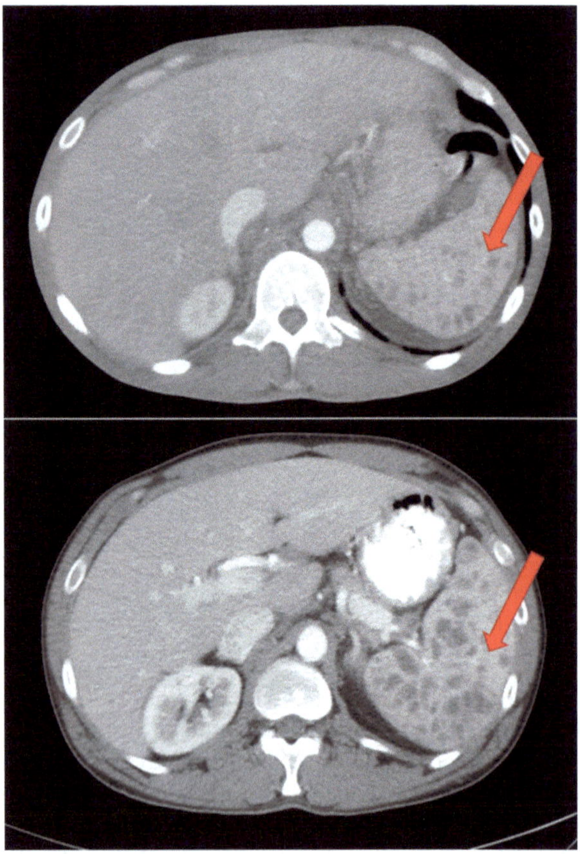

Fig. 6 45-year-old man with disseminated TB and suspected TB multiple splenic abscesses at the start of ART. He developed paradoxical TB IRIS after starting ART manifesting mostly with fevers and left upper quadrant abdominal as well as left shoulder pain. Computerized tomography (D) after ART initiation showed multiple new splenic abscesses. Pain resolved after months of continuing ART and corticosteroids for TB-IRIS treatment

Renal involvement may manifest with renal dysfunction (rising blood urea and creatinine) [13]. A renal biopsy is required to make a definitive diagnosis (a granulomatous infiltrate of the kidney is present) and exclude other causes of renal impairment (e.g. tenofovir nephrotoxicity).

The most severe presentations of TB-IRIS are when it involves the central nervous system. This may take the form of recurrent or new meningitis, formation or enlargement of cerebral tuberculomas or abscesses (often with associated cerebral oedema), and radiculomyelitis presenting with paraparesis [14]. The typical setting for this is a patient diagnosed with TB meningitis and started on TB treatment prior to starting ART, who initially has improvement in neurological symptoms and signs, but has neurologic deterioration after starting ART. However, CNS involvement is sometimes only recognized at the time of TB-IRIS presentation, for example as in a patient who is diagnosed and treated for pulmonary TB who subsequently develops new meningitis due to TB-IRIS after commencing ART or tuberculomas (undiagnosed at ART initiation) flaring with localizing symptoms [14, 15]. In such cases it is presumed that the dissemination of TB to the CNS was subclinical prior to ART, but becomes manifest due to recovering immune responses. One study found that patients who developed TB-IRIS had higher neutrophil numbers in the cerebrospinal fluid (CSF) at the time of TB meningitis diagnosis, and that the development of TBM-IRIS was associated with a recurrent increase in CSF neutrophils during early ART [16]. Neurologic TB-IRIS is discussed in more detail in chapter "Neurological Tuberculosis in HIV".

Patients with TB-IRIS frequently have symptoms and signs of systemic inflammation including fevers, night sweats, tachycardia and weight loss. Less common manifestations that have been reported include ureteric compression, epipidymo-orchitis, arthritis and osteitis (example in Fig. 7). Deep vein thromboses occurring at the time of TB-IRIS have also been reported.

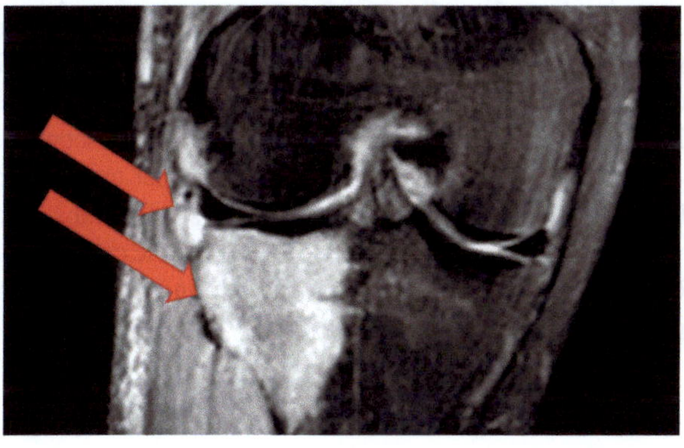

Fig. 7 41-year-old man with miliary TB and right knee TB arthritis with tibial osteomyelitis developed fevers and exacerbation of knee arthritis with swelling and pain 2 weeks after ART initiation. Magnetic resonance tomography (T2) showing bright enhancing lateral tibial lesion with cortical disruption. Further details of the clinical presentation and management have been previously reported [9]

Duration

Early cohort studies of TB-IRIS reported that the median duration of symptoms was 2–3 months [17–19]. A small proportion of cases had symptoms lasting over one year and a case with a recurrence 4 years after ART initiation was also described [20]. A more recent publication [21] that pooled data from 3 prospective studies conducted in South Africa, in which 216 patients with TB-IRIS were included and followed up, reported the median duration of TB-IRIS symptoms was 71 days (interquartile range, 41–113). In 73/181 patients (40.3%) with adequate follow-up data, IRIS duration was longer than 90 days. Six patients (3.3%), mainly with lymph node involvement, had IRIS duration of over 1 year. In multivariate analysis, lymph node TB-IRIS was one of the factors associated with prolonged TB-IRIS.

Diagnosis

The diagnosis of paradoxical TB-IRIS relies on a compatible clinical assessment and exclusion of relevant differential diagnoses pertinent to the individual patient, as there are no confirmatory laboratory tests. The typical scenario is that of a patient with HIV-associated TB who is started on TB treatment and experiences clinical improvement then starts ART and within the first 3 months, and most frequently within the first month, of ART develops new, recurrent or worsening symptoms and inflammatory signs of TB disease. The differential diagnoses to exclude will depend on the clinical presentation and the immunologic status of the patient (particularly CD4 count). For example, in patients who present with recurrent respiratory symptoms and new infiltrates on chest radiograph, the differential diagnosis may include bacterial, viral and pneumocystis pneumonia. In such cases, empiric treatment for bacterial pneumonia and diagnostic work-up for pneumocystis pneumonia may be considered before a diagnosis of TB-IRIS is made. An important differential diagnosis in all presentations is drug-resistant TB or non-adherence to medications, as patients may deteriorate clinically for these reasons and these scenarios may mimic or even co-exist with a TB-IRIS presentation [11]. In order to standardize the diagnosis of TB-IRIS across research studies and to assist clinicians in the diagnosis, INSHI published case definitions for TB-IRIS in 2008. The case definition for paradoxical TB-IRIS is shown in Table 1. This case definition has been validated in several subsequent studies [22–24].

Table 1 INSHI case definition for paradoxical TB-IRIS

There are 3 components to this case definition:
A. Antecedent requirements
Both of the 2 following requirements must be met:
1. Diagnosis of TB: the TB diagnosis was made before starting ART and this should fulfil WHO criteria for diagnosis of smear-positive PTB, smear-negative PTB or extrapulmonary TB
2. Initial response to TB treatment: the patient's condition should have stabilised or improved on appropriate TB treatment before ART initiation—e.g. cessation of night sweats, fevers, cough, weight loss. (Note: this does not apply to patients starting ART within 2 weeks of starting TB treatment since insufficient time may have elapsed for a clinical response to be reported)
B. Clinical criteria
The onset of TB-IRIS manifestations should be within 3 months of ART initiation, re-initiation, or regimen change because of treatment failure. Of the following, at least 1 major criterion or 2 minor clinical criteria are required:
Major criteria
1. New or enlarging lymph nodes, cold abscesses or other focal tissue involvement—e.g. tuberculous arthritis
2. New or worsening radiological features of TB (found by chest X-ray, abdominal US, CT or MRI)
3. New or worsening central nervous system TB (meningitis or focal neurological deficit—e.g. caused by tuberculoma)
4. New or worsening serositis (pleural effusion, ascites, or pericardial effusion)
Minor criteria
1. New or worsening constitutional symptoms such as fever, night sweats, or weight loss
2. New or worsening respiratory symptoms such as cough, dyspnoea, or stridor
3. New or worsening abdominal pain accompanied by peritonitis, hepatomegaly, splenomegaly, or abdominal adenopathy
C. Alternative explanations for clinical deterioration must be excluded if possible
1. Failure of TB treatment due to TB drug resistance
2. Poor adherence to TB treatment
3. Another opportunistic infection or neoplasm (it is particularly important to exclude an alternative diagnosis in patients with smear-negative PTB and extrapulmonary TB where the initial TB diagnosis has not been microbiologically confirmed)
4. Drug toxicity or reaction

Incidence

A wide range of incidence estimates for paradoxical TB-IRIS have been reported from retrospective and prospective cohort studies and randomized controlled trials of patients with HIV-associated TB starting ART. These were summarized in a systematic review and meta-analysis published in 2015 that included data from 40 studies [25]. Across the 40 studies that were included there were 7789 patients who were at risk for paradoxical TB-IRIS (i.e. with a diagnosis of HIV-associated TB and initiating ART); 1048 of these patients were diagnosed with TB-IRIS after initiating ART. The pooled incidence of paradoxical TB-IRIS across studies was 18% (95% CI: 16–21%), but estimates varied widely across individual studies from 0% to 54%. The pooled incidence reported from prospective observational studies was 23%, retrospective observational studies 16% and randomized controlled trials 16%.

The considerable heterogeneity in estimates across studies likely relates to two main factors: differences in methods of ascertainment of TB-IRIS cases and the underlying prevalence of risk factors for TB-IRIS in the studied population. Prospective studies that use a standardised case definition for TB-IRIS with regular clinical follow-up and questioning regarding symptoms of TB-IRIS would be more likely to ascertain cases that may be missed during routine clinical follow-up. The prevalence of baseline risk factors in the study population appears to be an important determinant of incidence in that cohort. Since the publication of the systematic review, two studies conducted in South Africa have reported incidence rates of around 50% when patients with low CD4 counts starting ART with a short delay from TB treatment are studied. In the PredART randomized controlled trial (discussed in detail in Prevention section below), patients in the placebo arm experienced TB-IRIS with an incidence of 46.7%; these patients had median CD4 count of 51 cells/mm^3 (all had a CD4 count < 100 cell/mm^3) and started ART a median of 16 days after TB treatment (interquartile range, 15–22) [26]. In another cohort study done in the same clinic, 57 ART-naive TB patients with CD4 counts <200 cells/mm^3 were enrolled and followed up after initiating ART: TB-IRIS was diagnosed in 29 of 49 (59.2%) patients who completed follow-up. In this cohort, the median duration from TB treatment to ART initiation was 15 days in patients that developed TB-IRIS and 22 days in those who did not [27].

Risk Factors

Across studies the risk factors most consistently associated with TB-IRIS include lower baseline CD4 count, higher baseline HIV viral load, short interval between TB treatment and ART initiation, and extra-pulmonary or disseminated TB. The association between a short interval from initiation of TB treatment to ART may reflect that a higher mycobacterial load and thus greater antigen stimulus increases TB-IRIS risk. Another risk factor, lower CD4 count, is a marker of immunosuppression and this in turn may be associated with a higher mycobacterial load due to TB dissemination.

Many studies have reported that a baseline CD4 count < 50 cells/mm^3 is associated with TB-IRIS [25]. The SAPIT, CAMELIA and STRIDE clinical trials all reported that lower baseline CD4 count and higher baseline HIV viral load were independent risk factors for paradoxical TB-IRIS in multivariate analyses [28–30]. Certain studies (including the CAMELIA trial [30]) have reported that a faster rise in CD4 count during early ART is associated with increased TB-IRIS risk, whereas other studies have reported a faster decline in HIV viral load to be associated [25].

Several early observational studies reported an association between the interval from initiating TB treatment to initiating ART and the risk of TB-IRIS [31–33]. An interval less than 60 days was a significant risk factor in a Spanish retrospective study [34]. In a UK retrospective study, an interval less than 6 weeks was associated with higher risk of TB-IRIS [35]. A prospective cohort study conducted in India reported that an interval of less than 30 days was a risk factor for TB-IRIS [31]. In a retrospective study conducted in Thailand, an interval of less than 60 days was found to be a significant risk factor after adjustment (odds ratio = 6.57; 95%CI: 1.61–26.86) [33]. In a retrospective observational study conducted in South Africa, an interval less than 30 days compared with greater than 90 days was a significant risk factor for TB-IRIS in multivariate analysis [14]. By contrast, in several studies the risk of paradoxical TB-IRIS was not associated with duration of TB treatment at ART initiation [36–38].

However, observational studies are prone to potential bias when addressing the association between ART timing and TB-IRIS; particularly confounding by indication given that clinicians are likely to initiate the sickest patients (those with low CD4 counts and disseminated TB) on ART more rapidly and this may contribute to the observed association. However, the association between earlier ART and TB-IRIS has subsequently been confirmed in several randomised controlled trials of ART timing in TB patients, which eliminates such biases. These trials were meta-analysed with the key finding with respect to TB-IRIS risk being that starting ART 1–4 weeks (early) versus 8–12 weeks (delayed) after starting TB treatment increases the risk for TB-IRIS 2.3-fold [39]. This meta-analysis is discussed in more detail in the Prevention section below. This provides strong evidence that earlier ART does indeed increase the risk of paradoxical TB-IRIS.

Extra-pulmonary or disseminated tuberculosis has been associated with an increased risk for paradoxical TB-IRIS; the magnitude of the increased risk varying from around 2- to 9-fold higher across studies [22, 30, 33, 38, 40, 41]. This likely reflects that a higher *Mycobacterium tuberculosis* antigen load is more likely to provoke the inflammatory response in tissues that results in the clinical presentation of TB-IRIS. Another line of evidence to support this is that studies have reported an association between paradoxical TB-IRIS and a positive urine lipo-arabinomannan (LAM) assay [27, 42, 43]. A positive urine LAM assay is thought to be a marker of dissemination of TB to the kidney in patients with advanced HIV [44]. Also, a cohort study of patients with TB meningitis starting ART, reported that having a TB culture positive in cerebrospinal fluid at the time of TB meningitis diagnosis was associated with a heightened risk of developing TB meningitis IRIS [16].

Concerns have been raised that the integrase strand transfer (InSTI) class of ART drugs may increase the risk for IRIS because, compared with other ART classes, they result in more rapid HIV viral load decline and CD4 cell count increase during initial ART [45]. Retrospective observational cohort studies have reported a 2 to 3-fold increase in IRIS in general with the InSTI class, but there were few TB-IRIS cases in these studies [46, 47]. In the INSPIRING trial that evaluated dolutegravir versus efavirenz-based ART in patients with HIV-associated TB there was no difference in the proportion of participants developing TB-IRIS in the dolutegravir (6%) compared with the efavirenz (9%) arms [48]. This finding is similar to that of the Reflate TB trial which compared two different doses of raltegravir to efavirenz in patients with HIV-associated TB: TB-IRIS occurred in 10% of patients in the efavirenz arm and in 6% in the raltegravir arms [49]. Therefore, based on prospective evidence, InSTIs do not appear to be associated with an increased risk for

TB-IRIS. An important consideration, however, is that patients at the highest risk for TB-IRIS (i.e. those with low CD4 counts) were under-represented in both the INSPIRING and the Reflate TB trials.

This question of whether InSTIs increase the risk for IRIS in general was also addressed in a secondary analysis from the REALITY trial (n = 1805, 15% were on TB treatment at study entry) [50]. One of the randomizations in this trial was to receive raltegravir plus standard ART versus standard ART alone (predominantly tenofovir, emtricitabine and efavirenz). Patients were followed for IRIS events (including those related to specific infections) and IRIS-related death. As expected, patients in the raltegravir arm had a significantly more rapid decline in plasma HIV viral load. However, there was no difference in IRIS-related mortality, overall IRIS events or IRIS events related to specific infections, including TB-IRIS. Cumulative incidence rates were similar between arms with IRIS occurring in 9.9% of patients in the raltegravir arm and 9.5% of patients in the no raltegravir arm. TB-IRIS was the most common IRIS event, occurring in 5.9% and 6.0% respectively. Future studies that include more patients at high risk for TB-IRIS will need to evaluate whether the InSTI class increases the risk for TB-IRIS.

Consequences

The consequences of paradoxical TB-IRIS include hospitalisation, the need for diagnostic and therapeutic procedures, and in some instances the use of antibiotics that are not required. This has health care resource implications in high burden settings. In a meta-analysis, 25% of patients who developed paradoxical TB-IRIS were hospitalized, with a range of 3–54% across studies, while 17%, with a range of 2–77%, required a therapeutic procedure [25]. Therapeutic procedures included aspiration or surgical drainage of lymph nodes and abscesses, aspiration of serous effusions and laparotomies.

In the past, clinicians interrupted ART in some patients who developed IRIS. This is seldom done currently unless IRIS is life-threatening where it may be considered. Interruption of ART may predispose to ART drug resistance and may result in worsening immunosuppression and risk of other opportunistic infections. IRIS may potentially negatively impact ART adherence, but in one study adherence was only slightly lower in patients who developed IRIS [51].

The all-cause mortality among paradoxical TB-IRIS cases was 7% (95% CI: 4–11%) and death attributable to paradoxical TB-IRIS occurred in 2% (95% CI: 1–3%) of TB-IRIS cases in a meta-analysis [25]. Although mortality due to TB-IRIS is uncommon overall, when the central nervous system is affected the condition is potentially life-threatening. In an Indian cohort study, 5 of 13 patients with paradoxical TB-IRIS died and TB-IRIS meningitis was the cause in 3 [52]. In a South African study, 4 of 16 patients diagnosed TB meningitis IRIS died and in 2, death was directly attributed to neurologic TB-IRIS [16]. In another Indian cohort study the 2 TB-IRIS deaths that occurred were also in patients with neurological involvement [31].

Recent studies suggest that acute increases in pulmonary inflammation are common in patients initiating ART with pulmonary TB, and that these increases are associated with decreases in lung function as assessed by spirometry. In one prospective cohort study from South Africa, clinically significant decreases in pulmonary function occurred in the first 4 weeks of ART in approximately half of all patients with TB who initiated ART, and decreases in lung function at week 4 were associated with worse lung function up to 24 weeks after TB treatment completion. Furthermore, in one study from Botswana, more rapid increases in CD4 cell count in the first 4 weeks on ART were associated with more limited lung function several months after TB was cured. These findings raise the possibility that immune restoration may trigger pulmonary damage that persists after TB cure. As has been noted from chest radiograph studies of HIV-infected patients with TB, low CD4 counts are associated with less lung infiltration at diagnosis. It is therefore plausible, and noted in the clinic not infrequently, that as CD4 cell counts recover lung involvement subsequently increases, usually in association with TB-IRIS symptoms [8].

Treatment

One randomised controlled trial has been conducted assessing treatment of paradoxical TB-IRIS. In this single-centre double-blind, placebo-controlled trial 110 participants with a clinical diagnosis of TB-IRIS were randomised to prednisone (1.5 mg/ kg per day for 2 weeks followed by 0.75 mg/kg per day for 2 weeks) or identical placebo [53]. Patients with immediately life-threatening TB-IRIS manifestations were excluded. The primary combined endpoint was days of hospitalization and outpatient therapeutic procedures. Procedures were counted as one hospital day. This primary endpoint was more frequent in the placebo arm (median hospital days 3 versus 0; p=0.04). There was more rapid improvement in symptoms, quality of life score and chest radiograph score in the prednisone arm. Infections occurred in more participants in the prednisone arm (mainly oral candida and uncomplicated herpes simplex), but there was no difference in severe infections. We suggest that patients with a clinical diagnosis of paradoxical TB-IRIS and without contra-indications to corticosteroids should be treated with a course of prednisone starting at 1.5mg/kg/day and weaning over 4 weeks. Some patients require longer courses of prednisone because their symptoms recur on weaning or stopping prednisone. Kaposi's sarcoma may worsen due to corticosteroids, so corticosteroids should be avoided in patients with this condition unless TB-IRIS is life-threatening [54].

In the Namale meta-analysis [25], 23 studies reported corticosteroid use, while 10 reported Non-steroidal anti-inflammatory drug (NSAID) use. Including only these studies, 28% of TB-IRIS cases were prescribed NSAID while 38% were prescribed corticosteroids, most frequently prednisone or prednisolone. In a retrospective cohort study conducted in France, of 34 patients managed with TB-IRIS, corticosteroids were prescribed in 61% and had no significant side effects apart from a trend towards a lower CD4 count at 12 months [55]. Such a trend was not observed in the randomised controlled trial of prednisone for TB-IRIS treatment [53].

NSAIDs have also been used to treat mild manifestations with anecdotal reports of benefit. In refractory cases, thalidomide, TNF-blockers and interleukin-6 blockers have been used [6, 9]. Aspiration procedures (e.g. lymph node aspiration, pericardiocentesis) may be required to relieve symptoms or mitigate complications. Interruption of ART is generally not advised. In the majority of patients with TB-IRIS, TB treatment does not need to be prolonged beyond 6 months. However, in patients with abscesses or tuberculomas that are present for longer than 6 months after starting TB treatment most clinicians would opt to prolong the treatment.

Prevention

There are two strategies that have been demonstrated in randomized controlled trials to reduce the incidence of paradoxical TB-IRIS: (1) deferring ART to 8–12 weeks after initiation of TB treatment; and (2) moderate dose prednisone to cover the first 4 weeks of ART.

A meta-analysis of trials that compared starting ART 1–4 weeks (early) versus 8–12 weeks (delayed) after starting TB treatment in patients with HIV-associated TB found that starting ART in the earlier time window was associated with a more than two-fold increase in the risk of paradoxical TB-IRIS. There were 1450 patients from 6 trials included in this component of the meta-analysis: in those who received early ART, 17.5% developed TB-IRIS compared with 8.3% in the delayed ART group (relative risk (RR) = 2.31 (95%CI= 1.87–2.86)). This heightened risk of TB-IRIS associated with early ART was observed for both patients with CD4 count <50 cells/mm^3 (RR = 2.50 (95%CI = 1.84–3.40)) and in those with a CD4 count >50 cells/mm^3 (RR = 2.21 (95%CI = 1.50–3.24)). Thus deferring ART to 8–12 weeks after starting TB treatment will reduce the incidence of TB-IRIS by more than 50%. However, deferring ART to reduce the risk of TB-IRIS can only be recommended in patients with CD4 counts > 50 cells/mm^3 and no urgent clinical indication to start ART (e.g. Kaposi's sarcoma). In the same meta-analysis, in patients with a CD4 count < 50 cells/mm^3 delaying ART until 8–12 weeks after starting TB treatment was associated with an increased risk of death; mortality in these patients was 29% lower in those who started early ART compared with deferred ART. There was no difference in mortality in the early versus delayed ART arms for patients with CD4 count > 50 cells/mm^3 (RR = 1.05 (95%CI = 0.68–1.61)). In line with this, the World Health Organisation guidelines recommend that TB treatment should be initiated first in patients with HIV-associated TB followed by ART within the first 8 weeks of treatment. Furthermore, they advise that patients with CD4 count < 50 cells/mm^3 should receive ART within 2 weeks of initiating TB treatment [56].

Given that patients with low CD4 counts have a higher risk of TB-IRIS that is further exacerbated by starting ART early after TB treatment (which is nonetheless indicated because it improves survival) it is important to consider adjunctive strategies for preventing TB-IRIS. The PredART trial investigated whether a moderate dose of prednisone to cover the first 4 weeks of ART could safely reduce the incidence of TB-IRIS in patients with HIV-associated TB who were at high risk for TB-IRIS when starting ART [26]. This was a randomized, double-blind, placebo-controlled trial that enrolled 240 HIV-positive ART-naïve patients initiating ART who had started TB treatment within 30 days before initiating ART, and had a CD4 count of 100 cells/mm^3 or less. Patients were randomised to receive either prednisone (at a dose of 40 mg/day for 14 days and then 20 mg/day for 14 days) or placebo. Paradoxical TB-IRIS was diagnosed in 39/120 patients (32.5%) in the prednisone arm and in 56/120 (46.7%) in the placebo arm (RR = 0.70; 95%CI = 0.51–0.96). Open-label corticosteroids were prescribed to treat TB-IRIS in 13% in the prednisone arm and in 28% in the placebo arm (RR = 0.47; 95%CI = 0.27–0.81). There was no difference in mortality between the two arms and prednisone was not associated with an excess risk of severe infections nor Kaposi's sarcoma. Whether higher doses of prednisone may have a more substantial impact on reducing the risk of TB-IRIS and whether higher doses are safe in this patient population could be investigated in future clinical trials. Based on the findings of the PredART trial, a 4-week course of prednisone could be used in clinical practice for preventing TB-IRIS in patients with TB and CD4 count less than or equal to 100 cells/mm^3 starting ART at the dose used in the trial. This recommendation does not apply to patients also diagnosed with conditions that could be worsened by corticosteroids (e.g. Kaposi's sarcoma, uncontrolled diabetes).

Other strategies proposed for preventing IRIS have been maraviroc and NSAIDs. Maraviroc is an ART drug that also potentially blocks immune cell migration to tissue by blocking the chemokine receptor CCR5. In a clinical trial of maraviroc versus placebo added to three drug ART, maraviroc was not associated with a reduced risk of IRIS. This trial did not specifically focus on TB patients and only 64 of the 276 participants (23%) had a diagnosis of TB at baseline. The baseline CD4 count was a median of 32 (placebo arm) and 36 (maraviroc arm) cells/mm^3. IRIS in general was not less common in participants who received maraviroc versus placebo (24% versus 23% respectively, $p = 0.74$). Specifically for paradoxical TB-IRIS, this was diagnosed in 8 (6% of all in maraviroc arm) versus 9 (7% of all in placebo arm) participants ($p = 0.80$) [57]. Neither NSAIDs nor any other immunomodulatory agents have been investigated for preventing TB-IRIS in a clinical trial.

Section 2: Clinical Aspects of Unmasking TB-IRIS

In settings with a high epidemiologic burden of TB, a substantial proportion of patients who present (or re-present) for HIV care do so with active TB, either overtly symptomatic or subclinical TB. Many of these patients will be diagnosed with TB and started on TB treatment before starting ART. However, because of the imperfect diagnostic sensitivity of TB diagnostics (particularly sputum smear), their limited availability, as well as health care workers occasionally missing symptoms of TB, a proportion of these patients will be started on ART with undiagnosed active TB. Such patients will then frequently be diagnosed with active TB during the first 3 months of ART when their symptoms manifest or worsen. High TB incidence rates (5.6–23 TB cases per 100 person years) in the first 3 months of ART have been reported from developing country ART programs [58–60]. Such presentations have been termed ART-associated TB [5]. A subset of these patients will present with unmasking TB-IRIS in the first 3 months of ART that the INSHI case definition for unmasking TB-IRIS [5] characterises as:

- Heightened intensity of clinical manifestations, particularly if there is evidence of a marked inflammatory component to the presentation.
- Once established on tuberculosis treatment, a clinical course that is complicated by a paradoxical reaction.

Examples of unmasking TB-IRIS include patients who present with pulmonary TB in the first few weeks after starting ART with rapid progression of symptoms and respiratory distress at presentation (a presentation that resembles a bacterial pneumonia). There is a case report of a patient who required mechanical ventilation for adult respiratory distress syndrome associated with miliary TB during early ART [61]. A fatal case of unmasking TB-IRIS presenting after 6 weeks on ART was shown at postmortem to have extensive infiltrate of the upper lobe of the right lung, with histology suggestive of bronchiolitis obliterans organizing pneumonia [62]. Another example is patients who present with cerebral tuberculomas and severe neurological deficits soon after starting ART. Also, patients who present with nodal TB after starting ART may manifest with paradoxical worsening even after being started on TB treatment, and may have a similar prolonged clinical course similar to patients with paradoxical TB-IRIS. Breen and colleagues reported 13 patients diagnosed with active TB in the first 3 months of ART. These patients developed paradoxical reactions more frequently (62%) than patients who were diagnosed with TB later on ART (0%) [63]. The clinical manifestations and incidence rate of unmasking TB-IRIS is less well characterised than paradoxical TB-IRIS with fewer cases reported in the literature. Incidence is likely influenced by baseline CD4 count of the cohort, screening and diagnostic practices for TB and local practices with respect to starting empiric TB treatment in patients with TB symptoms prior to ART.

It is important to note that not all patients with undiagnosed TB at the time of starting ART will develop unmasking TB-IRIS, but only a subgroup. Many will present with TB in a manner that is typical and without heightened inflammatory features [64]. In our experience, it is patients who were overtly symptomatic with TB where the diagnosis was overlooked by a health care worker prior to starting ART who develop severe unmasking TB-IRIS.

In order to prevent unmasking TB-IRIS it is key to screen for TB symptoms before ART and investigate those with symptoms. It has been reported that around 50–70% of patients diagnosed with TB in the first 3 months of ART had TB symptoms at the time of ART initiation [63, 65]. The use of the WHO symptom screen assessment [66] followed by investigation of those who are symptomatic (with sputum smear, Xpert and /or TB culture, chest radiograph and urinary LAM) will facilitate the diagnosis of many patients with TB prior to starting ART and the appropriate initiation of TB treatment. However, the diagnosis of TB may be difficult to prove in patients with advanced HIV given the insensitivity of sputum smears and difficulties with producing sputum and frequent extra-pulmonary TB. Empiric treatment, if there is a strong clinical suspicion and compatible imaging (chest radiograph or abdominal ultrasound), should be considered.

The management of unmasking TB-IRIS involves initiation of appropriate TB treatment and exclusion (and if needed treatment) of concomitant conditions contributing to clinical deterioration. There is no clinical trial data regarding the use of corticosteroids for this form of TB-IRIS, but some clinicians opt to use corticosteroids if there are severe inflammatory manifestations, for example tuberculomas with surrounding cerebral oedema or severe pulmonary TB with respiratory failure, or if there is paradoxical deterioration after starting TB treatment.

Section 3: Immunologic Mechanisms Involved In TB-IRIS

The immunopathogenesis of TB-IRIS involves both innate and adaptive immune activation resulting in a profound release of inflammatory cytokines that contribute to both the symptomatology of the syndrome and the tissue histopathology [6]. The immune mechanisms involved have been intensively studied shaping a better understanding of potential therapeutic targets. Most immunopathogenesis studies have focused on the paradoxical form of TB-IRIS.

Role of T Lymphocytes

The most notable immunologic change after ART initiation in all patients starting therapy is the recovery of CD4 T lymphocytes. As such, in the early years that TB-IRIS was recognized, the focus of pathogenesis research was the T lymphocytes and specifically the CD4 T cells [67, 68]. Recovery of CD4 T lymphocytes after ART initiation is both quantitative, with increases in circulating numbers, and qualitative, manifesting as changes in phenotype and improvement in function.

Although the overall numbers of CD4 counts in people with HIV-TB and paradoxical TB-IRIS are not always higher than in those without IRIS, evaluation of phenotype, by assessing differentiation and maturation markers of memory versus naïve T cells as well as activation markers signifying exposure to antigen and cycling of T cells, has suggested a more robust recovery of effector memory CD4 T cells in people who develop the syndrome [69, 70]. More recently this was further characterized by looking at expression of chemokine receptors namely CCR6 and CXCR3, showing higher expression of CXCR3+ CD4 T cells in people with emerging TB-IRIS [71]. CXCR3 T cells are typically enriched in Th1 cells, which are cells with higher potential for production of cytokines like interferon gamma (IFNγ), the same cytokine measured in the diagnostic whole blood IFN-γ release assays (IGRAs), and tumor necrosis factor (TNF).

Patients with paradoxical TB-IRIS have also been shown to have robust TB-specific responses, measured by assessing production of IFNγ after stimulation with PPD or TB peptides, during IRIS events, which also usually represent striking increases from pre-ART baseline values. These responses to *M. tuberculosis* antigens have been demonstrated using enzyme-linked immunospot (ELISPOT) or whole blood IGRAs or flow cytometric analysis of intracellular cytokine production in T cells [67, 72, 73]. In addition, although the majority of patients with advanced HIV-associated TB experience increases in TB-specific immune responses on ART, increases in TB-specific CD4 responses are often significantly higher in those who develop TB-IRIS when compared to those who do not. Since the initial early reports documenting this association [67], several subsequent studies have concluded that there is a significant expansion of TB-specific effector polyfunctional CD4 T cells during IRIS that produce cytokines including INFγ, TNF, IL-2 and IL-17 [73, 74]. Studies have also suggested that specific types of effector cells may be involved. For example, Bourgarit et al. [75] found that higher proportions of TCRγδ+ T cells not expressing CD94/NKG2 inhibitory receptors were observed pre-ART in those patients who ended up developing paradoxical TB-IRIS. Wilkinson KA et al., studied the effector function of TB-specific CD4 T cells and demonstrated an increase in IFNγ response as well as in perforin 1 and granzyme B expression in heat-killed H37Rv stimulated human PBMCs from paradoxical TB-IRIS patients compared to non-IRIS patients with HIV-associated TB [76]. Notably, this observation is in agreement with a recent study of HIV patients with *Mycobacterium avium* complex IRIS [77]. Furthermore, recent studies from sub-Saharan African adults with HIV-associated TB suggest that greater recovery of TB-specific CD4 cell function on ART is associated with greater inflammation in the lung and worse lung function after TB cure, indicating that cellular immune recovery may promote clinically-relevant inflammation at the site of disease (Fig. 3). Taken together, existing research supports an association between more robust recovery of pathogen-specific cellular immune responses on ART and development of paradoxical TB-IRIS. In this model, pathogen-specific immune cells unleashed by ART-mediated virologic suppression promote immunopathology at anatomic sites where antigen burden is high. Activation of myeloid cells appear to also directly mediate tissue damage. Consistent with this model, greater antigen burden, assumed by a shorter interval between TB treatment and ART initiation or measured directly by higher acid fast bacilli status or by detection of LAM in urine at baseline, increases the risk of paradoxical TB-IRIS.

This association has not been universally observed in all cohorts [68], however, and controversy remains as to whether increases in the numbers and function of antigen-specific CD4+ T cells are as important as other immunologic mediators that may more directly drive IRIS risk. Further confusion stems from limited baseline data time points in many studies, complicating the ability to definitively conclude that pathogen-specific cellular responses were indeed not actually already higher prior to ART initiation compared to persons with HIV-associated TB who did not develop IRIS. Moreover, despite the fact that expansion of polyfunctional and potentially cytotoxic CD4 T cells after ART is a fairly consistent finding in TB-IRIS, the presence of these cells after ex vivo stimulation with TB antigens persists after treatment of IRIS with corticosteroids or even cytokine blockade. In addition, some patients without TB-IRIS have very robust increases in pathogen-specific, polyfunctional CD4+ T cell responses on ART [73]. While the majority of TB-IRIS investigations have measured cellular immune responses in blood and not at the site of disease, these apparent inconsistencies make it possible that although they seem involved in IRIS pathogenesis, they may not be mediating the inflammation driving the clinical symptoms of TB-IRIS.

The magnitude of CD4 cellular immune recovery on ART is a key correlate of survival, including among those with HIV-associated TB, where the lack of recovery of TB-specific CD4 cellular immune function on ART has been associated with death within 6 months after ART initiation, despite virologic suppression. One prospective cohort study of HIV-infected adults initiating ART in the setting of active pulmonary TB indicated that there were three general groups of patients: those who developed TB-IRIS, those who died, and those who survived without IRIS. Those with rapid recovery of pathogen-specific CD4 cell function had a higher risk of TB-IRIS, whereas those who had poor recovery of these cells had a high risk of death. These data suggest that patients who die of TB despite ART initiation have a biomarker profile distinct from patients who go on to develop TB-IRIS and may not benefit from strategies to immunosuppress HIV-associated TB patients but may in fact benefit from immune boosting strategies [78].

Role of NK Cells

The role of NK cells has been investigated in two studies of TB-IRIS. The presence of activated NK cells expressing CD69 and HLA-DR was found in one cross-sectional study to be associated with unmasking TB-IRIS events at the time of IRIS events [79]. In the second study of patients with fairly advanced disease participating in a randomized controlled trial of timing of ART initiation in TB showed that a higher proportion of cytotoxic NK cells was associated with eventual development of paradoxical TB-IRIS [30]. It remains unclear how the NK cell activation ties in with the myeloid cell activation that may represent, possibly along with inflammasome activation, the cornerstone of IRIS pathogenesis.

Role of Myeloid Cells

The role of myeloid cells had long been hypothesised after biomarker investigation revealed a prominent signature of pro-inflammatory cytokines of predominantly myeloid cell origin such as IL-6, TNF, CXCL10, IL-6, IL-1β and IL-18 in peripheral blood associated with TB-IRIS onset [80, 81]. When biomarkers were assessed prior to ART, after short (<4 weeks) versus longer (>4 weeks) anti-tuberculous treatment, a clear predominance of myeloid origin cytokines and chemokines was observed after short TB treatment which was also highly associated with development of paradoxical TB-IRIS [80]. Further studies looking specifically at subsets of monocytes identified an expansion of inflammatory monocytes ($CD14^{high}CD16^{-}$) in people developing TB-IRIS that was observed prior to ART initiation. The expansion of this monocyte subset preceded ART initiation and was independently associated with IRIS emergence [80]. Inflammatory monocytes not only produce inflammatory cytokines such as IL-6, TNF and IL-1β but also express CCR2 which can facilitate tissue migration in response to CCL2 or MCP-1, a chemokine that is also elevated in paradoxical TB-IRIS. It is hypothesized that when these inflammatory monocytes migrate into tissue they mature into inflammatory macrophages which are important cells in the formation of TB granulomas.

Inflammasone activation has also been implicated in IRIS and was investigated after elevated concentrations of IL-1β and IL-18, two signature infammasome cytokines, were observed in patients developing TB-IRIS [80–82]. Inflammasomes are cytosolic protein aggregates assembled to coordinate distinct immune responses to infectious agents or physiological perturbations. In addition to its role in the immune response against pathogens, inflammasome activation has also been shown to be dysregulated in cancer, cardiovascular and neurodegenerative disorders, autoinflammatory syndromes and diabetes. In a study assessing gene expression in blood at the start of ART in patients with HIV-associated TB and then at regular intervals during early ART, differences were observed not at baseline but after ART initiation, when genes related to inflammasome activation were more upregulated in patients who developed TB-IRIS [83]. This was further evaluated by PBMC cultures ex vivo showing that blocking MyD88 reduced cytokine production. These data support a significant role of inflammasome activation in IRIS pathogenesis. Whole blood transcripts from patients with TB meningitis (TBM) who were starting ART showed prominent inflammasome gene upregulation in those who went on to develop TBM-IRIS as well as an IL-8 signature [84], which corroborated the finding of increased neutrophils in CSF in those TBM patients who develop TBM-IRIS [85].

A previous study had evaluated the transciptome of isolated monocytes in patients with TB-IRIS and controls (HIV-associated TB patients without IRIS) prior to ART initiation and after two weeks on ART, which was close to IRIS symptoms presentation [86, 87]. Upregulated C1Q and C1-INH was observed before and around the IRIS event in monocytes but the respective complement protein measurements were essentially not different therefore not providing serological evidence of the gene upregulation observation. It thus remains unclear if complement may be playing a role in initiating the aberrant inflammatory cascade of TB-IRIS.

In a case-control study, Torrado et al. [88] found that in vitro *Mycobacterium tuberculosis* stimulated PBMCs from patients with paradoxical TB-IRIS transcribe more IL-27p28 than do PBMCs from HIV-associated TB patients without IRIS. Plasma IL-27p28 subunit level was higher in those who developed IRIS compared with those who did not before ART initiation, suggesting also a potential role of IL-27, another myeloid derived cytokine, in development of IRIS which has not been further explored to date.

In summary, myeloid cells are important in IRIS pathogenesis and tissue pathology. Data supportive of their role in the syndrome emanate from multiple lines of evidence ranging from neutrophils that dominate in necrotic lymph nodes and TBM-IRIS to monocytes with inflammasome activation and to tissue macrophages with their prominent role in granuloma formation in various affected organs including lymph nodes, bone marrow, liver, spleen, kidneys and genitourinary tract.

Role of Tissue Damage

Evidence of tissue destruction with neutrophil infiltration and granule production and increased production of matrix metalloproteinases (MMPs) including MMP 3, 7, 8, 9 and 10 in TB-IRIS seems to further support an important role of myeloid cells in IRIS pathogenesis including macrophages which also represent relevant cells for granuloma formation [27, 89]. Importantly, MMPs can degrade lung matrix and lead to lung function abnormalities despite microbiologic clearance. Consistent with this hypothesis, increased MMP-8 levels were independently associated with TB-IRIS development and with decreases in lung function after treatment [8].

In addition, imaging with FDG-PET scans has shown a high uptake of glucose in more extensive areas of involvement in TB patients who developed IRIS versus those who did not [90]. In the lung, higher uptake of radiolabeled glucose has been shown to be associated with worsening lung function. These observations are in agreement with the hypothesis that inflammatory myeloid cells as well as effector memory CD4 T cells rely on glycolysis as a fast energy source, and also with the clinical observation that patients with more disseminated extensive disease and higher antigen burden are more prone to develop paradoxical IRIS. High levels of glycolysis are typically accompanied by higher expression of Glut-1 which was observed on both CD4 T cells and monocytes in patients with paradoxical TB-IRIS. This observation highlights the possible dependence of inflammatory cells on glycolytic pathways in the IRIS immune response.

Conclusion

No single specific cellular population or immunologic mechanism perfectly predicts TB-IRIS risk or identifies all patients who have the syndrome. Rather, over a decade of research has implicated a variety of inter-related adaptive and innate immune mechanisms that can drive development of pathologic inflammation in patients with active TB who virologically respond to ART. The key immunologic components of TB-IRIS appear to be severe lymphopenia with immune suppression and quantitative and qualitative myeloid and T cell recovery in the setting of a high mycobacterial antigen load and HIV viral suppression. It remains unclear if genetics play a role in either predisposition to the syndrome or response to treatment and resolution. It also remains unclear what, if any, the long-term consequences for IRIS may be in overall status of residual immune activation, or other immunologic and virologic outcomes. Further research on the immunologic mechanisms of TB-IRIS will ideally lead to new treatments that may decrease the morbidity of the syndrome. These research efforts may also yield insights into mechanisms of TB-associated inflammation more broadly (i.e., beyond HIV and TB co-infection).

References

1. Battegay M, Nuesch R, Hirschel B, Kaufmann GR (2006) Immunological recovery and antiretroviral therapy in HIV-1 infection. Lancet Infect Dis 6(5):280–287
2. Autran B, Carcelain G, Li TS, Blanc C, Mathez D, Tubiana R, Katlama C, Debre P, Leibowitch J (1997) Positive effects of combined antiretroviral therapy on CD4+ T cell homeostasis and function in advanced HIV disease. Science 277(5322):112–116
3. Guihot A, Bourgarit A, Carcelain G, Autran B (2011) Immune reconstitution after a decade of combined antiretroviral therapies for human immunodeficiency virus. Trends Immunol 32(3):131–137
4. Gorochov G, Neumann AU, Kereveur A, Parizot C, Li T, Katlama C, Karmochkine M, Raguin G, Autran B, Debre P (1998) Perturbation of CD4+ and CD8+ T-cell repertoires during progression to AIDS and regulation of the CD4+ repertoire during antiviral therapy. Nat Med 4(2):215–221
5. Meintjes G, Lawn SD, Scano F, Maartens G, French MA, Worodria W, Elliott JH, Murdoch D, Wilkinson RJ, Seyler C et al (2008) Tuberculosis-associated immune reconstitution inflammatory syndrome: case definitions for use in resource-limited settings. Lancet Infect Dis 8(8):516–523
6. Walker NF, Stek C, Wasserman S, Wilkinson RJ, Meintjes G (2018) The tuberculosis-associated immune reconstitution inflammatory syndrome: recent advances in clinical and pathogenesis research. Curr Opin HIV AIDS 13(6):512–521
7. Walker NF, Scriven J, Meintjes G, Wilkinson RJ (2015) Immune reconstitution inflammatory syndrome in HIV-infected patients. HIV AIDS (Auckl) 7:49–64
8. Ravimohan S, Tamuhla N, Kung SJ, Nfanyana K, Steenhoff AP, Gross R, Weissman D, Bisson GP (2016) Matrix metalloproteinases in tuberculosis-immune reconstitution inflammatory syndrome and impaired lung function among advanced HIV/TB co-infected patients initiating antiretroviral therapy. EBioMedicine 3:100–107
9. Hsu DC, Faldetta KF, Pei L, Sheikh V, Utay NS, Roby G, Rupert A, Fauci AS, Sereti I (2016) A paradoxical treatment for a paradoxical condition: infliximab use in three cases of mycobacterial IRIS. Clin Infect Dis 62(2):258–261
10. Lawn SD, Wood R (2007) Hepatic involvement with tuberculosis-associated immune reconstitution disease. AIDS 21(17):2362–2363
11. Meintjes G, Rangaka MX, Maartens G, Rebe K, Morroni C, Pepper DJ, Wilkinson KA, Wilkinson RJ (2009) Novel relationship between tuberculosis immune reconstitution inflammatory syndrome and antitubercular drug resistance. Clin Infect Dis 48(5):667–676
12. Weber E, Gunthard HF, Schertler T, Seebach JD (2009) Spontaneous splenic rupture as manifestation of the immune reconstitution inflammatory syndrome in an HIV type 1 infected patient with tuberculosis. Infection 37(2):163–165
13. Salliot C, Guichard I, Daugas E, Lagrange M, Verine J, Molina JM (2008) Acute kidney disease due to immune reconstitution inflammatory syndrome in an HIV-infected patient with tuberculosis. J Int Assoc Physicians AIDS Care (Chic) 7(4):178–181
14. Pepper DJ, Marais S, Maartens G, Rebe K, Morroni C, Rangaka MX, Oni T, Wilkinson RJ, Meintjes G (2009) Neurologic manifestations of paradoxical tuberculosis-associated immune reconstitution inflammatory syndrome: a case series. Clin Infect Dis 48(11):e96–e107

15. Coleman BW, Sereti I, Bishop R, Smith BR (2018) Upbeat nystagmus in an HIV-positive patient with a tuberculoma in the medulla. Lancet Infect Dis 18(2):225
16. Marais S, Meintjes G, Pepper DJ, Dodd LE, Schutz C, Ismail Z, Wilkinson KA, Wilkinson RJ (2013) Frequency, severity, and prediction of tuberculous meningitis immune reconstitution inflammatory syndrome. Clin Infect Dis 56(3):450–460
17. Burman W, Weis S, Vernon A, Khan A, Benator D, Jones B, Silva C, King B, LaHart C, Mangura B et al (2007) Frequency, severity and duration of immune reconstitution events in HIV-related tuberculosis. Int J Tuberc Lung Dis 11(12):1282–1289
18. Olalla J, Pulido F, Rubio R, Costa MA, Monsalvo R, Palenque E, Costa JR, Del PA (2002) Paradoxical responses in a cohort of HIV-1-infected patients with mycobacterial disease. Int J Tuberc Lung Dis 6(1):71–75
19. Michailidis C, Pozniak AL, Mandalia S, Basnayake S, Nelson MR, Gazzard BG (2005) Clinical characteristics of IRIS syndrome in patients with HIV and tuberculosis. Antivir Ther 10(3):417–422
20. Huyst V, Lynen L, Bottieau E, Zolfo M, Kestens L, Colebunders R (2007) Immune reconstitution inflammatory syndrome in an HIV/TB co-infected patient four years after starting antiretroviral therapy. Acta Clin Belg 62(2):126–129
21. Bana TM, Lesosky M, Pepper DJ, van der Plas H, Schutz C, Goliath R, Morroni C, Mendelson M, Maartens G, Wilkinson RJ et al (2016) Prolonged tuberculosis-associated immune reconstitution inflammatory syndrome: characteristics and risk factors. BMC Infect Dis 16(1):518
22. Manosuthi W, Van Tieu H, Mankatitham W, Lueangniyomkul A, Ananworanich J, Avihingsanon A, Siangphoe U, Klongugkara S, Likanonsakul S, Thawornwan U et al (2009) Clinical case definition and manifestations of paradoxical tuberculosis-associated immune reconstitution inflammatory syndrome. AIDS 23(18):2467–2471
23. Eshun-Wilson I, Havers F, Nachega JB, Prozesky HW, Taljaard JJ, Zeier MD, Cotton M, Simon G, Soentjens P (2010) Evaluation of paradoxical TB-associated IRIS with the use of standardized case definitions for resource-limited settings. J Int Assoc Physicians AIDS Care (Chic) 9(2):104–108
24. Haddow LJ, Moosa MY, Easterbrook PJ (2010) Validation of a published case definition for tuberculosis-associated immune reconstitution inflammatory syndrome. AIDS 24(1):103–108
25. Namale PE, Abdullahi LH, Fine S, Kamkuemah M, Wilkinson RJ, Meintjes G (2015) Paradoxical TB-IRIS in HIV-infected adults: a systematic review and meta-analysis. Future Microbiol 10(6):1077–1099
26. Meintjes G, Stek C, Blumenthal L, Thienemann F, Schutz C, Buyze J, Ravinetto R, van Loen H, Nair A, Jackson A et al (2018) Prednisone for the prevention of paradoxical tuberculosis-associated IRIS. N Engl J Med 379(20):1915–1925
27. Walker NF, Wilkinson KA, Meintjes G, Tezera LB, Goliath R, Peyper JM, Tadokera R, Opondo C, Coussens AK, Wilkinson RJ et al (2017) Matrix degradation in human immunodeficiency virus type 1-associated tuberculosis and tuberculosis immune reconstitution inflammatory syndrome: a prospective observational study. Clin Infect Dis 65(1):121–132
28. Naidoo K, Yende-Zuma N, Padayatchi N, Naidoo K, Jithoo N, Nair G, Bamber S, Gengiah S, El-Sadr WM, Friedland G et al (2012) The immune reconstitution inflammatory syndrome after antiretroviral therapy initiation in patients with tuberculosis: findings from the SAPiT trial. Ann Intern Med 157(5):313–324
29. Luetkemeyer AF, Kendall MA, Nyirenda M, Wu X, Ive P, Benson CA, Andersen JW, Swindells S, Sanne IM, Havlir DV et al (2014) Tuberculosis immune reconstitution inflammatory syndrome in A5221 STRIDE: timing, severity, and implications for HIV-TB programs. J Acquir Immune Defic Syndr 65(4):423–428
30. Laureillard D, Marcy O, Madec Y, Chea S, Chan S, Borand L, Fernandez M, Prak N, Kim C, Dim B et al (2013) Paradoxical tuberculosis-associated immune reconstitution inflammatory syndrome after early initiation of antiretroviral therapy in a randomized clinical trial. AIDS 27(16):2577–2586
31. Narendran G, Andrade BB, Porter BO, Chandrasekhar C, Venkatesan P, Menon PA, Subramanian S, Anbalagan S, Bhavani KP, Sekar S et al (2013) Paradoxical tuberculosis immune reconstitution inflammatory syndrome (TB-IRIS) in HIV patients with culture con-

firmed pulmonary tuberculosis in India and the potential role of IL-6 in prediction. PLoS One 8(5):e63541
32. Lawn SD, Myer L, Bekker LG, Wood R (2007) Tuberculosis-associated immune reconstitution disease: incidence, risk factors and impact in an antiretroviral treatment service in South Africa. AIDS 21(3):335–341
33. Limmahakhun S, Chaiwarith R, Nuntachit N, Sirisanthana T, Supparatpinyo K (2012) Treatment outcomes of patients co-infected with tuberculosis and HIV at Chiang Mai University Hospital, Thailand. Int J STD AIDS 23(6):414–418
34. Navas E, Martin-Davila P, Moreno L, Pintado V, Casado JL, Fortun J, Perez-Elias MJ, Gomez-Mampaso E, Moreno S (2002) Paradoxical reactions of tuberculosis in patients with the acquired immunodeficiency syndrome who are treated with highly active antiretroviral therapy. Arch Intern Med 162(1):97–99
35. Breen RA, Smith CJ, Bettinson H, Dart S, Bannister B, Johnson MA, Lipman MC (2004) Paradoxical reactions during tuberculosis treatment in patients with and without HIV co-infection. Thorax 59(8):704–707
36. Kumarasamy N, Chaguturu S, Mayer KH, Solomon S, Yepthomi HT, Balakrishnan P, Flanigan TP (2004) Incidence of immune reconstitution syndrome in HIV/tuberculosis-coinfected patients after initiation of generic antiretroviral therapy in India. J Acquir Immune Defic Syndr 37(5):1574–1576
37. Kumarasamy N, Venkatesh KK, Vignesh R, Devaleenal B, Poongulali S, Yepthomi T, Flanigan TP, Benson C, Mayer KH (2013) Clinical outcomes among HIV/tuberculosis-coinfected patients developing immune reconstitution inflammatory syndrome after HAART initiation in South India. J Int Assoc Provid AIDS Care 12(1):28–31
38. Worodria W, Menten J, Massinga-Loembe M, Mazakpwe D, Bagenda D, Koole O, Mayanja-Kizza H, Kestens L, Mugerwa R, Reiss P et al (2012) Clinical spectrum, risk factors and outcome of immune reconstitution inflammatory syndrome in patients with tuberculosis-HIV coinfection. Antivir Ther 17(5):841–848
39. Uthman OA, Okwundu C, Gbenga K, Volmink J, Dowdy D, Zumla A, Nachega JB (2015) Optimal timing of antiretroviral therapy initiation for HIV-infected adults with newly diagnosed pulmonary tuberculosis: a systematic review and meta-analysis. Ann Intern Med 163(1):32–39
40. Manosuthi W, Kiertiburanakul S, Phoorisri T, Sungkanuparph S (2006) Immune reconstitution inflammatory syndrome of tuberculosis among HIV-infected patients receiving antituberculous and antiretroviral therapy. J Infect 53(6):357–363
41. Tieu HV, Ananworanich J, Avihingsanon A, Apateerapong W, Sirivichayakul S, Siangphoe U, Klongugkara S, Boonchokchai B, Hammer SM, Manosuthi W (2009) Immunologic markers as predictors of tuberculosis-associated immune reconstitution inflammatory syndrome in HIV and tuberculosis coinfected persons in Thailand. AIDS Res Hum Retrovir 25(11):1083–1089
42. Conesa-Botella A, Loembe MM, Manabe YC, Worodria W, Mazakpwe D, Luzinda K, Mayanja-Kizza H, Miri M, Mbabazi O, Koole O et al (2011) Urinary lipoarabinomannan as predictor for the tuberculosis immune reconstitution inflammatory syndrome. J Acquir Immune Defic Syndr 58(5):463–468
43. Lawn SD, Edwards DJ, Kranzer K, Vogt M, Bekker LG, Wood R (2009) Urine lipoarabinomannan assay for tuberculosis screening before antiretroviral therapy diagnostic yield and association with immune reconstitution disease. AIDS 23(14):1875–1880
44. Lawn SD, Gupta-Wright A (2016) Detection of lipoarabinomannan (LAM) in urine is indicative of disseminated TB with renal involvement in patients living with HIV and advanced immunodeficiency: evidence and implications. Trans R Soc Trop Med Hyg 110(3):180–185
45. Walmsley SL, Antela A, Clumeck N, Duiculescu D, Eberhard A, Gutierrez F, Hocqueloux L, Maggiolo F, Sandkovsky U, Granier C et al (2013) Dolutegravir plus abacavir-lamivudine for the treatment of HIV-1 infection. N Engl J Med 369(19):1807–1818
46. Psichogiou M, Basoulis D, Tsikala-Vafea M, Vlachos S, Kapelios CJ, Daikos GL (2017) Integrase Strand Transfer Inhibitors and the Emergence of Immune Reconstitution Inflammatory Syndrome (IRIS). Curr HIV Res 15(6):405–410

47. Dutertre M, Cuzin L, Demonchy E, Pugliese P, Joly V, Valantin MA, Cotte L, Huleux T, Delobel P, Martin-Blondel G et al (2017) Initiation of antiretroviral therapy containing integrase inhibitors increases the risk of IRIS requiring hospitalization. J Acquir Immune Defic Syndr 76(1):e23–e26
48. Dooley KE, Kaplan R, Mwelase N, Grinsztejn B, Ticona E, Lacerda M, Sued O, Belonosova E, Ait-Khaled M, Angelis K et al (2019) Dolutegravir-based antiretroviral therapy for patients co-infected with tuberculosis and HIV: a multicenter, noncomparative, open-label, randomized trial. Clin Infect Dis. https://doi.org/10.1093/cid/ciz256
49. Grinsztejn B, De Castro N, Arnold V, Veloso VG, Morgado M, Pilotto JH, Brites C, Madruga JV, Barcellos NT, Santos BR et al (2014) Raltegravir for the treatment of patients co-infected with HIV and tuberculosis (ANRS 12 180 Reflate TB): a multicentre, phase 2, non-comparative, open-label, randomised trial. Lancet Infect Dis 14(6):459–467
50. Gibb D, Szuber AJ, Chidziva E, Lugemwa A, Mwaringa S, Siika A, Mallewa JE, Bwakura-Dangarembizi M, Kabahenda S, Reid A, Baleeta K, Walker S, Pett S Impact of raltegravir intensification of first-line ART on IRIS in the REALITY trial. Abstract 23. In: Conference on Retroviruses and Opportunistic Infections, March 4–7, 2018, Boston, Massachusetts
51. Nachega JB, Morroni C, Chaisson RE, Goliath R, Efron A, Ram M, Maartens G (2012) Impact of immune reconstitution inflammatory syndrome on antiretroviral therapy adherence. Patient Prefer Adherence 6:887–891
52. Agarwal U, Kumar A, Behera D, French MA, Price P (2012) Tuberculosis associated immune reconstitution inflammatory syndrome in patients infected with HIV: meningitis a potentially life threatening manifestation. AIDS Res Therapy 9(1):17
53. Meintjes G, Wilkinson RJ, Morroni C, Pepper DJ, Rebe K, Rangaka MX, Oni T, Maartens G (2010) Randomized placebo-controlled trial of prednisone for paradoxical tuberculosis-associated immune reconstitution inflammatory syndrome. AIDS 24(15):2381–2390
54. Manion M, Uldrick T, Polizzotto MN, Sheikh V, Roby G, Lurain K, Metzger D, Mican JM, Pau A, Lisco A et al (2018) Emergence of Kaposi's sarcoma herpesvirus-associated complications following corticosteroid use in TB-IRIS. Open Forum Infect Dis 5(10):ofy217
55. Breton G, Bourgarit A, Pavy S, Bonnet D, Martinez V, Duval X, Longuet P, Abgrall S, Simon A, Leport C et al (2012) Treatment for tuberculosis-associated immune reconstitution inflammatory syndrome in 34 HIV-infected patients. Int J Tuberc Lung Dis 16(10):1365–1370
56. World Health Organisation (2016) Consolidated guidelines on the use of antiretroviral drugs for treating and preventing HIV infection. Recommendations for a public health approach, 2nd edn. WHO, Geneva
57. Sierra-Madero JG, Ellenberg SS, Rassool MS, Tierney A, Belaunzaran-Zamudio PF, Lopez-Martinez A, Pineirua-Menendez A, Montaner LJ, Azzoni L, Benitez CR et al (2014) Effect of the CCR5 antagonist maraviroc on the occurrence of immune reconstitution inflammatory syndrome in HIV (CADIRIS): a double-blind, randomised, placebo-controlled trial. Lancet HIV 1(2):e60–e67
58. Lawn SD, Myer L, Bekker LG, Wood R (2006) Burden of tuberculosis in an antiretroviral treatment programme in sub-Saharan Africa: impact on treatment outcomes and implications for tuberculosis control. AIDS 20(12):1605–1612
59. Antiretroviral Therapy in Low-Income Countries Collaboration of the International epidemiological Databases to Evaluate A, Collaboration ARTC, Brinkhof MW, Egger M, Boulle A, May M, Hosseinipour M, Sprinz E, Braitstein P, Dabis F et al (2007) Tuberculosis after initiation of antiretroviral therapy in low-income and high-income countries. Clin Infect Dis 45(11):1518–1521
60. Moh R, Danel C, Messou E, Ouassa T, Gabillard D, Anzian A, Abo Y, Salamon R, Bissagnene E, Seyler C et al (2007) Incidence and determinants of mortality and morbidity following early antiretroviral therapy initiation in HIV-infected adults in West Africa. AIDS 21(18):2483–2491
61. Goldsack NR, Allen S, Lipman MC (2003) Adult respiratory distress syndrome as a severe immune reconstitution disease following the commencement of highly active antiretroviral therapy. Sex Transm Infect 79(4):337–338

62. Lawn SD, Wainwright H, Orrell C (2009) Fatal unmasking tuberculosis immune reconstitution disease with bronchiolitis obliterans organizing pneumonia: the role of macrophages. AIDS 23(1):143–145
63. Breen RA, Smith CJ, Cropley I, Johnson MA, Lipman MC (2005) Does immune reconstitution syndrome promote active tuberculosis in patients receiving highly active antiretroviral therapy? AIDS 19(11):1201–1206
64. Kerkhoff AD, Wood R, Lawn SD (2011) Optimum time to start antiretroviral therapy in patients with HIV-associated tuberculosis: before or after tuberculosis diagnosis? AIDS 25(7):1003–1006
65. Koenig SP, Riviere C, Leger P, Joseph P, Severe P, Parker K, Collins S, Lee E, Pape JW, Fitzgerald DW (2009) High mortality among patients with AIDS who received a diagnosis of tuberculosis in the first 3 months of antiretroviral therapy. Clin Infect Dis 48(6):829–831
66. Getahun H, Kittikraisak W, Heilig CM, Corbett EL, Ayles H, Cain KP, Grant AD, Churchyard GJ, Kimerling M, Shah S et al (2011) Development of a standardized screening rule for tuberculosis in people living with HIV in resource-constrained settings: individual participant data meta-analysis of observational studies. PLoS Med 8(1):e1000391
67. Bourgarit A, Carcelain G, Martinez V, Lascoux C, Delcey V, Gicquel B, Vicaut E, Lagrange PH, Sereni D, Autran B (2006) Explosion of tuberculin-specific Th1-responses induces immune restoration syndrome in tuberculosis and HIV co-infected patients. AIDS 20(2):F1–F7
68. Meintjes G, Wilkinson KA, Rangaka MX, Skolimowska K, van Veen K, Abrahams M, Seldon R, Pepper DJ, Rebe K, Mouton P et al (2008) Type 1 helper T cells and FoxP3-positive T cells in HIV-tuberculosis-associated immune reconstitution inflammatory syndrome. Am J Respir Crit Care Med 178(10):1083–1089
69. Haridas V, Pean P, Jasenosky LD, Madec Y, Laureillard D, Sok T, Sath S, Borand L, Marcy O, Chan S et al (2015) TB-IRIS, T-cell activation, and remodeling of the T-cell compartment in highly immunosuppressed HIV-infected patients with TB. AIDS 29(3):263–273
70. Antonelli LR, Mahnke Y, Hodge JN, Porter BO, Barber DL, DerSimonian R, Greenwald JH, Roby G, Mican J, Sher A et al (2010) Elevated frequencies of highly activated CD4+ T cells in HIV+ patients developing immune reconstitution inflammatory syndrome. Blood 116(19):3818–3827
71. Silveira-Mattos PS, Narendran G, Akrami K, Fukutani KF, Anbalagan S, Nayak K, Subramanyam S, Subramani R, Vinhaes CL, Souza DO et al (1502) Differential expression of CXCR3 and CCR6 on CD4(+) T-lymphocytes with distinct memory phenotypes characterizes tuberculosis-associated immune reconstitution inflammatory syndrome. Sci Rep 9(1):2019
72. Elliott JH, Vohith K, Saramony S, Savuth C, Dara C, Sarim C, Huffam S, Oelrichs R, Sophea P, Saphonn V et al (2009) Immunopathogenesis and diagnosis of tuberculosis and tuberculosis-associated immune reconstitution inflammatory syndrome during early antiretroviral therapy. J Infect Dis 200(11):1736–1745
73. Ravimohan S, Tamuhla N, Nfanyana K, Steenhoff AP, Letlhogile R, Frank I, MacGregor RR, Gross R, Weissman D, Bisson GP (2016) Robust Reconstitution of Tuberculosis-Specific Polyfunctional CD4+ T-Cell Responses and Rising Systemic Interleukin 6 in Paradoxical Tuberculosis-Associated Immune Reconstitution Inflammatory Syndrome. Clin Infect Dis 62(6):795–803
74. Mahnke YD, Greenwald JH, DerSimonian R, Roby G, Antonelli LR, Sher A, Roederer M, Sereti I (2012) Selective expansion of polyfunctional pathogen-specific CD4(+) T cells in HIV-1-infected patients with immune reconstitution inflammatory syndrome. Blood 119(13):3105–3112
75. Bourgarit A, Carcelain G, Samri A, Parizot C, Lafaurie M, Abgrall S, Delcey V, Vicaut E, Sereni D, Autran B et al (2009) Tuberculosis-associated immune restoration syndrome in HIV-1-infected patients involves tuberculin-specific CD4 Th1 cells and KIR-negative gammadelta T cells. J Immunol 183(6):3915–3923
76. Wilkinson KA, Walker NF, Meintjes G, Deffur A, Nicol MP, Skolimowska KH, Matthews K, Tadokera R, Seldon R, Maartens G et al (2015) Cytotoxic mediators in paradoxical HIV-tuberculosis immune reconstitution inflammatory syndrome. J Immunol 194(4):1748–1754

77. Hsu DC, Breglio KF, Pei L, Wong CS, Andrade BB, Sheikh V, Smelkinson M, Petrovas C, Rupert A, Gil-Santana L et al (2018) Emergence of Polyfunctional Cytotoxic CD4+ T Cells in Mycobacterium avium Immune Reconstitution Inflammatory Syndrome in Human Immunodeficiency Virus-Infected Patients. Clin Infect Dis 67(3):437–446
78. Ravimohan S, Tamuhla N, Steenhoff AP, Letlhogile R, Nfanyana K, Bellamy SL, MacGregor RR, Gross R, Weissman D, Bisson GP (2015) Immunological profiling of tuberculosis-associated immune reconstitution inflammatory syndrome and non-immune reconstitution inflammatory syndrome death in HIV-infected adults with pulmonary tuberculosis starting anti-retroviral therapy: a prospective observational cohort study. Lancet Infect Dis 15(4):429–438
79. Conradie F, Foulkes AS, Ive P, Yin X, Roussos K, Glencross DK, Lawrie D, Stevens W, Montaner LJ, Sanne I et al (2011) Natural killer cell activation distinguishes Mycobacterium tuberculosis-mediated immune reconstitution syndrome from chronic HIV and HIV/MTB coinfection. J Acquir Immune Defic Syndr 58(3):309–318
80. Andrade BB, Singh A, Narendran G, Schechter ME, Nayak K, Subramanian S, Anbalagan S, Jensen SM, Porter BO, Antonelli LR et al (2014) Mycobacterial antigen driven activation of CD14++CD16- monocytes is a predictor of tuberculosis-associated immune reconstitution inflammatory syndrome. PLoS Pathog 10(10):e1004433
81. Tan HY, Yong YK, Andrade BB, Shankar EM, Ponnampalavanar S, Omar SF, Narendran G, Kamarulzaman A, Swaminathan S, Sereti I et al (2015) Plasma interleukin-18 levels are a biomarker of innate immune responses that predict and characterize tuberculosis-associated immune reconstitution inflammatory syndrome. AIDS 29(4):421–431
82. Tan HY, Yong YK, Shankar EM, Paukovics G, Ellegard R, Larsson M, Kamarulzaman A, French MA, Crowe SM (2016) Aberrant Inflammasome Activation Characterizes Tuberculosis-Associated Immune Reconstitution Inflammatory Syndrome. J Immunol 196(10):4052–4063
83. Lai RP, Meintjes G, Wilkinson KA, Graham CM, Marais S, Van der Plas H, Deffur A, Schutz C, Bloom C, Munagala I et al (2015) HIV-tuberculosis-associated immune reconstitution inflammatory syndrome is characterized by Toll-like receptor and inflammasome signalling. Nat Commun 6:8451
84. Marais S, Lai RPJ, Wilkinson KA, Meintjes G, O'Garra A, Wilkinson RJ (2017) Inflammasome Activation Underlying Central Nervous System Deterioration in HIV-Associated Tuberculosis. J Infect Dis 215(5):677–686
85. Marais S, Wilkinson KA, Lesosky M, Coussens AK, Deffur A, Pepper DJ, Schutz C, Ismail Z, Meintjes G, Wilkinson RJ (2014) Neutrophil-associated central nervous system inflammation in tuberculous meningitis immune reconstitution inflammatory syndrome. Clin Infect Dis 59(11):1638–1647
86. Tran HT, Van den Bergh R, Loembe MM, Worodria W, Mayanja-Kizza H, Colebunders R, Mascart F, Stordeur P, Kestens L, De Baetselier P et al (2013) Modulation of the complement system in monocytes contributes to tuberculosis-associated immune reconstitution inflammatory syndrome. AIDS 27(11):1725–1734
87. Tran HT, Van den Bergh R, Vu TN, Laukens K, Worodria W, Loembe MM, Colebunders R, Kestens L, De Baetselier P, Raes G et al (2014) The role of monocytes in the development of Tuberculosis-associated Immune Reconstitution Inflammatory Syndrome. Immunobiology 219(1):37–44
88. Torrado E, Fountain JJ, Liao M, Tighe M, Reiley WW, Lai RP, Meintjes G, Pearl JE, Chen X, Zak DE et al (2015) Interleukin 27R regulates CD4+ T cell phenotype and impacts protective immunity during Mycobacterium tuberculosis infection. J Exp Med 212(9):1449–1463
89. Tadokera R, Meintjes GA, Wilkinson KA, Skolimowska KH, Walker N, Friedland JS, Maartens G, Elkington PT, Wilkinson RJ (2014) Matrix metalloproteinases and tissue damage in HIV-tuberculosis immune reconstitution inflammatory syndrome. Eur J Immunol 44(1):127–136
90. Hammoud DA, Boulougoura A, Papadakis GZ, Wang J, Dodd LE, Rupert A, Higgins J, Roby G, Metzger D, Laidlaw E et al (2019) Increased metabolic activity on 18F-Fluorodeoxyglucose positron emission tomography-computed tomography in human immunodeficiency virus-associated immune reconstitution inflammatory syndrome. Clin Infect Dis 68(2):229–238

Diagnosis of HIV-Associated Tuberculosis

Andrew D. Kerkhoff and Adithya Cattamanchi

Abstract Of the estimated 1.2 million tuberculosis (TB) cases among people living with HIV (PLHIV), less than half are diagnosed and reported to health authorities. This is a key reason why TB remains the leading cause of death among PLHIV. Systematic screening approaches coupled with improved diagnostics are critical to reducing the gap and have begun to emerge over recent years. This chapter reviews current approaches to screening for and diagnosing HIV-associated TB, including drug-resistant TB, in adults. The chapter is organized into three parts: Part I provides an overview of World Health Organization (WHO)-recommended tools to facilitate TB screening and diagnosis among PLHIV, Part II provides a selective overview of tools and tests currently in the later stages of the TB diagnostic pipeline and Part III provides a clinically-oriented, step-wise approach for diagnosing TB in PLHIV in resource-limited settings.

Keywords HIV · Tuberculosis · Screening · Diagnosis · Point-of-care · Drug susceptibility testing · Microscopy · Culture · Xpert · LAM

A. D. Kerkhoff (✉)
Division of HIV, Infectious Diseases and Global Medicine, University of California San Francisco, San Francisco, CA, USA
e-mail: Andrew.Kerkhoff@ucsf.edu

A. Cattamanchi
Division of Pulmonary and Critical Care Medicine and Center for Tuberculosis, University of California San Francisco, San Francisco, CA, USA
e-mail: Adithya.Cattamanchi@ucsf.edu

I. Sereti et al. (eds.), *HIV and Tuberculosis*,
https://doi.org/10.1007/978-3-030-29108-2_7

Introduction

Of the estimated 1.2 million tuberculosis (TB) cases among people living with HIV (PLHIV), less than half are diagnosed and reported to health authorities [1]; this is a key factor that contributes to why TB remains the leading cause of death among PLHIV. Better diagnostics are critical to reducing the gap and, after more than 150 years, smear microscopy is finally starting to be eclipsed as the primary diagnostic method for TB diagnosis in high burden countries. Since 2010, the World Health Organization (WHO) has endorsed several new diagnostic tools including (1) Xpert MTB/RIF, a semi-automated molecular assay that has higher sensitivity than smear microscopy and can identify rifampin resistance; (2) Determine TB-LAM, a lateral flow assay that can detect lipoarabinomannan (LAM) in urine of the sickest HIV/AIDS patients in less than 30 min at the bedside [2, 3]; and (3) line probe assays (LPAs) that rapidly identify mutations conferring resistance to first and second line anti-TB drugs in reference laboratories [4]. The devastating toll of TB on PLHIV has also led to guidelines emphasizing the need for systematic screening rather than reliance on passive case detection alone.

This chapter will review current approaches to screening and diagnosis of HIV-associated TB, including drug-resistant TB, in adults. The chapter is organized into three parts: **Part I** provides an overview of WHO-recommended tools to facilitate screening for and diagnosis of HIV-associated TB, **Part II** provides a selective overview of tools and tests currently in the later stages of the TB diagnostic pipeline and **Part III** provides a clinically-oriented, step-wise approach for diagnosing TB in PLHIV in resource-limited settings. Of note, the diagnosis of latent tuberculosis infection (LTBI) is covered separately in the chapter "Recent Advances in the Treatment of Latent Tuberculosis Infection Among Adults Living with HIV Infection", the diagnosis of TB immune reconstitution inflammatory syndrome (IRIS) is covered in the chapter "The Tuberculosis-Associated Immune Reconstitution Inflammatory Syndrome (TB-IRIS)" and the diagnosis of pediatric TB disease is covered in the chapter "HIV and Tuberculosis in Children".

Part I: Overview of Screening Tools and Diagnostic Tests for HIV-Associated TB

Types of Available Tests for HIV-Associated TB and Desired Characteristics

Tools for identifying patients with HIV-associated TB can be broadly organized into one of two categories: screening (typically non-microbiological assays) and diagnostic (typically microbiological assays) tools.

Screening tools are ideally simple, low-cost and can be used at the point-of-care to differentiate between people living with HIV (PLHIV) with a low probability of having active TB who can be safely started on TB preventative therapy and PLHIV with an increased likelihood of having active TB who should undergo further microbiological testing. A positive result, however, does not provide confirmation of TB disease. The WHO has proposed that a screening tool/test for TB should be at least 90% sensitive (to make it very unlikely that those screening negative have TB and can therefore safely start TB preventative therapy) and at least 70% specific (to reduce the number of unnecessary confirmatory tests by limiting false-positive results) [5].

Microbiological tests directly detect the presence of *Mycobacterium tuberculosis* (MTB) in a clinical specimen, providing confirmation of a TB diagnosis in the correct clinical setting. An ideal microbiological test would be rapid, inexpensive, have minimal infrastructure requirements and be available for use (and provide results) at the point-of-care [5]. Traditionally, microbiological assays for TB have included acid fast bacilli (AFB) smear microscopy and culture-based methods. However, rapid tests based on molecular methods (Xpert and Xpert Ultra) or detection of TB antigens (lipoarabinomannan) have emerged from the pipeline. The WHO has proposed that new diagnostic tests for TB should have excellent specificity (>98%) to minimize false-positive results and that the sensitivity should be >80% [5].

Below we outline and discuss WHO-recommended screening and diagnostic tools for TB, highlighting their performance among PLHIV. A discussion of tests available for monitoring response to TB therapy is beyond the scope of this chapter.

Tools and Tests for TB Screening

Symptom-Based Screening Rules

Because of the non-specific symptoms of TB in PLHIV, often including an absence of cough, many HIV-associated TB diagnoses are missed. Standardized, symptom-based screening can help maximize case detection. In 2011, a meta-analysis evaluating different symptom screening rules for HIV-associated TB found that the presence of any one of four symptoms—cough, night sweats, fevers, or weight loss (of any duration)—had a sensitivity of ~79% and specificity of ~50%. This corresponded to a negative predictive value of >90% when TB prevalence ranged from 5% to 20%. On the basis of this study, in 2011, the WHO recommended screening all PLHIV for TB using this screening rule at every clinical encounter, regardless of reason for presentation [6]. PLHIV who screen positive should undergo further microbiological testing, ideally with sputum Xpert, while those testing negative should be evaluated for initiation of TB preventative therapy [7]. More recently, a meta-analysis found that the symptom screen was associated with poor sensitivity among PLHIV receiving antiretroviral therapy (ART) (51%) compared to those who were ART-naive (89%) [8]. It also found that specificity among ART-naive patients was only 28%. These data highlight the urgent need for improved TB screening tools.

The clinical application of this screening rule within a TB diagnosis algorithm as well as its limitations are further described in **Part III**, step 2.

Radiologic Screening Tools

Chest X-Ray

Chest X-ray has long been a mainstay of TB diagnostic algorithms. There is no single chest radiographic pattern that is pathognomonic for TB, especially in PLHIV where significant variation in radiographic patterns across CD4 strata are observed. This is because more advanced immunodeficiency is associated with an impaired local tissue inflammatory response and results in reduced consolidation, fibrosis and cavitation [9]. PLHIV with a greater degree of immunosuppression are more likely to demonstrate a lower lobe and miliary pattern; however, those on ART and with well-controlled disease may manifest more typical patterns (as seen in HIV-negative persons), such as upper lobe infiltrates with or without cavitation. The diagnostic performance of chest X-ray for detecting HIV-associated TB is dependent on the definition applied to determine an 'abnormal chest X-ray' as well as the average CD4 count of the population in which a study is being conducted. It is well-recognized that those with pulmonary TB (PTB) may have completely normal chest imaging (up to 30%) [10–13]. Thus, a normal chest-X-ray does not exclude the diagnosis of active TB disease. Chest X-rays are non-specific as a patient may have alternative lung pathology accounting for radiographic lesions and they are also subject to both intra- and inter-reader variability. Radiographic findings in PLHIV with TB are discussed in greater depth in the chapter "Clinical Manifestations of HIV-Associated Tuberculosis in Adults".

In PLHIV, chest-X-rays may be complementary to symptom-based screening and serve as an important screening tool for active TB disease. Notably, a meta-analysis found that among patients receiving ART, the addition of chest radiography to the WHO symptom screen increased sensitivity for active TB from 52% (95% CI 38–66) to 85% (95% CI 70–93) [8]; because this results in a substantial improvement in the negative predictive value, TB preventive therapy can be initiated with greater confidence in such patients. Additionally, chest X-rays may provide rapid clues towards a diagnosis in those in whom TB is suspected and Xpert testing (or sputum microscopy) is negative, unavailable or result turnaround time may delay initiation of possibly life-saving therapy (i.e., severely ill patients) [7]. The use of chest X-rays within the TB diagnosis algorithm is discussed in **Part III**, steps 2 and 3.

There have been several recent advances in chest radiography. Digital chest X-rays are now available that may be associated with lower radiation doses, more immediate results without the requirement for film, improved image quality, while also allowing for the transmission and storage of images. They are also associated with lower operational costs when compared to film-based X-rays [14]; however, substantial upfront costs have limited their uptake. Furthermore, there are now portable digital X-ray machines that can allow the technology to be decentralized and integrated into mobile screening units/programs. Computer-aided algorithms have been developed to systematically read digital chest X-rays and detect abnormalities that may be compatible with PTB. A systematic review found that, while available evidence was limited, new computer-aided algorithms are likely as good as novice readers and likely approach the diagnostic accuracy of expert radiologists [15]. HIV prevalence among patients included in the meta-analysis ranged from 33% to 68%, however only one study explicitly reported the sensitivity and specificity of a computer-aided diagnosis (CAD) program for scoring chest X-rays in PLHIV [16]. Among 57 PLHIV in Zambia with Xpert-confirmed pulmonary TB, a CAD score >60 was associated with a sensitivity of 100%, but specificity was only 18% [16]. Automated, computer-aided algorithms are not recommended by the WHO at this time due to insufficient evidence [14], but are due to be formally evaluated by the WHO in the near future.

Ultrasound for Extra-Pulmonary TB (EPTB)

Ultrasonography is available as a portable, hand-held device with a number of clinical applications. It can rapidly identify abnormal signs that in high incidence settings may suggest EPTB. There is a standardized protocol for the assessment of HIV-associated TB called FASH (focused assessment with sonography for HIV-associated TB). FASH includes two different types of assessments [17]. The FASH basic assessment attempts to identify the presence of a pericardial effusion (possible pericardial TB), a pleural effusion (possible pleural TB) or ascites (possible abdominal TB). The FASH-plus examine requires greater skill and user experience, but looks for the presence of periportal/para-aortic lymphadenopathy (possible abdominal TB), focal liver lesions (possible liver abscesses due to TB) and focal splenic lesions (possible splenic abscesses due to TB). Several studies have demonstrated the utility of ultrasound to improve and expedite the diagnosis of EPTB, especially abdominal and pericardial disease [10, 18, 19]. One important limitation of ultrasound is its lack of specificity, as findings may be mimicked by other opportunistic infections, Kaposi sarcoma and lymphoma [10].

Microbiological Assays (Confirmatory Tests) for TB

Smear Microscopy

AFB smear microscopy remains the most commonly available microbiological test for TB in most low-resource settings as it is simple, rapid and relatively inexpensive. There are two different staining techniques that can be utilized to evaluate for AFB – Ziehl-Neelsen (ZN) staining is used with light microscopy and auramine fluorochrome staining is used with fluorescence microscopy. When available, fluorescence microscopy is preferred over light microscopy as it allows for more rapid scanning of sputum smears at low magnification and has improved sensitivity when compared to light microscopy [20]. Traditional fluorescence microscopy requires dark room isolation and expensive equipment with ongoing need for replacement bulbs. However, light-emitting diode (LED) fluorescence microscopes are less expensive and have fewer technology requirements. LED fluorescence microscopy is being increasingly utilized in resource-limited settings.

Although widely available, smear microscopy has low and variable sensitivity, particularly for HIV-associated TB. One systematic review found that sensitivity of sputum smear microscopy for HIV-associated PTB ranged from 39% to 76% [21]. The sensitivity of smear microscopy for EPTB varies by sample type, however is generally poor (0–40%) [22, 23], given the often paucibacillary nature of disease. Other disadvantages of smear microscopy include that results are operator-dependent, it cannot differentiate MTB from non-tuberculous mycobacteria (NTM) and it is unable to identify drug resistance.

Culture

Growth-based detection of MTB remains the gold-standard for the diagnosis of all forms of HIV-associated TB (pulmonary and extra-pulmonary) as it has the highest sensitivity and specificity. Culture can be performed using solid or liquid media. Solid media culture is typically less sensitive and takes longer than liquid media culture, but is less expensive. However, both methods require weeks to provide results, substantial laboratory infrastructure (including biosafety requirements) and highly trained staff. Liquid culture is also prone to contamination and thus rapid specimen transport and quality assurance protocols are crucial. These requirements typically preclude the use of culture-based methods for routine diagnosis of TB in poorly resourced, high burden countries. However, culture-based methods are commonly available at referral laboratories and remain the primary method for drug susceptibility testing, particularly for second-line anti-TB drugs.

Xpert MTB/RIF Assay

The Xpert MTB/RIF assay (Cepheid Inc., Sunnyvale, CA, USA) is a nucleic acid amplification test (NAAT) that utilizes a semi-automated, cartridge-based system to detect MTB and the presence of RIF resistance within 2.5 h [24]. Single-use plastic cartridges that contain the necessary buffers and reagents for sample processing, DNA extraction and real-time PCR are loaded with a clinical specimen that has been treated with a sample reagent. The cartridge is then loaded into the GeneXpert PCR platform. Five overlapping molecular probes (A-E) that span the entire rpoB core region (81 base pairs) are used to detect the presence of MTB. The probes bind to a matching sequence in the clinical specimen producing a fluorescence signal, indicating the presence of one of the gene sequences. The number of PCR cycles required to detect a minimum fluorescence signal is called a 'cycle threshold (C_T)' and the assay will terminate after 38 cycles [24]. When at least two of the five probes produce a positive signal in less than 38 cycles, MTB is detected. The assay provides one of the following results for TB diagnosis: (1) MTB not detected, (2) MTB detected (high, medium, low or very low), or (3) 'error', 'invalid' or 'no result.' In addition, when MTB is detected, RIF resistance results are also reported as (1) RIF resistance detected, (2) RIF resistance not detected, or (3) RIF resistance indeterminate. The C_T is also reported with a positive Xpert result and provides an approximation of bacillary burden. Studies have found that a C_T value cutoff of ≤28 corresponds to a high bacillary burden and predicts sputum smear-status [25, 26].

Among PLHIV, Xpert has a pooled sensitivity of 97% (95% CI 90–99) for smear-positive PTB and a sensitivity of 61% (95% CI 40–81) for smear-negative PTB [27]; its overall pooled sensitivity is 79% (95% CI 70–86) and pooled specificity is 98% (95% CI 96–99). The sensitivity for EPTB ranges dramatically by sample type (corresponding to disease site) [28]. It performs best on bone/joint, lymph node and urine samples (sensitivity 82–88%), moderately for TB meningitis (sensitivity 71%) and less favorably on pericardial, pleural and peritoneal fluid samples (<31–66%). It should be noted that the sensitivity of urine Xpert (pooled estimate 83%) is among those with genitourinary disease; it has decreased performance when used for testing all PLHIV regardless of symptoms [29–31]. Table 1 summarizes the diagnostic accuracy of Xpert for important non-respiratory samples.

In 2010, the WHO recommended that Xpert replace sputum microscopy as the initial test for the microbiological evaluation of PTB in PLHIV. Subsequent WHO recommendations also endorsed Xpert MTB/RIF as the first line assay for EPTB in PLHIV as well as the first-line diagnostic for the rapid detection of RIF resistance in those with confirmed TB [7].

Table 1 Pooled estimates of sensitivity and specificity of Xpert MTB/RIF for different forms of EPTB (adapted from Kohli et al.) [28]

	Number of patients	Number of specimens with culture-confirmed TB	Pooled sensitivity (95% CI)	Pooled specificity (95% CI)
TB of blood (Disseminated TB)				
Blood	266	23	(Numbers insufficient)	(Numbers insufficient)
TB of genitourinary tract (renal TB)				
Urine	1199	73	82.7 (69.6–91.1)	98.7 (94.8–99.7)
TB of lymph node (TB lymphadenitis)				
Lymph node aspirate	1710	671	87.6 (81.7–92.0)	86.0 (78.4–91.5)
Lymph node tissue	484	147	84.4 (74.7–91.0)	78.9 (52.6–91.5)
TB meningitis				
Cerebrospinal fluid	3774	433	71.1 (60.9–80.4)	98.0 (97.0–98.8)

(continued)

Table 1 (continued)

	Number of patients	Number of specimens with culture-confirmed TB	Pooled sensitivity (95% CI)	Pooled specificity (95% CI)
TB of musculoskeletal system				
Bone or joint fluid	385	58	97.2 (89.5–99.6)	90.2 (55.6–98.5)
Bone or joint tissue	618	179	82.0 (56.6–94.9)	91.8 (70.1–98.4)
TB of pericardium (pericardial TB)				
Pericardial fluid	324	76	65.7 (46.3–81.4)	96.0 (85.8–99.3)
TB of peritoneum (peritoneal TB)				
Peritoneal fluid	712	115	59.2 (45.2–73.5)	97.9 (96.2–99.1)
TB pleurisy (pleural TB)				
Pleural fluid	4006	607	50.9 (39.7–62.8)	99.2 (98.2–99.7)
Pleural tissue	207	71	30.5 (3.5–77.8)	97.4 (92.1–99.3)

For all forms of EPTB except pleural TB, solid or liquid mycobacterial culture was used as the reference standard. For pleural TB, either culture or the presence of granulomatous inflammation on histopathological examination defined the reference standard

Unfortunately, cost remains an issue even with subsidized pricing for the GeneXpert platform and Xpert MTB/RIF cartridges (~$10/cartridge). The GeneXpert platform is also sensitive to heat and dust, requires a continuous power supply to operate as well as ongoing maintenance [32]. For these reasons, Xpert testing has mainly been available in higher-level health facilities in high burden countries. Several studies have shown that implementation of Xpert has resulted in increased detection of mycobacteriologically-confirmed TB, reduced time to diagnosis and reduced time to TB treatment. The implementation of Xpert has been associated with a mortality reduction in some settings [33, 34], however, this has not been a universal finding, as the majority of trials did not find a survival benefit associated with its use [35–42].

Xpert MTB/RIF Ultra (Xpert Ultra) Assay

The Xpert Ultra cartridge utilizes the existing GeneXpert platform, but incorporates two new multi-copy amplification targets (IS6110 and IS1081) and a larger DNA amplification reaction chamber than the original Xpert cartridge. This contributes to an improved lower limit of detection compared to the original Xpert cartridge (16 vs 114 bacterial colony forming units per milliliter), and increased sensitivity [43].

The Xpert Ultra test adds a new result category, 'trace-positive', which corresponds to the lowest bacillary burden for MTB detection. A large multi-country evaluation found that among PLHIV, Xpert Ultra increased the sensitivity for the detection of PTB by 13% (95% CI 6–21) compared to Xpert (90% versus 77%) [44]. However Xpert Ultra was also associated with a small decrease in specificity (2.7%). Specificity was higher when not considering trace results to be positive, and when excluding patients previously treated for TB [44]. Evaluations of Xpert Ultra for EPTB are limited among PLHIV, however a study evaluating its utility for detecting TB meningitis (TBM) found that the sensitivity for probable or definite TBM in PLHIV was 70% (95% CI 47–87), compared to 43% (95% CI 23–66) using either Xpert or culture [45]. On the basis of these early, but highly encouraging results, in 2017 the WHO recommended that the Xpert Ultra cartridge replace the original Xpert cartridge as the first line test for HIV-associated TB (pulmonary and extra-pulmonary samples) [46].

Lipoarabinomannan (LAM)

LAM comprises a group of lipopolysaccharides within the cell wall of MTB [3]. A commercially available lateral-flow urine assay, called 'Determine TB-LAM' (Alere Inc. Waltham, Massachusetts, USA), was the first truly rapid, inexpensive, point-of-care assay available for the diagnosis of HIV-associated-TB. The assay is a lateral-flow, urine-based, dip-stick assay (henceforth known as 'LF-LAM') that does not have any storage requirements, has minimal training requirements and is capable of providing results within 30 min at the point-of-care [3]. The assay currently costs between $2.50 and $3.00 a test. The sensitivity of LF-LAM strongly correlates with the immune status of HIV-patients as demonstrated by a meta-analysis that showed sensitivity in patients with CD4 count <100 cells/μL was 56% (95% CI 41–70) compared to 26% (95% CI 16–46) in patients with CD4 count >100 cells/μL [47]. Similarly, sensitivity was greater among hospitalized patients than among ambulatory outpatients (53% versus ~20%). Pooled specificity was found to be 92%, but approaches 99% when a rigorous reference standard is utilized [48, 49].

While the LF-LAM assay has only moderate sensitivity among immunocompromised HIV patients, it rapidly detects TB in the sickest patients at the highest risk for poor outcomes [50]. For example, one study found that LF-LAM detected TB in two-thirds of all patients with evidence of mycobacteremia, including all patients dying within 90 days [51]. Furthermore, a meta-analysis among HIV patients found that mortality was 2.5-fold higher among those with a positive versus a negative LF-LAM result [2]. Notably, two randomized trials have evaluated the addition of LF-LAM to the local diagnostic standard of care in sub-Saharan Africa and have demonstrated a mortality reduction associated with its use among those with a CD4 count <100 cells/μL [52, 53]. This mortality benefit likely reflects the ability to more rapidly detect TB and start potentially life-saving anti-TB therapy. LF-LAM was conditionally recommended by the WHO in 2015 for use in PLHIV with signs and symptoms of TB (pulmonary and/or extra-pulmonary) who either have a CD4 count ≤100 cells/μL or who are seriously ill with any 'danger signs' as defined by the presence of respiratory rate >30, temperature > 39.0° C, heart rate > 120 beats per minute, or inability to ambulate unassisted (independent of CD4 count) [54]. Since 2015, a number of additional studies have reported on the diagnostic performance of LF-LAM among PLHIV; in 2019 the WHO is expected to reappraise the available evidence and issue updated guidance on the use of LF-LAM.

Loop-Mediated Isothermal Amplification (LAMP)

TB LAMP (Eiken Chemical Company Ltd. Tokyo, Japan) is a rapid assay that can provide results in less than 1 h. It uses a temperature-independent method for DNA amplification that is easy to use, requires minimal laboratory infrastructure and that can be read using the naked eye under ultraviolet light. However, the assay has several steps and requires trained laboratory personnel. A systematic review was undertaken in 2016 to evaluate its diagnostic performance against smear microscopy as well as Xpert [55]. There was limited data available among PLHIV. Overall, the sensitivity of TB LAMP for pulmonary TB ranged from 64% to 73% and its specificity from 95% to 99% depending on the reference standard used. On the basis of these results, TB LAMP was recommended by the WHO as a replacement for sputum smear microscopy or as a follow-on test after a negative sputum smear result [55]. However, the WHO advised that TB LAMP should not replace Xpert where available, and felt that there was insufficient evidence to recommend the use of TB LAMP for non-respiratory samples or for testing for TB among PLHIV.

Detection of TB Drug-Resistance (Drug-Susceptibility Testing)

Only one-quarter of RIF-resistant (RR) and multi-drug resistant (MDR)-TB cases worldwide are detected each year. The rapid and accurate detection of drug resistance is important to the individual and to public health alike. For the individual, rapid drug susceptibility testing (DST) allows for initiation of the most effective anti-TB regimen as soon as possible, which allows for the highest likelihood of cure. For the community, rapid DST can help to minimize the transmission of drug-resistant TB, help to guide appropriate care for contacts and help prevent the spread of drug-resistant TB. The END TB strategy rightfully calls for universal access to DST [56].

DST is broadly comprised by two major methodologic categories – growth-based (phenotypic) and molecular-based (genotypic). Generally, culture-based DST is thought to be more reliable than molecular methods because an MTB isolate is grown on a culture media containing the critical concentration of a given anti-TB agent. It is therefore typically assumed that if growth of MTB is inhibited by that agent on DST, that same agent should be reliably effective for the patient's isolate in vivo; however, up to 5% of wild-type strains may be classified as resistant, in part likely due to limitations of critical-concentration methods [57, 58]. Additionally, there are reports of specific rpoB mutations that confer rifampicin resistance not being detected on liquid culture DST [59]. This is compared to molecular methods that detect known mutations for drug resistance. If all resistance mutations are not known or included in the probe, drug-resistance using molecular techniques may be underdiagnosed in a proportion of patients.

One important difference between growth-based and molecular-based DST is the requirement for a pure MTB isolate to be obtained from either solid or liquid culture media before culture-based DST can be performed. When coupled with the further requirement to monitor growth (or lack of growth) in the setting of agar or liquid culture media containing a specific drug, the overall process can take several weeks to months. Molecular methods not only provide for more rapid results, but also offer standardized testing with fewer biosafety requirements; both of which may allow for increased throughput. On this basis, the WHO recommends that molecular methods for TB DST be performed in addition to culture-based DST whenever available [60, 61].

Culture-Based Methods for DST

Phenotypic methods, or culture (growth)-based DST remain the gold standard for DST. There are multiple methods available and in clinical use. Critical concentrations, not minimum inhibitory concentrations (MIC), are used to determine the susceptibility or resistance of anti-TB agents for a given culture isolate. The critical concentration is defined as the lowest concentration that reliably inhibits >99% of wild-type MTB complex strains in vivo, while also not inhibiting strains considered to be resistant [62]. The critical concentration varies slightly between culture media and in 2018 the WHO published standard critical concentrations for most first-, second- and third-line agents [62].

Solid media-based DST (the indirect agar proportion method) most commonly utilizes Lowenstein-Jensen, Middlebrook 7H10 or 7H11 agar. Using this technique, a culture isolate is directly inoculated into a quadrant of the plate. Three quadrants contain a specific anti-TB agent at its critical concentration, while one quadrant without a drug serves as a control. After 21 days colony counts are taken and if the number of colonies in a drug-containing quadrant is >1% of the colonies in the control quadrant, the isolate is considered to be resistant to that drug.

Liquid media-based DST has faster turnaround time when compared to solid media-based techniques, with results available in as little as 7–10 days after inoculation. There are several commercially available platforms, but WHO critical concentrations are only available for the MGIT 960 platform (Becton Dickinson, Sparks, MD). The MGIT 960 platform can provide DST for first- and second-line agents. The method is based on fluorescence that is produced from the MGIT medium when bacterial growth results in reduced oxygen. The amount of fluorescence generated is then converted to growth units (GU), where greater GU corresponds to more growth. If a drug containing tube yields a GU < 100 at the end of incubation then the organism is considered susceptible, while a GU ≥100 is considered resistant.

Molecular Methods for DST

There are several benefits associated with molecular methods compared to growth-based methods. The most important is the short turnaround time for DST results, which may be as few as 1–2 days as compared to at least several weeks associated with culture-based methods. Additionally, unlike culture-based methods, molecular methods can be run on smear-positive/ culture-negative specimens, as well as fixed pathology specimens. In general, there are two broad categories of molecular methods available for DST: sequencing and non-sequencing based techniques. Currently, non-sequencing methods predominate especially in low- and middle-income settings. However, sequencing-based methods are expected to become more simplified and increasingly affordable, which will likely translate to increased availability over the next several years.

Xpert and Xpert Ultra

As noted previously, the Xpert assay is able to rapidly detect RIF resistance in clinical specimens in which MTB is confirmed and results are provided within 2.5 h. It is recommended by the WHO as the first-line assay for the rapid detection of RIF-resistance and has become the most widely available assay for TB DST globally. Its pooled sensitivity and specificity for the detection of RIF-resistance in patients with HIV-associated TB is 95% (95% CI 90–97) and 98% (95% CI 97–99), respectively [27].

The Xpert Ultra cartridge utilizes a new melt curve analysis to detect RIF-resistance and data to date suggest that the Xpert Ultra cartridge provides similar (non-inferior) diagnostic accuracy for the detection of RIF resistance compared to the traditional Xpert cartridge [44]. As described in **Part III**, the detection of RIF resistance by Xpert and Xpert Ultra testing should prompt further DST for first and second line anti-TB agents (injectable agents and fluoroquinolones).

Line Probe Assays

Line probe assays (LPA) are a type of molecular test that permit the detection of *M. tuberculosis* complex, as well as mutations associated with TB drug resistance. LPA involves a multi-step process that includes: (1) DNA extraction, (2) PCR-based amplification of known resistance determining regions using primers, (3) reverse hybridization of amplicons to probes affixed on the assay strip and (4) colorimetric detection of captured hybrids allowing for visualization of bands. LPAs can be performed on DNA extracted from clinical specimens (direct method) or from culture isolates (indirect method). LPAs can detect specific mutations known to be associated with drug resistance but can also indirectly indicate drug resistance when a mutation is present in one of the target regions, resulting in the amplicon not hybridizing with a wild-type probe [63].

The WHO has made formal recommendations for two commercially available LPAs that detect drug resistance associated with RIF and INH (first-line agents). These include the GenoType MTBDRplusv2.0 (Hain Lifescience, Nehren, Germany) and the Nipro NTM + MDRTB Detection Kit 2 (Nipro, Tokyo, Japan). The diagnostic accuracy of both assays for the detection of RIF and INH resistance directly on smear-positive sputum samples was evaluated and found to be comparable (~97–98%, ~95% sensitive for RIF and INH resistance, respectively; ~98% specific for RIF and INH resistance) [60]. However, the sensitivity of both assays for indirect testing of MTB culture isolates was lower, ~90–91% [60]. On the basis of these results, the WHO recommended that for persons with sputum smear-positive disease or any culture-isolate positive for MTB complex, either LPA (MTBDRplusv2.0 or NTM+MDRTB Detection Kit 2) may be used as the initial test for the rapid detection of RIF and INH resistance in addition to conventional culture-based DST [60]. There is limited data available that specifically evaluate the performance of these LPAs among PLHIV, however one study suggested that the MTBDRplusv2.0 had excellent sensitivity for the detection of RIF resistance (>90%), but only moderate sensitivity for the detection of INH resistance (~70%) [64]. Incomplete sensitivity for INH resistance likely reflects the fact that additional resistance conferring mutations are not included in the assay.

The GenoType MTBDRsl (Hain Lifescience, Nehren, Germany) version 1.0 was the first commercially available LPA able to rapidly detect mutations associated with resistance to second-line agents, thus allowing for the diagnosis of MDR-, pre-extensively drug resistant- (XDR) and XDR-TB [61]. The assay can detect the presence of MTB complex, mutations associated with fluoroquinolones (ofloxacin, levofloxacin, moxifloxacin, gatifloxacin) and second-line injectable agents (kanamycin, amikacin, capreomycin). The pooled sensitivities and specificities of the version 1.0 assay for second-line TB drugs are shown in Table 2.

The manufacturer has subsequently introduced a newer generation of the GenoType MTBDRsl assay (version 2.0) that detects additional resistance mutations as well as all identified by the version 1.0 assay. There is limited published data on its diagnostic accuracy specifically among PLHIV, however, in one study testing 268 respiratory isolates from a high burden HIV-associated TB setting, the sensitivity of the version 2.0 assay for fluoroquinolones (100%; 95% CI 96–100) and second-line injectable agents (89%; 95% CI 79–96) was excellent and was associated with a specificity >98.5% for all agents with the exception of capreomycin (95.9%) [65].

In patients with either confirmed RR-TB or MDR-TB (detected using Xpert, LPA or culture-based methods), the WHO recommends that the MTBDRsl assay may be used as the initial test (in addition to culture-based DST) to rapidly detect

Table 2 Pooled sensitivity and specificity estimates of GenoType MTBDRsl v1.0 for fluoroquinolones and second-line injectable agents using conventional culture-based DST reference standard[a]

	Number of patients	Pooled sensitivity (95% CI)	Pooled specificity (95% CI)
Fluoroquinolones, direct testing	1771	86.2 (74.6–93.0)	98.6 (96.9–99.4)
Ofloxacin	1667	90.9 (84.7–94.7)	98.9 (97.8–99.4)
Moxifloxacin	821	95.0 (92.1–96.9)	99.0 (97.5–99.6)
Fluoroquinolones, indirect testing	2223	85.6 (79.2–90.4)	98.5 (95.7–99.5)
Levofloxacin[b]	169	80.0–100[b]	96–100[b]
Ofloxacin	1927	85.2 (78.5–90.1)	98.5 (95.6–99.5)
Moxifloxacin	419	94.0 (82.2–98.1)	96.6 (85.2–99.3)
Second-line injectable agents, direct testing	1639	87.0 (38.1–98.6)	99.5 (93.6–100)
Amikacin	1491	91.9 (71.5–98.1)	99.9 (95.2–100)
Capreomycin	1027	76.6 (61.1–87.3)	98.2 (92.5–99.6)
Kanamycin	1020	78.7 (11.9–99.0)	99.7 (93.8–100)
Second-line injectable agents, indirect testing	1921	76.5 (63.3–86.0)	99.1 (97.1–99.7)
Amikacin	1301	84.9 (79.2–89.1)	99.1 (97.6–99.6)
Capreomycin	1406	79.5 (58.4–91.4)	95.6 (93.4–97.3)
Kanamycin	1342	66.9 (44.1–83.8)	98.6 (96.1–99.5)

[a]For the MTBDRsl v2.0 there was insufficient data to undertake a meta-analysis or compare direct and indirect testing
[b]Insufficient data precluded pooled estimates; numbers represent ranges from study point estimates

resistance associated with fluoroquinolones or second line injectable agents on (1) sputum samples (irrespective of smear status—*direct testing*) or (2) cultured isolates of MTB complex from any respiratory or non-respiratory samples (*indirect testing*) [61].

Part II: Novel Approaches to Diagnosis of HIV-Associated TB

Overview

There are considerable ongoing efforts to develop TB tests that are faster, cheaper, simpler and can be performed on samples that are easier to collect than sputum. These range from discovery phase studies that seek to identify and validate novel biomarkers in blood, urine and breath, to the development and evaluation of new technologies to facilitate sample processing and analysis [4, 66]. In this section, we will provide a selective overview of tools in the TB diagnostic pipeline that are either at the later stages of development or have later phase clinical data published and focus on tests and platforms that we anticipate will improve the diagnosis of HIV-associated TB in the near future.

For the most up-to-date information on the TB diagnostics pipeline, the Foundation for New Innovative Diagnostics (FIND) has developed an online, interactive diagnostics pipeline that shows the current status and estimated release dates of various diagnostic tools and assays. Please visit: **https://www.finddx.org/tb/pipeline/**

Tools and Tests for TB Screening

Clinical Prediction Scores

Clinical prediction scores may combine symptoms as well as easily obtained clinical information (body mass index, vital signs, ART status) with routinely available laboratory tests (hemoglobin, CD4 cell counts) to direct diagnostic testing for HIV-associated TB. Notably, two clinical prediction scores have been studied among ambulatory HIV patients screening positive using the WHO symptom screen [67, 68]. Both studies propose that a defined cutoff could be used to safely reduce the overall number of patients requiring further TB testing without missing a large number of TB cases. One of the clinical scores utilized ART status (ART >3 months vs. pre-ART or ART <3 months), body mass index, CD4 count and the number of WHO symptoms present (1 versus >1 symptom). When used among those with a positive WHO symptom screen, a cutoff score of 3 had a sensitivity and specificity for HIV-associated TB that was 92% and 34%, respectively and would have resulted in a >30% reduction in need for further TB testing while missing <10% of all TB diagnoses (predominantly among those on ART and with higher CD4 cell counts) [68].

C-Reactive Protein (CRP)

CRP is an acute phase reactant that is detectable in serum and can be rapidly measured at the point-of-care [69]. A systematic review among predominantly ambulatory PLHIV found that the sensitivity and specificity of CRP (cutoff: 10 mg/L) for the detection of confirmed pulmonary TB was 93% (95% CI 88–98) and 60% (95% CI 40–75), respectively [70]. Prospective studies in Uganda and South Africa have demonstrated that point-of-care CRP testing has similar sensitivity when compared to the WHO symptom screen (~90%), but has significantly improved specificity (59–72%) [69, 71]. Dependent on the CRP cut-off level used, the specificity associated with CRP is 21–58% higher than that of symptom screening. While these results must be further validated, the results suggest that use of CRP in place of the WHO symptom screen as part of intensified case finding for PLHIV would detect a similar number of HIV-associated TB cases, while significantly reducing the number of patients requiring further TB investigations by >50%. Indeed, one study among ambulatory PLHIV showed that CRP-based TB screening followed by confirmatory testing with LF-LAM (if CD4 count < 100), Xpert and a single liquid culture, would increase case detection relative to the currently recommended strategy [72].

Microbiological Assays (Confirmatory Tests) for TB

Next-Generation LAM Assays

Several urine-based assays that detect the presence of LAM are undergoing development and evaluation. They aim to retain the point-of-care quality of the currently available LF-LAM assay while improving upon sensitivity that would expand utility beyond only the sickest HIV patients [4]. One test, the SILVAMP TB LAM assay (FujiFilm Global, Tokyo, Japan), had a sensitivity of 70.4% compared to 42.3% using the LF-LAM assay without a significant difference in specificity, when retrospectively testing 968 urine samples from PLHIV in South Africa. Among those with a CD4 count ≤100 cells/μL, the SILVAMP TB LAM had a sensitivity of 84.2% versus 57.3% using the LF-LAM assay [73]; it also demonstrated useful sensitivity in those with CD4 counts 101–200 cells/μL—60.6% compared to 26.4% using LF-LAM. prospective evaluations of its performance will be undertaken in 2019.

Xpert Omni

In 2019, a new Xpert platform called Xpert Omni is expected to be introduced that may allow for truly point-of-care detection of MTB and the presence of RIF resistance within 2 h. The single module unit is lightweight (~1 kg), portable, battery-powered (up to 12 h rechargeable battery life) and is designed to allow for testing in more extreme clinical settings. Its initial cost is expected to be ~$5,000 per device and it will require special cartridges which will be ~$1.50 more expensive than traditional Xpert cartridges (to allow for the incorporation of wireless near-field communication) [74]. It is currently undergoing feasibility studies and is expected to become commercially available in 2019.

Nucleic Acid Amplification Tests (NAAT) Other Than GeneXpert

Since the introduction of GeneXpert, there have been many companies that have sought to develop competing rapid NAAT-based assays for the diagnosis of TB. Some of the assays furthest along in development and evaluation include the

Genedrive MTB/RIF assay (Epistem Ltd., UK), TrueNat MTB RIF assay (Molbio, Goa, India), TRCReady (Tosoh Bioscience, Tokyo, Japan), EasyNAT TB assay (Ustar Biotechnologies Ltd., Hangzhou, China), RealTime MTB (and MTB RIF/INH) assay (Abbott, Chicago, USA), and the FluoroType MTB assay (Hain Lifescience, Nehren, Germany). Of these, only the TRCReady assay represents a stand-alone, semi-automated NAAT similar to Xpert; however, it does not provide simultaneous RIF resistance detection. While some of these assays are already commercially available and even in use in countries such as India and China, there are minimal published data to recommend their routine use in PLHIV [4].

Molecular Methods for DST

Xpert Xtend XDR

A new cartridge utilizing the GeneXpert platform called the Xpert Xtend XDR will test for resistance associated with isoniazid (INH) as well as fluoroquinolones and injectable aminoglycosides. The Xtend XDR cartridge is expected in 2019 and may potentially allow for decentralized, rapid detection (results available within 90 min) of resistance associated with second-line agents. An initial prototype demonstrated promising results [58].

Sequencing

Next generation sequencing (NGS) is the latest advance in the rapid detection of TB-associated drug resistance. It can be used to perform targeted and whole genome sequencing. Non-sequencing, molecular methods such as Xpert and LPAs may miss important resistance-conferring mutations if not encapsulated within the target probe(s) or may detect mutations that do not confer resistance, resulting in false-negative and false-positive results, respectively. One major advantage of NGS is its ability to identify all known mutations simultaneously.

Studies have demonstrated that NGS has good concordance with culture-based methods and NGS can be performed directly on smear-positive clinical specimens [75–78]. NGS may ultimately one day allow for more individualized treatment regimens based on knowledge of the most effective anti-TB drugs for each person. However, a number of challenges face the implementation and scale-up of NGS, especially in resource-limited settings. These include the ability to reliably extract sufficient mycobacterial DNA from clinical samples for sequencing, the cost of sequencing platforms and laboratory infrastructure requirements, as well as the need for improved means to process and analyze large amounts of raw data [4].

Part III: A Suggested Step-Wise Approach to Diagnosing HIV-Associated TB for Clinicians with a Focus on Resource-Limited Settings

Overview

Recent WHO guidelines highlight the shift towards active case finding among PLHIV as well as new diagnostic tools for rapid TB detection and DST. In the subsequent sections, we present a suggested step-wise approach for the diagnosis of HIV-associated TB using current WHO recommendations and the revised 2018 Global Laboratory Initiative (GLI) model TB diagnostic algorithms for PLHIV as a framework (Figs. 1–3) [79].

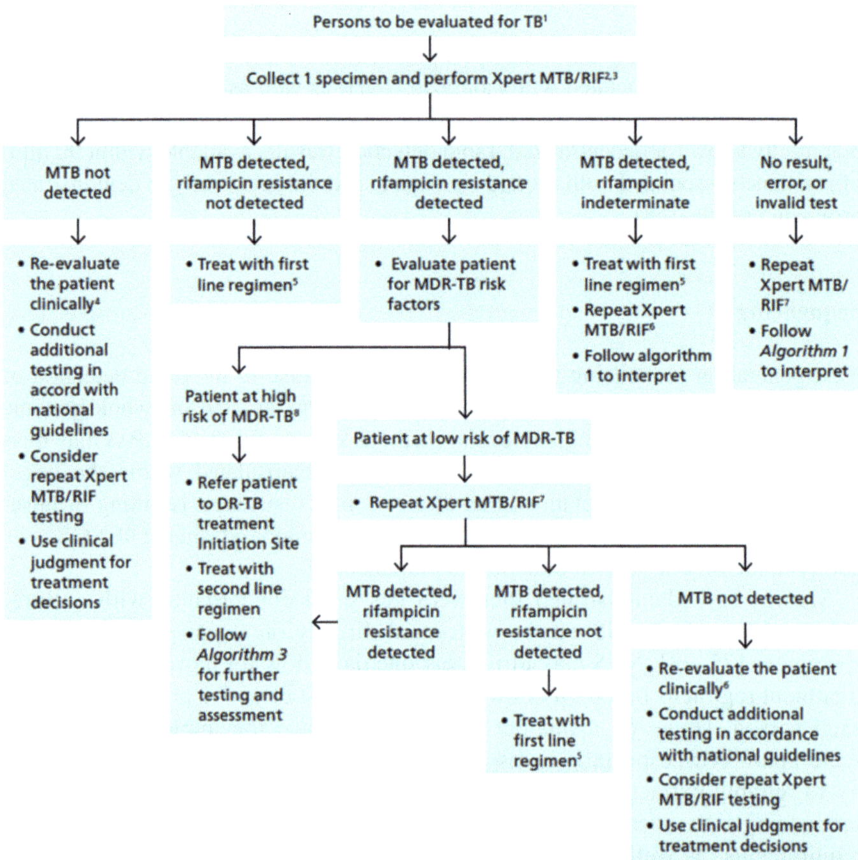

Fig. 1 WHO recommended algorithm for evaluating persons for TB (Xpert as the initial test)

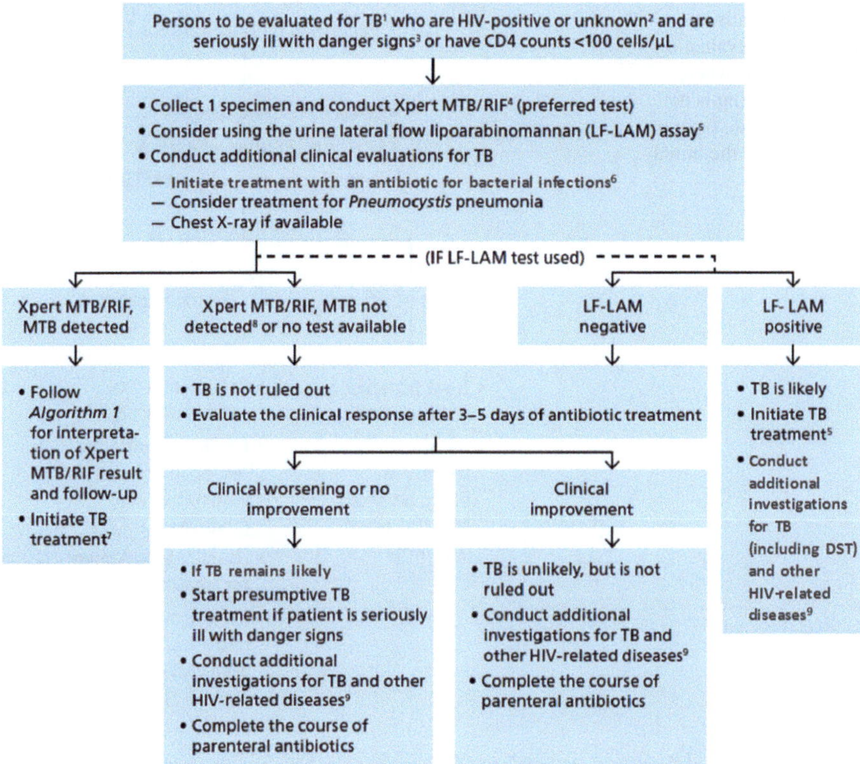

Fig. 2 Algorithm for evaluating PLHIV for TB among those who are seriously ill with danger signs or have CD4 count ≤100 cells/μL

Step 1. Who Should I Screen for HIV-Associated TB?

Current WHO recommendation: All patients with confirmed HIV (or an unknown HIV status) should be screened for TB at each health care encounter.

Further information: There are typically two broad approaches to identifying people with HIV-associated TB - passive and active case finding. Passive case finding is reliant upon symptomatic TB patients to self-present to a health-care setting followed by a health worker recognizing that their symptoms may be due to TB and ordering TB testing [80]. This approach on its own has led to substantial underdiagnosis of HIV-associated TB globally for several reasons. These include that patients with early TB disease may not be symptomatic (or symptoms may be nonspecific) and that health workers often fail to order TB testing even when indicated. In contrast, active or intensified case finding (ICF) in either facility- or community-based settings involves screening everyone within a high-risk group, such as PLHIV,

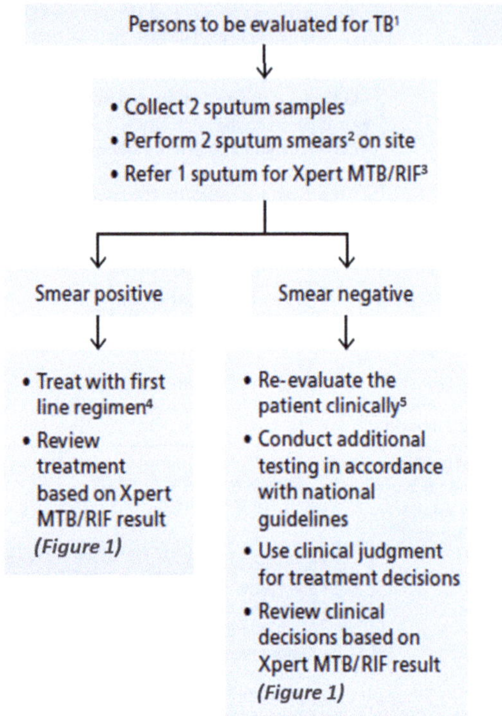

Fig. 3 Alternative algorithm for evaluating persons for TB where molecular testing is not readily available (sputum microscopy as the initial test)

followed by confirmatory diagnostic testing for those who screen positive. The goals of ICF are not only to identify more people with TB but also to identify them earlier in order to reduce morbidity and community transmission [80].

Relevant guidelines:

- Consolidated guidelines on the use of antiretroviral drugs for treating and preventing HIV infection. Geneva: WHO; 2016.
- Guidelines for intensified tuberculosis case-finding and isoniazid preventative therapy for people living with HIV in resource-constrained settings. Geneva: WHO; 2011.
- Systematic screening for active tuberculosis: Principles and recommendations. Geneva: WHO; 2013.

Step 2. How Should I Screen for HIV-Associated TB?

Current WHO recommendation: A four-part symptom screen should be used: current cough, fever, weight loss or night sweats. For PLHIV on ART, chest radiography may be considered in addition to symptom screening. Chest radiography (when available) is also recommended as a screening tool in addition to symptom-based screening in all PLHIV with a CD4 count <100 cells/μL or those presenting with 'danger signs,' regardless of symptoms (Fig. 2).

Further information: PLHIV who have a negative symptom screen with or without a negative chest X-ray are unlikely to have active TB and should be offered TB preventive therapy, regardless of ART status. In addition to the above symptom screen (with or without chest radiography), all PLHIV should have a careful history and vital signs obtained and physical exam undertaken to determine: 1) if there are 'danger signs' present (defined as any one of the following: respiratory rate > 30, temperature > 39.0C, heart rate > 120 beats per minute, or unable to ambulate unassisted) that would suggest a need for referral to a higher level of clinical care and 2) if there are any signs or symptoms that might suggest EPTB (Step 5).

Limitations of currently recommended strategy: There are several limitations of the WHO standard symptom-screening rule that has kept it from being widely implemented in high burden settings. It is not objective in that it relies on patients' self-reported symptoms. Additionally, and more pragmatically challenging, it has low overall specificity (~50%) [8], and even poorer specificity among ART-naive PLHIV (~28%) [8]. As many clinicians and policy makers point out, universal application of this recommendation would result in a large proportion of PLHIV requiring additional TB investigations, of whom only a small number might have TB. This may stretch the resources of HIV/AIDS programs as well as delay and reduce the number of patients initiated on TB preventative therapy. Furthermore, it demonstrates poor sensitivity among those receiving ART (~50%) [8]. The WHO symptom screen therefore falls short of the WHO proposed cutoffs for a screening tool - >90% sensitivity and > 70% specificity [5]. Thus, there is significant interest in developing improved screening strategies that might help better identify PLHIV who should be prioritized for TB testing.

Relevant guidelines:

- Chest radiography in tuberculosis detection – summary of current WHO recommendations and guidance on programmatic approaches. Geneva: WHO; 2016.
- Consolidated guidelines on the use of antiretroviral drugs for treating and preventing HIV infection. Geneva: WHO; 2016.
- Guidelines for intensified tuberculosis case-finding and isoniazid preventative therapy for people living with HIV in resource-constrained settings. Geneva: WHO; 2011.
- Latent tuberculosis infection: updated and consolidated guidelines for programmatic management. Geneva: WHO; 2018.
- Systematic screening for active tuberculosis: Principles and recommendations. Geneva: WHO; 2013.

Step 3: Whom Should I Investigate Further for Pulmonary TB?

Current WHO recommendation: All PLHIV who screen positive using four-part symptom screen should be investigated for active TB. Furthermore, anyone with clinical exam findings or radiology (chest X-ray, ultrasound [when undertaken]) findings potentially consistent with PTB should also be further investigated for active TB, regardless of symptoms.

Relevant guidelines:

- Chest radiography in tuberculosis detection – summary of current WHO recommendations and guidance on programmatic approaches. Geneva: WHO; 2016.
- Consolidated guidelines on the use of antiretroviral drugs for treating and preventing HIV infection. Geneva: WHO; 2016.
- Guidelines for intensified tuberculosis case-finding and isoniazid preventative therapy for people living with HIV in resource-constrained settings. Geneva: WHO; 2011.
- Improving the diagnosis and treatment of smear-negative pulmonary and extrapulmonary tuberculosis among adults and adolescents. Geneva: WHO; 2007.
- Latent tuberculosis infection: updated and consolidated guidelines for programmatic management. Geneva: WHO; 2018.
- Systematic screening for active tuberculosis: Principles and recommendations. Geneva: WHO; 2013.

Step 4. How Should I Test for PTB?

Current WHO recommendation: Xpert MTB/RIF (Xpert Ultra if available) should be used as the initial diagnostic test for PTB (Figs. 1 and 2). In addition, the LF-LAM assay should be performed in all PLHIV who are severely ill or have CD4 count <100 cells/mm^3 to enable rapid diagnosis and treatment initiation (Fig. 2). Where Xpert MTB/RIF is not readily available, sputum microscopy should be used as the initial diagnostic test for PTB (Fig. 3).

Further information:

Xpert (Ultra) for PTB: For patients with suspected PTB, one fresh sputum sample should be collected and tested using Xpert (or preferably Xpert Ultra) (Fig. 1). If the initial Xpert test result is negative, but the clinical suspicion for PTB remains high, undertaking repeat Xpert testing on a newly collected, fresh sputum specimen may be considered as this has been associated with up to a 20% increase in diagnostic sensitivity for smear-negative disease [81]. It is not yet clear if there is increased diagnostic yield associated with undertaking repeat sputum Xpert Ultra testing if the first Xpert Ultra test is negative. When Xpert Ultra testing is utilized, the WHO recommends that for PLHIV a trace positive result be regarded as a true positive result and that these patients be initiated on anti-TB therapy [46].

Smear microscopy for PTB: Where Xpert testing is not available for the investigation of PTB, it is recommended that microscopy (LED fluorescent microscopy preferred) be performed on two sputum samples to evaluate for the presence of acid-fast bacilli (Fig. 3). Same-day microscopy involves collecting two spot sputum samples at the initial health center visit and is the recommended approach as it is more patient-friendly and retains similar sensitivity and specificity when compared to multiple day sputum collection [82, 83]. Use of a concentrated sputum sample does not appear to increase sensitivity and is not recommended because it increases resource requirements [84]. A positive sputum AFB microscopy result should be confirmed as MTB (when possible) as this may represent non-tuberculous mycobacteria (NTM); however, this should not delay treatment, especially if the patient is at risk for further clinical deterioration.

Culture-based methods: When the results of rapid tests are negative, culture-based methods should be considered where resources permit. A recent study demonstrated considerable incremental yield with the addition of a single liquid culture when Xpert results are negative [72].

TB-LAMP: Where available, TB-LAMP may be used as a replacement test for sputum-smear microscopy for the diagnosis of PTB only, or may be considered as a follow-on test in those testing sputum-smear negative (see Fig. 3).

Relevant guidelines:

- Consolidated guidelines on the use of antiretroviral drugs for treating and preventing HIV infection. Geneva: WHO; 2016.
- Fluorescent light-emitting diode (LED) microscopy for diagnosis of tuberculosis. Geneva: WHO; 2011.
- GLI model TB diagnostic algorithms. Geneva: WHO; 2018.
- Same-day diagnosis of tuberculosis by microscopy. Geneva: WHO; 2011.
- Xpert MTB/RIF implementation manual – technical and operational "how-to". Practical considerations. Geneva: WHO; 2014.

Step 5: Whom Should I Investigate Further for EPTB and How Should I Test for EPTB?

Current WHO recommendation: All PLHIV with signs or symptoms of EPTB should be investigated for TB using microbiological tests. If PLHIV have respiratory symptoms or chest radiograph abnormalities, sputum-based testing with Xpert should be performed (Fig. 1). Even in PLHIV without respiratory symptoms, sputum-based testing will yield some TB diagnoses. If the results of rapid sputum-based testing (Xpert or microscopy) are negative, or PLHIV are unable to produce sputum, microbiological testing should be undertaken on non-respiratory samples corresponding to the extra-pulmonary manifestation most strongly suspected using Xpert (or Xpert Ultra). In PLHIV with signs and symptoms of TB who either have a CD4 count ≤100 cells/μL or are seriously ill (independent of CD4 count), the LF-LAM assay should be performed in parallel with sputum Xpert testing for the diagnosis of disseminated TB; its use should especially be considered in those unable to produce sputum (Fig. 2) [54].

Further Information:

Overview of EPTB: EPTB is defined as any case of TB that involves an organ or anatomic site other than the lungs. EPTB is common among PLHIV, especially those with severe immunosuppression (present in up to 90%). Disseminated and extra-pulmonary disease is associated with significant morbidity and mortality [1]; therefore, timely diagnosis is crucial. Unfortunately, the diagnosis of EPTB remains challenging given its non-specific presentations and traditional difficulty in obtaining non-respiratory samples.

Clinical and radiological features of EPTB: TB can involve almost any anatomic site, but patients will often have local signs and symptoms related to the site of their disease with or without constitutional symptoms. EPTB clinical manifestations are reviewed in greater detail in the chapter "Clinical Manifestations of HIV-Associated Tuberculosis in Adults". Clinicians should have heightened suspicion for EPTB in PLHIV presenting with a positive symptom screen as well as dyspnea (possible TB pleural effusion and/or TB pericarditis), enlarged cervical/axillary lymph nodes (possible TB lymphadenitis), headache or altered mental status (possible TB meningitis). The WHO has previously outlined a pragmatic clinical approach to help identify cases of EPTB by "looking and listening" for signs of four common forms of EPTB, including TB lymphadenitis, pleural TB, TB pericarditis and TB meningitis (Table 3) [85].

As discussed in **Part I**, ultrasonography may also help rapidly and inexpensively detect pleural or pericardial effusions suggesting pleural and pericardial TB, respectively [17]. The abdomen is the most frequent site of TB disease dissemination beyond the chest cavity and almost any structure (i.e., peritoneum, gastrointestinal tract, lymph nodes) or solid organ (spleen, liver, pancreas) may be involved. Intra-abdominal findings on ultrasonography, especially ascites, diffuse lymphadenopathy or splenic or liver micro-abscesses should result in microbiological TB investigations.

Overview of diagnosing EPTB: When EPTB is suspected on the basis of clinical or radiologic features (chest X-ray or ultrasound), rapid investigations to confirm a TB diagnosis should be undertaken to allow for the prompt initiation of TB therapy. If empiric TB treatment is initiated on the basis of high clinical suspicion (for example: the patient is symptomatic, has compatible ultrasound findings and is at high risk for clinical deterioration), clinical specimens should still be obtained for TB confirmation and DST.

Table 3 'Look and listen for' signs of possible EPTB

• **Possible TB lymphadenitis**
– Cervical (typically unilateral) or axillary lymphadenopathy
• **Possible pleural TB**
– Absent breath sounds
– Reduced chest wall movement
– Dullness to percussion
• **Possible TB pericarditis**
– Distant heart sounds
– Peripheral edema and/or abdominal distension
– Jugular venous distension
• **Possible TB meningitis**
– Neck stiffness
– Confusion
– Atypical eye movements

The WHO recommends that Xpert should be the initial test for the investigation of all forms of EPTB. The most common forms of EPTB are listed in Table 4 along with associated clinical samples that might be obtained and submitted for further microbiological testing when those forms of EPTB are clinically suspected. The approach to microbiological testing for EPTB is described below.

Sputum-based testing for those able to produce sputum: A large proportion of patients with extra-pulmonary disease also have concomitant pulmonary disease [29]. For PLHIV with suspected EPTB, those who are able to produce a sputum sample should still undergo initial testing with sputum Xpert testing (Figs. 1 and 2) or sputum AFB microscopy (and culture) testing where Xpert testing is unavailable (Fig. 3).

Obtaining non-respiratory clinical specimens: When the diagnosis of EPTB is suspected but cannot be made via sputum-based methods (either sputum testing negative, or patient is too sick/unable to provide a sputum sample) then further non-respiratory samples should be obtained and submitted for rapid microbiological testing. The obtainment of clinical samples should be guided by which clinical site/organ is suspected to be involved, as well as what investigations are locally available (Table 4). When multiple anatomic sites are thought to be involved, the least invasive clinical specimen that can be obtained for microbiological testing should be prioritized.

Xpert MTB/RIF (and Xpert Ultra) for EPTB: The diagnostic sensitivity of Xpert for non-respiratory samples is summarized in Table 1. While evaluations to-date are limited, Xpert Ultra is expected to improve detection of EPTB in PLHIV. As above, a 'trace-positive' Xpert Ultra result in PLHIV should be regarded as a true-positive result [46].

LF-LAM assay: When LF-LAM results are positive, an additional microbiological test that provides drug susceptibility testing results should be performed if possible (steps 6 and 7).

Microscopy and culture for EPTB: In PLHIV with suspected EPTB for which Xpert and LF-LAM testing is either negative or unavailable, smear microscopy and

Table 4 Forms of EPTB and associated clinical samples for TB testing

EPTB form	Sample
Bacteremia	Blood
Genitourinary TB	Urine, semen (men), organ biopsy
Lymphadenitis	Fine needle aspirate of affected tissue, excisional biopsy
Meningitis	Cerebrospinal fluid, tuberculoma biopsy
Pericarditis	Pericardial fluid, pericardial biopsy
Peritonitis	Ascitic fluid (paracentesis), peritoneal biopsy
Pleurisy (pleural TB)	Pleural fluid (thoracentesis), pleural biopsy
Skeletal (bone/joint)	Synovial fluid (arthrocentesis), bone biopsy

culture may be considered on EPTB samples (Table 4). Smear microscopy is often of limited value given the pauci-bacillary nature of most EPTB samples and the clinical utility of culture-based methods is greatly diminished, especially among sick hospitalized patients given prolonged time-to-positivity and these patients' predisposition to rapid clinical deterioration without appropriate treatment.

Relevant guidelines:
- Consolidated guidelines on the use of antiretroviral drugs for treating and preventing HIV infection. Geneva: WHO; 2016.
- GLI model TB diagnostic algorithms. Geneva: WHO; 2018.
- Improving the diagnosis and treatment of smear-negative pulmonary and extrapulmonary tuberculosis among adults and adolescents. Geneva: WHO; 2007.
- The use of lateral flow urine lipoarabinomannan assay (LF-LAM) for the diagnosis and screening of active tuberculosis in people living with HIV. Geneva: WHO; 2015.
- Xpert MTB/RIF implementation manual – technical and operational "how-to". Practical considerations. Geneva: WHO; 2014.

Step 6: Whom Should I Test for Drug Resistance?

Current WHO recommendation: All PLHIV with confirmed TB should undergo rapid DST for RIF. Patients with HIV-associated TB and evidence of RIF resistance should have further DST undertaken for other first-line drugs and at least for fluoroquinolones and second-line injectable agents.

Relevant guidelines:
- Framework of indicators and targets for laboratory strengthening under the End TB Strategy. Geneva: WHO; 2016.
- WHO treatment guidelines for drug-resistant tuberculosis, 2016 update. October 2016 revision. Geneva: WHO; 2016.

Step 7: How Should I Test for Drug Resistance?

Current WHO recommendations: Xpert (or Xpert Ultra) should be used for first-line DST to evaluate for RIF-resistance. In PLHIV with confirmed RR-TB or MDR-TB, further DST should be undertaken for other first-line drugs and at least fluoroquinolones and second-line injectable agents using LPA or other molecular methods (where available), in addition to culture-based methods.

Further information: For PLHIV, the WHO recommends universal access to rapid drug-susceptibility testing (DST) for at least RIF and if RIF-resistance is present, further DST for fluoroquinolones and second-line injectable agents. This allows for the prompt identification of RR-TB, MDR-TB and XDR/pre-XDR TB.

Relevant guidelines:

- Framework of indicators and targets for laboratory strengthening under the End TB Strategy. Geneva: WHO; 2016.
- The use of molecular line probe assays for the detection of mutations associated with resistance to fluoroquinolones (FQs) and second-line injectable drugs (SLIDs). Policy guidance. Geneva: WHO; 2016.
- The use of molecular line probe assays for the detection of resistance to isoniazid and rifampicin. Geneva: WHO; 2016.
- WHO treatment guidelines for drug-resistant tuberculosis

Step 8: For Whom Should I Consider Initiation of Empiric TB Therapy?

Current WHO recommendations: Every effort should be made to confirm the diagnosis of TB. When sputum Xpert (or smear microscopy) testing is negative, or in settings where TB investigations are limited, empiric TB therapy should be considered in those who are seriously ill due to suspected TB.

Further information: Whenever possible, all attempts should be made to make a microbiological diagnosis of TB as outlined in steps 4 and 5, before initiating empiric TB therapy. However, there are circumstances when empiric therapy (the administration of TB therapy without microbiological confirmation of TB) might be warranted. According to the current WHO algorithm for ambulatory HIV patients [7], empiric therapy might be considered in those for which TB is still felt to be likely despite negative Xpert testing (on respiratory and/or non-respiratory samples) or negative sputum microscopy (if Xpert testing is unavailable). However, if the patient's clinical stability will allow for further TB investigations (i.e., repeat sputum testing, repeat chest imaging, abdominal ultrasound and extra-pulmonary sampling) these should be preferentially pursued before initiating empiric therapy. A multi-country trial among ambulatory PLHIV with CD4 counts <50 cells/μL randomized patients to either ART plus isoniazid preventive therapy (IPT) or ART plus active TB treatment after systematic TB screening and further TB investigations were negative [86]. No difference in 24-week mortality between the two arms was found; this suggests that empiric TB treatment does not improve outcomes in ambulatory PLHIV if TB investigations are negative and that IPT can be safely initiated even in those with severe immunodeficiency if TB symptom screening and/or subsequent TB investigations are negative.

In hospitalized PLHIV or those who are seriously ill as defined by the presence of one or more danger signs (Fig. 2), if one or more Xpert tests (or sputum smear microscopy where Xpert is not available) and a LF-LAM test (where available) are negative, empiric therapy should be started when the patient fails to clinically improve on broad-spectrum antibiotics within 3–5 days and TB remains clinically suspected. This approach is supported by studies demonstrating that among seriously ill hospitalized patients with smear-negative, but suspected TB, early empiric therapy was associated with reduced hospitalization and improved survival at 8 weeks [87, 88].

The initiation of empiric TB therapy may be necessary in settings where there are limited or no TB investigations routinely available. In patients who are seriously ill due to suspected TB (based on compatible clinical history, exam and/or imaging findings) a clinician at their discretion may choose to start empiric TB-therapy. In such settings, clinical prediction scores may be helpful in assessing which patients should be started on empiric TB therapy (see **Part II**). For example, one study among HIV inpatients with cough (of any duration) and at least one WHO danger sign found that a clinical prediction rule using only clinical, laboratory and radiographic characteristics might have utility for determining who may benefit from empiric TB initiation [89]; a cutoff score of 3 or 4 was associated with a sensitivity of 87–90% and a specificity of 45–59% for culture-confirmed TB and thus might be used to guide initiation of empiric therapy.

Relevant guidelines:
- Consolidated guidelines on the use of antiretroviral drugs for treating and preventing HIV infection. Geneva: WHO; 2016.

Conclusions

As more than half of incident TB cases in PLHIV remain undiagnosed and unreported, significant challenges remain in the diagnosis of HIV-associated TB. However, there is much reason to be excited about new and imminent diagnostic tools and tests. With improved implementation of currently recommended WHO universal screening and testing strategies for HIV-associated TB, the diagnosis gap can be greatly reduced allowing for significant progress to be achieved towards improved individual patient outcomes among PLHIV and enhanced TB control.

References

1. Gupta RK, Lucas SB, Fielding KL, Lawn SD (2015) Prevalence of tuberculosis in post-mortem studies of HIV-infected adults and children in resource-limited settings: a systematic review and meta-analysis. AIDS 29:1987–2002
2. Gupta-Wright A, Peters JA, Flach C, Lawn SD (2016) Detection of lipoarabinomannan (LAM) in urine is an independent predictor of mortality risk in patients receiving treatment for HIV-associated tuberculosis in sub-Saharan Africa: a systematic review and meta-analysis. BMC Med 14:53
3. Lawn SD (2012) Point-of-care detection of lipoarabinomannan (LAM) in urine for diagnosis of HIV-associated tuberculosis: a state of the art review. BMC Infect Dis 12:103
4. World Health Organization (ed) (2017) Tuberculosis-diagnostics technology landscape, 5th edn. World Health Organization, Geneva
5. World Health Organization (2014) High-priority target product profiles for new tuberculosis diagnostics: report of a consensus meeting. World Health Organization, Geneva
6. World Health Organization (2014) Guidelines for intensified tuberculosis case-finding and isoniazid preventative therapy for people living with HIV in resource-constrained settings. World Health Organization, Geneva
7. World Health Organization (2016) Consolidated guidelines on the use of antiretroviral drugs for treating and preventing HIV infection. World Health Organization, Geneva
8. Hamada Y, Lujan J, Schenkel K, Ford N, Getahun H (2018) Sensitivity and specificity of WHO's recommended four-symptom screening rule for tuberculosis in people living with HIV: a systematic review and meta-analysis. Lancet HIV 5:e515–e523
9. Lawn SD, Wood R (2011) Tuberculosis in antiretroviral treatment services in resource-limited settings: addressing the challenges of screening and diagnosis. J Infect Dis 204(Suppl 4):S1159–S1167
10. Heller T, Goblirsch S, Bahlas S et al (2013) Diagnostic value of FASH ultrasound and chest X-ray in HIV-co-infected patients with abdominal tuberculosis [Notes from the field]. Int J Tuberc Lung Dis 17:342–344
11. Dawson R, Masuka P, Edwards DJ et al (2010) Chest radiograph reading and recording system: evaluation for tuberculosis screening in patients with advanced HIV. Int J Tuberc Lung Dis 14:52–58
12. Davis JL, Worodria W, Kisembo H et al (2010) Clinical and radiographic factors do not accurately diagnose smear-negative tuberculosis in HIV-infected inpatients in Uganda: a cross-sectional study. PLoS One 5:e9859–e9858
13. Cain KP, McCarthy KD, Heilig CM et al (2010) An algorithm for tuberculosis screening and diagnosis in people with HIV. N Engl J Med 362:707–716
14. World Health Organization (2016) Chest radiography in tuberculosis detection—summary of current WHO recommendations and guidance on programmatic approaches. World Health Organization, Geneva
15. Pande T, Cohen C, Pai M, Ahmad Khan F (2016) Computer-aided detection of pulmonary tuberculosis on digital chest radiographs: a systematic review. Int J Tuberc Lung Dis 20:1226–1230
16. Muyoyeta M, Maduskar P, Moyo M et al (2014) The sensitivity and specificity of using a computer aided diagnosis program for automatically scoring chest X-rays of presumptive TB patients compared with Xpert MTB/RIF in Lusaka Zambia. PLoS One 9:e93757–e93759
17. Heller T, Wallrauch C, Goblirsch S, Brunetti E (2012) Focused assessment with sonography for HIV-associated tuberculosis (FASH): a short protocol and a pictorial review. Crit Ultrasound J 4:21–29
18. Spalgais S, Jaiswal A, Puri M, Sarin R, Agarwal U (2013) Clinical profile and diagnosis of extrapulmonary tb in HIV infected patients: routine abdominal ultrasonography increases detection of abdominal tuberculosis. Indian J Rheumatol 60:147–153

19. Patel MN, Beningfield S, Burch V (2011) Abdominal and pericardial ultrasound in suspected extrapulmonary or disseminated tuberculosis. S Afr Med J 101:39–42
20. World Health Organization (2011) Fluorescent light-emitting diode (LED) microscopy for diagnosis of tuberculosis. World Health Organization, Geneva
21. Getahun H, Harrington M, O'Brien R, Nunn P (2007) Diagnosis of smear-negative pulmonary tuberculosis in people with HIV infection or AIDS in resource-constrained settings: informing urgent policy changes. Lancet 369:2042–2049
22. Chakravorty S, Sen MK, Tyagi JS (2005) Diagnosis of extrapulmonary tuberculosis by smear, culture, and PCR using universal sample processing technology. J Clin Microbiol 43:4357–4362
23. Lee JY (2015) Diagnosis and treatment of extrapulmonary tuberculosis. Tuberc Respir Dis 78:47–55
24. Lawn SD, Nicol MP (2011) Xpert® MTB/RIF assay: development, evaluation and implementation of a new rapid molecular diagnostic for tuberculosis and rifampicin resistance. Future Microbiol 6:1067–1082
25. Hanrahan CF, Theron G, Bassett J et al (2014) Xpert MTB/RIF as a measure of sputum bacillary Burden. Variation by HIV status and immunosuppression. Am J Respir Crit Care Med 189:1426–1434
26. Beynon F, Theron G, Respeito D et al (2018) Correlation of Xpert MTB/RIF with measures to assess *Mycobacterium tuberculosis* bacillary burden in high HIV burden areas of Southern Africa. Sci Rep 8:5201
27. Steingart KR, Schiller I, Horne DJ, Pai M, Boehme CC, Dendukuri N (2014) Xpert® MTB/RIF assay for pulmonary tuberculosis and rifampicin resistance in adults. Cochrane Database Syst Rev (1):CD009593
28. Kohli M, Schiller I, Dendukuri N et al (2018) Xpert®MTB/RIF assay for extrapulmonary tuberculosis and rifampicin resistance. Cochrane Database Syst Rev 124:1382–1247
29. Lawn SD, Kerkhoff AD, Burton R et al (2015) Rapid microbiological screening for tuberculosis in HIV-positive patients on the first day of acute hospital admission by systematic testing of urine samples using Xpert MTB/RIF: a prospective cohort in South Africa. BMC Med 13:192
30. Peter JG, Theron G, Muchinga TE, Govender U, Dheda K (2012) The diagnostic accuracy of urine-based Xpert MTB/RIF in HIV-infected hospitalized patients who are smear-negative or sputum scarce. PLoS One 7:e39966–e39968
31. Lawn SD, Kerkhoff AD, Vogt M, Wood R (2012) High diagnostic yield of tuberculosis from screening urine samples from HIV-infected patients with advanced immunodeficiency using the Xpert MTB/RIF assay. J Acquir Immune Defic Syndr 60:289–294
32. Albert H, Nathavitharana RR, Isaacs C, Pai M, Denkinger CM, Boehme CC (2016) Development, roll-out and impact of Xpert MTB/RIF for tuberculosis: what lessons have we learnt and how can we do better? Eur Respir J 48:516–525
33. Ngwira LG, Corbett EL, Khundi M et al (2018) Screening for tuberculosis with Xpert MTB/RIF versus fluorescent microscopy among adults newly diagnosed with HIV in rural Malawi: a cluster randomized trial (CHEPETSA). Clin Infect Dis 363:1005
34. Trajman A, Durovni B, Saraceni V et al (2015) Impact on patients' treatment outcomes of XpertMTB/RIF implementation for the diagnosis of tuberculosis: follow-up of a stepped-wedge randomized clinical trial. PLoS One 10:e0123252–e0123211
35. Auld AF, Fielding KL, Gupta-Wright A, Lawn SD (2016) Xpert MTB/RIF—why the lack of morbidity and mortality impact in intervention trials? Trans R Soc Trop Med Hyg 110:432–444
36. Theron G, Zijenah L, Chanda D et al (2014) Feasibility, accuracy, and clinical effect of point-of-care Xpert MTB/RIF testing for tuberculosis in primary-care settings in Africa: a multicentre, randomised, controlled trial. Lancet 383:424–435
37. Churchyard GJ, Stevens WS, Mametja LD et al (2015) Xpert MTB/RIF versus sputum microscopy as the initialdiagnostic test for tuberculosis: a cluster-randomised trialembedded in South African roll-out of Xpert MTB/RIF. Lancet Glob Health 3:e450–e457
38. Cox HS, Mbhele S, Mohess N et al (2014) Impact of Xpert MTB/RIF for TB diagnosis in a primary care clinic with high TB and HIV prevalence in South Africa: a pragmatic randomised trial. PLoS Med 11:e1001760–e1001712

39. Yoon C, Cattamanchi A, Davis JL et al (2012) Impact of Xpert MTB/RIF testing on tuberculosis management and outcomes in hospitalized patients in Uganda. PLoS One 7:e48599
40. Mupfumi L, Makamure B, Chirehwa M et al (2014) Impact of Xpert MTB/RIF on antiretroviral therapy-associated tuberculosis and mortality: a pragmatic randomized controlled trial. Open Forum Infect Dis 1:1897–1898
41. Calligaro GL, Theron G, Khalfey H et al (2015) Burden of tuberculosis in intensive care units in Cape Town, South Africa, and assessment of the accuracy and effect on patient outcomes of the Xpert MTB/RIF test on tracheal aspirate samples for diagnosis of pulmonary tuberculosis: a prospective burden of disease study with a nested randomised controlled trial. Lancet Respir Med 3:621–630
42. van Kampen SC, Susanto NH, Simon S et al (2015) Effects of introducing Xpert MTB/RIF on diagnosis and treatment of drug-resistant tuberculosis patients in Indonesia: a pre-post intervention study. PLoS One 10:e0123536–e0123511
43. Chakravorty S, Simmons AM, Rowneki M et al (2017) The new Xpert MTB/RIF ultra: improving detection of mycobacterium tuberculosis and resistance to rifampin in an assay suitable for point-of-care testing. MBio 8:e00812–e00817
44. Dorman SE, Schumacher SG, Alland D et al (2018) Xpert MTB/RIF Ultra for detection of Mycobacterium tuberculosis and rifampicin resistance: a prospective multicentre diagnostic accuracy study. Lancet Infect Dis 18:76–84
45. Bahr NC, Nuwagira E, Evans EE et al (2018) Diagnostic accuracy of Xpert MTB/RIF Ultra for tuberculous meningitis in HIV-infected adults: a prospective cohort study. Lancet Infect Dis 18:68–75
46. World Health Organization (2017) WHO meeting report of a technical expert consultation: non-inferiority analysis of Xpert MTB/RIF ultra compared to Xpert MTB/RIF. World Health Organization, Geneva
47. Shah M, Hanrahan C, Wang ZY et al (2016) Lateral flow urine lipoarabinomannan assay for detecting active tuberculosis in HIV-positive adults. Cochrane Database Syst Rev (5):CD011420
48. Lawn SD, Kerkhoff AD, Burton R et al (2017) Diagnostic accuracy, incremental yield and prognostic value of determine TB-LAM for routine diagnostic testing for tuberculosis in HIV-infected patients requiring acute hospital admission in South Africa: a prospective cohort. BMC Med 15:67
49. Lawn SD, Kerkhoff AD, Vogt M, Wood R (2012) Diagnostic accuracy of a low-cost, urine antigen, point-of-care screening assay for HIV-associated pulmonary tuberculosis before antiretroviral therapy: a descriptive study. Lancet Infect Dis 12:201–209
50. Lawn SD, Kerkhoff AD, Vogt M, Wood R (2013) HIV-associated tuberculosis: relationship between disease severity and the sensitivity of new sputum-based and urine-based diagnostic assays. BMC Med 11:231
51. Kerkhoff AD, Barr DA, Burton R et al (2017) Disseminated tuberculosis among hospitalised HIV patients in South Africa: a common condition that can be rapidly diagnosed using urine-based assays. Sci Rep 7:10931
52. Peter JG, Zijenah LS, Chanda D et al (2016) Effect on mortality of point-of-care, urine-based lipoarabinomannan testing to guide tuberculosis treatment initiation in HIV-positive hospital inpatients: a pragmatic, parallel-group, multicountry, open-label, randomised controlled trial. Lancet 387:1187–1197
53. Gupta-Wright A, Corbett EL, van Oosterhout JJ et al (2018) Rapid urine-based screening for tuberculosis in HIV-positive patients admitted to hospital in Africa (STAMP): a pragmatic, multicentre, parallel-group, double-blind, randomised controlled trial. Lancet 392:292–301
54. World Health Organization (2015) The use of lateral flow urine lipoarabinomannan assay (LF-LAM) for the diagnosis and screening of active tuberculosis in people living with HIV. World Health Organization, Geneva
55. World Health Organization (2016) The use of loop-mediated isothermal ampli cation (TB-LAMP) for the diagnosis of pulmonary tuberculosis. Policy guidance. World Health Organization, Geneva

56. World Health Organization (2015) The end TB strategy. World Health Organization, Geneva
57. Angeby K, Jureen P, Kahlmeter G, Hoffner SE, Schön T (2012) Challenging a dogma: antimicrobial susceptibility testing breakpoints for *Mycobacterium tuberculosis*. Bull World Health Organ 90:693–698
58. Xie YL, Chakravorty S, Armstrong DT et al (2017) Evaluation of a rapid molecular drug-susceptibility test for tuberculosis. N Engl J Med 377:1043–1054
59. Rigouts L, Gumusboga M, de Rijk WB et al (2013) Rifampin resistance missed in automated liquid culture system for *Mycobacterium tuberculosis* isolates with specific rpoB mutations. J Clin Microbiol 51:2641–2645
60. World Health Organization (2016) The use of molecular line probe assays for the detection of resistance to isoniazid and rifampicin. World Health Organization, Geneva
61. World Health Organization (2016) The use of molecular line probe assays for the detection of resistance to second-line anti-tuberculosis drugs. Policy guidance. World Health Organization, Geneva
62. World Health Organization (2018) Critical concentrations for drug susceptibility testing of medicines used in the treatment of drug-resistant tuberculosis. World Health Organization, Geneva
63. World Health Organization (2008) Molecular line probe assays for rapid screening of patients at risk of multidrug-resistant tuberculosis (MDR-TB)—policy statement. World Health Organization, Geneva
64. Luetkemeyer AF, Charlebois ED, Flores LL et al (2007) Comparison of an interferon-gamma release assay with tuberculin skin testing in HIV-infected individuals. Am J Respir Crit Care Med 175:737–742
65. Gardee Y, Dreyer AW, Koornhof HJ et al (2017) Evaluation of the genotype MTBDRslVersion 2.0 assay for second-line drug resistance detection of Mycobacterium tuberculosis isolates in South Africa. J Clin Microbiol 55:791–800
66. DiNardo A, Saavedra B, Silva DR et al (2018) Point of care diagnostics for tuberculosis. Rev Port Pneumol 24:73–85
67. Balcha TT, Skogmar S, Sturegård E et al (2014) A clinical scoring algorithm for determination of the risk of tuberculosis in HIV-infected adults: a cohort study performed at Ethiopian Health Centers. Open Forum Infect Dis 1:1906–1909
68. Hanifa Y, Fielding KL, Chihota VN et al (2017) A clinical scoring system to prioritise investigation for tuberculosis among adults attending HIV clinics in South Africa. PLoS One 12:e0181519–e0181520
69. Yoon C, Semitala FC, Atuhumuza E et al (2017) Point-of-care C-reactive protein-based tuberculosis screening for people living with HIV: a diagnostic accuracy study. Lancet Infect Dis 17:1285–1292
70. Yoon C, Chaisson LH, Patel SM et al (2017) Diagnostic accuracy of C-reactive protein for active pulmonary tuberculosis: a meta-analysis. Int J Tuberc Lung Dis 21:1013–1019
71. Shapiro AE, Hong T, Govere S et al (2018) C-reactive protein as a screening test for HIV-associated pulmonary tuberculosis prior to antiretroviral therapy in South Africa. AIDS 32:1811–1820
72. Yoon C, Semitala FC, Asege L et al (2019) Yield and efficiency of novel intensified tuberculosis case-finding algorithms for people living with HIV. Am J Respir Crit Care Med 199(5):643–650
73. Broger T, Sossen B, Toit du E et al (2019) Novel lipoarabinomannan point-of-care tuberculosis test for people living with HIV with superior sensitivity: a diagnostic accuracy study. Lancet Infect Dis 19(8):852–861
74. Médecins Sans Frontières MSF (2018) Xpert Omni Factsheet. MSF, Geneva
75. Pankhurst LJ, del Ojo Elias C, Votintseva AA et al (2016) Rapid, comprehensive, and affordable mycobacterial diagnosis with whole-genome sequencing: a prospective study. Lancet Respir Med 4:49–58
76. Votintseva AA, Bradley P, Pankhurst L et al (2017) Same-day diagnostic and surveillance data for tuberculosis via whole-genome sequencing of direct respiratory samples. J Clin Microbiol 55:1285–1298

77. Doughty EL, Sergeant MJ, Adetifa I, Antonio M, Pallen MJ (2014) Culture-independent detection and characterisation of *Mycobacterium tuberculosis* and *M. africanumin* sputum samples using shotgun metagenomics on a benchtop sequencer. PeerJ 2:e585–e518
78. Brown AC, Bryant JM, Einer-Jensen K et al (2015) Rapid whole-genome sequencing of Mycobacterium tuberculosis isolates directly from clinical samples. J Clin Microbiol 53:2230–2237
79. World Health Organization (2018) GLI model TB diagnostic algorithms. World Health Organization, Geneva
80. Ho J, Fox GJ, Marais BJ (2016) Passive case finding for tuberculosis is not enough. Int J Mycobacteriol 5:374–378
81. Lawn SD, Brooks SV, Kranzer K et al (2011) Screening for HIV-associated tuberculosis and rifampicin resistance before antiretroviral therapy using the Xpert MTB/RIF assay: a prospective study. PLoS Med 8:e1001067
82. World Health Organization (2011) Same-day diagnosis of tuberculosis by microscopy: policy statement. World Health Organization, Geneva
83. Davis JL, Cattamanchi A, Cuevas LE, Hopewell PC, Steingart KR (2013) Diagnostic accuracy of same-day microscopy versus standard microscopy for pulmonary tuberculosis: a systematic review and meta-analysis. Lancet Infect Dis 13:147–154
84. Cattamanchi A, Dowdy DW, Davis JL et al (2009) Sensitivity of direct versus concentrated sputum smear microscopy in HIV-infected patients suspected of having pulmonary tuberculosis. BMC Infect Dis 9:240–249
85. World Health Organization (2007) Improving the diagnosis and treatment of smear-negative pulmonary and extrapulmonary tuberculosis among adults and adolescents. World Health Organization, Geneva
86. Hosseinipour MC, Bisson GP, Miyahara S et al (2016) Empirical tuberculosis therapy versus isoniazid in adult outpatients with advanced HIV initiating antiretroviral therapy (REMEMBER): a multicountry open-label randomised controlled trial. Lancet 387:1198–1209
87. Katagira W, Walter ND, Boon den S et al (2016) Empiric TB treatment of severely ill patients with HIV and presumed pulmonary TB improves survival. J Acquir Immune Defic Syndr 72:297–303
88. Holz TH, Kabera G, Mthiyane T et al (2011) Use of a WHO-recommended algorithm to reduce mortality in seriously ill patients with HIV infection and smear-negative pulmonary tuberculosis in South Africa: an observational cohort study. Lancet Infect Dis 11:533–540
89. Griesel R, Stewart A, van der Plas H et al (2017) Optimizing tuberculosis diagnosis in human immunodeficiency virus-infected inpatients meeting the criteria of seriously Ill in the World Health Organization algorithm. Clin Infect Dis 66:1419–1426

Recent Advances in the Treatment of Latent Tuberculosis Infection Among Adults Living with HIV Infection

April C. Pettit and Timothy R. Sterling

Abstract In this chapter, we review evidence supporting the tuberculosis (TB) disease spectrum, rather than dichotomous categories of latent *M. tuberculosis* infection (LTBI) and active disease, as TB pathophysiology is important when considering TB prevention and treatment. We also describe the impact of immunocompromising conditions, specifically human immunodeficiency virus (HIV) infection, along the TB disease spectrum. We review the indications for treatment of LTBI among people living with HIV (PLWH) and the evidence behind the LTBI treatment regimens currently recommended by the World Health Organization and the United States Centers for Disease Control and Prevention. Lastly, we discuss the importance of antiretroviral therapy (ART) in addition to anti-mycobacterial therapy for the prevention of TB among PLWH, as well as the drug-drug interactions with concomitant antiretroviral therapy and LTBI treatment.

Keywords Tuberculosis disease spectrum · Latent *M. tuberculosis* infection · Tuberculin skin test · Interferon gamma release assay · Human immunodeficiency virus · Antiretroviral therapy · Isoniazid · Rifampicin · Rifabutin · Rifapentine · Directly observed therapy

A. C. Pettit (✉) · T. R. Sterling
Department of Medicine, Division of Infectious Diseases, Vanderbilt University Medical Center, Nashville, TN, USA

Vanderbilt Tuberculosis Center, Nashville, TN, USA
e-mail: april.pettit@vumc.org

I. Sereti et al. (eds.), *HIV and Tuberculosis*,
https://doi.org/10.1007/978-3-030-29108-2_8

Latent TB Infection and TB Disease: A Spectrum

Traditionally, tuberculosis (TB) has been viewed as a dichotomous disease entity in which a person with TB has either latent TB infection (LTBI) or active TB disease. LTBI is classically defined as evidence of an immunologic response to *M. tuberculosis* (Mtb) antigens via an *in vivo* tuberculin skin test (TST) or an *in vitro* interferon-gamma release assay (IGRA) in the absence of other signs and/or symptoms of TB disease. Active TB disease has been distinguished from LTBI by the presence of objective signs and/or symptoms of TB disease, such as cough, fever, night sweats, hemoptysis, and weight loss. However, evidence of an immunologic response to Mtb antigens via either a TST or IGRA is not required for the diagnosis of active TB disease. Importantly, the TST and IGRA are unable to distinguish between LTBI and active TB disease.

More recent evidence suggests that TB is a spectrum of disease pathology, and not simply a dichotomous disease entity. In 2009, Young and colleagues proposed a novel paradigm that involved a spectrum of responses to and control of Mtb infection (Fig. 1): (1) an innate immune response in which Mtb infection is eliminated without priming of antigen-specific T-cells; (2) an acquired immune response in

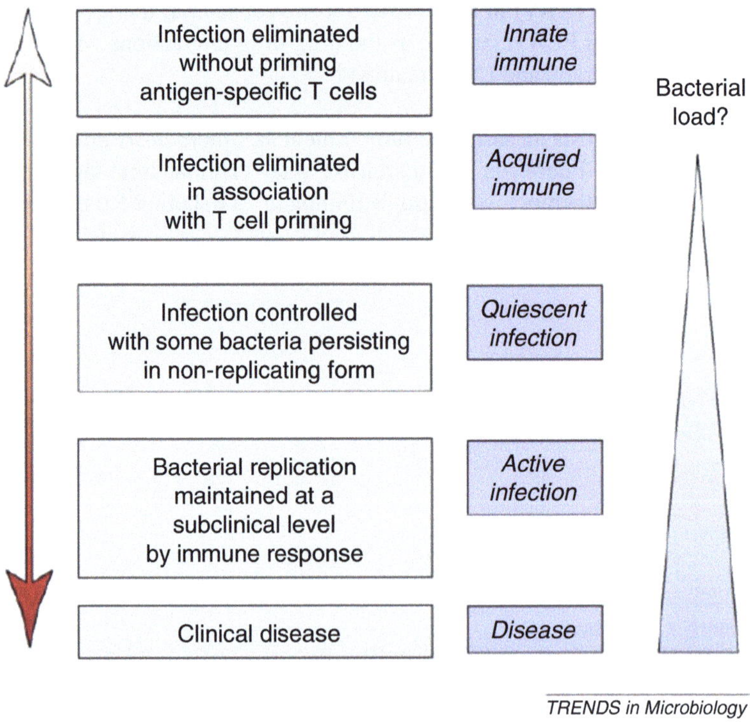

Fig. 1 A spectrum of responses to tuberculosis infection (Adapted from Young et al. Trends in Microbiology 2009) [1]

which infection is eliminated with T-cell priming; (3) quiescent infection during which bacteria can persist in a non-replicating form and infection is controlled; (4) active infection in which bacteria are replicating at a sub-clinical level (i.e. asymptomatic) that is controlled by the immune response and; (5) clinical disease in which overt signs and symptoms are observed due to bacterial replication despite immune responses [1]. There is *in vitro*, non-human primate, and clinical human evidence to support this concept of a spectrum of TB disease pathology, as discussed below.

First, the elimination of Mtb infection by the innate immune response, without T-cell priming via an acquired immune response, is supported by the finding that some healthy people who are repeatedly exposed to TB do not develop a positive TST or IGRA [2–5]. In addition to never becoming infected, another explanation for this finding is that current tests are unable to detect the host's acquired immune response. The latter could be due in part to the imperfect sensitivity of the TST or the IGRA, particularly when the host's immune system is compromised (e.g. human immunodeficiency virus [HIV]) [6–9]. Another possible explanation is that some people exposed to TB do not need an acquired immune response for bacterial elimination. For example, a person might expectorate Mtb bacilli before the immune system is exposed. Alternatively, the innate immune response (via neutrophils or macrophages) could eliminate Mtb without necessitating an acquired immune response [10, 11].

Second, elimination of Mtb infection by the acquired immune response is supported by the finding that some people with a positive IGRA revert to negative on subsequent testing, and never develop a positive TST. Although some reversions may be related to the sub-optimal reproducibility of the IGRA using current cutoffs for a positive result [12, 13], elimination of Mtb infection via transient activation of the acquired immune response may also revert an IGRA from positive to negative [14].

While *in vivo* evidence is lacking, there is *in vitro* evidence that Mtb can persist in a non-replicating form, as seen in quiescent infection. This dormant state has been described using the Wayne model, in which bacilli are deprived of oxygen [15]. Oxygen-deprived bacilli upregulate genes controlled by two sensor kinases (*dos*S and *dos*T) and a response regulator (*dos*R) [16, 17]. The *dos*R response regulator includes genes involved in triglyceride synthesis, which are required when the bacilli shift their carbon source from glucose to fatty acids [18, 19]. With more prolonged oxygen deprivation, many genes responsible for transcriptional regulation are induced [20]. Dormant bacteria may be resuscitated, either randomly or due to unknown signals, and serve as "scouts" which test the conditions of the local environment. If these scouts are not cleared by the immune system (e.g., immunocompromised states, including HIV-infection), they will continue to replicate and signal the remainder of the dormant bacilli to replicate which, in turn, leads to sub-clinical and clinical TB disease. These scout signals may take the form of a protein called Rpf (resuscitation promoting factor), which has been shown to increase the recovery of Mtb from the sputum of patients with active TB [21].

Sub-clinical TB disease occurs in the macaque TB model, which may most closely resemble the human TB disease spectrum. Monkeys with sub-clinical TB, termed "percolators", are clinically normal but may have positive cultures for Mtb months and even up to 1 year after infection with Mtb [22]. Sub-clinical TB is also well-described in humans, particularly among people living with HIV [23]. Perhaps the best human example of the transition from subclinical to clinical TB disease is provided by treatment of people living with HIV with antiretroviral therapy (ART). A vigorous recovery of T cell responses on ART has been associated with unmasking of sub-clinical TB in the setting of immune reconstitution inflammatory syndrome (IRIS) [24, 25].

Understanding this spectrum of TB disease is important when considering treatment for latent Mtb infection. Based on this paradigm, persons diagnosed with LTBI represent a heterogenous group of individuals—those who have already cleared the infection, are infected with bacilli in a non-replicating form, or infected with actively replicating bacilli but have no overt signs or symptoms of clinical disease. Therefore, anti-tuberculous drugs may be effective because non-replicating bacilli can be targeted by affecting cell wall maintenance and repair, or because Mtb bacilli periodically replicate (e.g. "scouts") and this is inhibited by the drugs. An improved understanding of the TB disease spectrum is critical for future TB drug development.

Immunocompromise in the setting of HIV-infection likely has effects all along this disease spectrum. First, PLWH may be unable to mount an adequate innate or acquired immune response, leaving them less likely to clear infection following exposure compared to HIV-negative persons. Second, PLWH may also be unable to clear "scouts" that signal other dormant bacilli to replicate and cause sub-clinical or clinical TB disease following infection. Thirdly, PLWH and severe immunocompromise may not progress from subclinical to clinical TB for long periods of time. In contrast, the progression from subclinical to clinical TB may be accelerated among PLWH in the setting of antiretroviral therapy initiation and immune reconstitution.

Treatment of Latent TB Infection with Anti-TB Therapy

There are several regimens available for the treatment of LTBI among people living with HIV (Table 1). Currently in the US, guidelines recommend testing for latent TB infection using TST or IGRA at the time of HIV diagnosis, regardless of epidemiological risk factors for TB exposure. PLWH, CD4+ counts <200 cells/μL, and negative tests for LTBI should be retested once they have started antiretroviral therapy (ART) and CD4+ counts are over 200 cells/μL. Annual testing (using TST) is only recommended for PLWH if they are at high risk for repeated/ongoing exposure to people with active TB [31]. In the US, treatment for LTBI is only recommended for PLWH who test positive for LTBI [28–30]. According to the WHO guidelines, PLWH who have an unknown or positive TST (and are unlikely to have active TB) are recommended to receive treatment for latent TB infection. HIV-negative persons are recommended to receive treatment if they are household contacts of a person with pulmonary TB or if they belong to other at-risk groups [26, 27].

Table 1 Treatment regimens for LTBI among PLWH Endorsed by the WHO and CDC

	WHO Guidelines [26, 27]	CDC Guidelines [28–30]
Isoniazid Dose: 5 mg/kg/day Maximum dose 300 mg daily	6 months given daily in settings with high or low TB incidence (strong recommendation with high-quality evidence) 36 months given daily in settings with high TB incidence and high rates of TB transmission as determined by national authorities (conditional recommendation with low-quality evidence)	9 months given daily (AII evidence rating) Alternatives: 6 months given daily (CI evidence rating); 6–9 months given biweekly by directly observed therapy (CI evidence rating for 6 months and BII evidence rating for 9 months)
Rifampicin plus isoniazid Isoniazid dosing: 5 mg/kg/day Maximum dose 300 mg daily Rifampicin dosing: 10 mg/kg/day Maximum dose 600 mg daily	3–4 months given daily in settings with low TB incidence (strong recommendation with moderate to high quality evidence)	Not recommended
Rifampicin Dose 10 mg/kg/day Maximum dose 300 mg daily	3–4 months given daily in settings with low TB incidence (strong recommendation with moderate to high quality evidence)	4 months given daily Rifabutin may be used as a substitute among PLWH on ART with significant drug interactions (BIII evidence rating)
Rifapentine plus isoniazid Isoniazid dosing: 15 mg/kg/week Maximum dose 900 mg weekly Rifapentine dosing: 10–14 kg = 300 mg/week 14.1–25 kg = 450 mg/week 25.1–32 kg = 600 mg/week 32.1–50 kg = 750 mg/week >50 kg = 900 mg/week Maximum dose 900 mg weekly	12 weeks given weekly by directly observed or self-administered therapy in settings with low TB incidence (strong recommendation with moderate to high quality evidence) or high TB incidence (strong recommendation with moderate quality evidence)	12 weeks given weekly by directly observed or self-administered therapy (no evidence rating provided yet)

Definitions: High TB incidence, estimated annual TB incidence rate ≥ 100 per 100,000 population; Low TB incidence, estimated annual TB incidence rate < 100 per 100,000

WHO quality of evidence rating: High-We are very confident that the true effect lies close to that of the estimate of the effect; Moderate-We are moderately confident that the true effect is likely to be close to the estimate of the effect, but there is a possibility that it is substantially different; Low-Our confidence in the effect estimate is limited (the true effect may be substantially different); Very low-We have very little confidence in the effect estimate (the true effect is likely to be substantially different)

WHO strength of recommendation definitions: Strong-the GDG was confident that the desirable effects of adherence would outweigh the undesirable effects (either in favor of or against an intervention); Conditional-the GDG concluded that the desirable effects of adherence would probably outweigh the undesirable effects, but the GDG was not confident about the trade-off (i.e. absence of high-quality evidence, imprecise estimates of benefits/harm, uncertainty/variation in the value of the outcomes for different individuals, and small benefits or benefits that might not be worth the cost)

CDC quality of evidence definitions: (I)-randomized clinical trial data; (II)-data from clinical trials that are not randomized or were conducted in other populations; (III)-expert opinion

CDC strength of recommendation definitions: (A)-preferred; (B)-acceptable alternative; (C)-offer when A and B cannot be given

Isoniazid

Isoniazid (INH) monotherapy is endorsed as an LTBI treatment option by both the World Health Organization (WHO) [26, 27] and the United States Centers for Disease Control and Prevention (CDC) [28–30]. The WHO guidelines recommend INH daily for 6 months regardless of HIV-status, in countries with either high or low TB incidence, as a strong recommendation. In contrast, the CDC recommends INH daily for 9 months, regardless of HIV-status. The CDC endorses isoniazid for 6 months as an alternative regimen. There are no special antiretroviral therapy considerations during INH monotherapy for LTBI.

Randomized trials have shown that both 6 and 12 months of INH, when compared to placebo, significantly reduce the incidence of TB disease in both HIV-negative [32–34] and HIV-positive populations [35, 36]. Only one trial, conducted by the International Union Against Tuberculosis (IUAT) among an HIV-negative population, was designed to compare 6 vs. 12 months of INH. This study showed a 65% efficacy of 6 months of INH and 75% efficacy of 12 months of INH during 5 years of follow-up [33]. The 6-month INH regimen has not been directly compared to 12 months duration in clinical trials among HIV-positive populations.

Data on direct comparisons of treatment pairs for the outcome of active TB are lacking for 9-months compared to either 6 or 12 months. Through a network meta-analysis, indirect evidence regarding these comparisons is available. For example, although 9 months of INH has not been directly compared to either 6 or 12 months, all three have been compared to no treatment and both 6 and 12 months have been compared to placebo, allowing indirect comparisons to be made. Using network meta-analysis, the odds ratio for active TB was 0.65 (95% Credible Interval [Crl] 0.50–0.83) for 6 months INH, 0.75 (95% Crl 0.35–1.62) for 9 months INH, and 0.50 (95% Crl 0.41–0.62) for 12–72 months INH compared to placebo. The 95% credible intervals for all INH durations are overlapping, indicating no significant differences in the prevention of active TB. Stratifying results by HIV-status revealed no statistically significant difference in effect estimates for all INH durations [37]. In addition to these data, the WHO Guideline Development Group (GDG) considered programmatic feasibility, resource requirements, and patient acceptability in their preference for 6 over 9 months of INH. The 6-month regimen is more cost-effective than the 12-month regimen [38].

Data supporting the CDC's recommendation for 9-month INH duration over the 6-month duration are largely extrapolated from two studies including only HIV-negative persons, neither of which were designed to directly compare various INH durations. A post-hoc analysis of a randomized trial conducted by the US Public Health Service [34] showed that among persons believed to have taken at least 80% of INH therapy, TB rates dropped 68% if INH was taken for 10–12 months and only 16% when taken for less than 10 months when compared to placebo [32]. Another post-hoc analysis was conducted using data from the Bethel Isoniazid Studies [39, 40] and included participants who had taken a wide range of INH durations, from none up to 24 months. TB case rates per 100 persons were plotted against the duration of INH in months and the observed points were fitted with a simple curve. This curve approached horizontal at approximately 9–10 months duration [41]. However, the CDC also notes that cost-effectiveness may be a consideration for the 6-month duration depending on the local health department conditions.

Continuous Isoniazid

A treatment duration of at least 36 months (or continuous isoniazid) is recommended by the WHO for PLWH who have either a positive TST, or unknown TST result in settings with high TB incidence and high rates of Mtb transmission, as defined by national authorities. It was recognized by the GDG that PLWH with a positive TST (or unknown TST result) as well as those receiving antiretroviral therapy (ART) are more likely to benefit [42]. This is a conditional recommendation with low-quality evidence based largely on the results of a systematic review including three studies conducted in Botswana, South Africa, and India [43]. Compared to 6 months of INH, patients receiving continuous isoniazid had a 38% lower risk of active TB (relative risk [RR] 0.62, 95% CI 0.42–0.89) [43]. Among those with a positive TST there was a 49% lower risk of active TB (RR 0.51, 95% CI 0.30–0.86) and a 50% lower risk of mortality (RR 0.50, 95% CI 0.27–0.91) [43]. Among those with a negative TST there was no significant decrease in active TB or mortality, although the point estimate indicated a reduction in TB incidence of 27% [43]. All but one of these studies excluded participants eligible for ART [44]. In this study, those PLWH with a positive TST receiving continuous isoniazid and 360 days of ART had a 96% lower risk of active TB than those PLWH with a positive TST who received only 6 months of isoniazid and no ART (adjusted hazard ratio [aHR] 0.04, 95% CI 0.005–0.35). Among those with a positive TST receiving 6-months of isoniazid and 360 days of ART, the reduction in the risk of active TB was only 50% (aHR 0.50, 95% CI 0.26–0.97) compared to TST-positive participants receiving 6-months of isoniazid but not receiving ART [44].

Two of the studies found no statistically significant increase in adverse events in the continuous isoniazid group [44, 45]; a third study reported an increase in grade 3 or 4 adverse events (32 vs. 9.5%, RR 3.41, 95% CI 2.28–5.09) [46], but meta-analysis was not preformed due to study heterogeneity [43]. The pooled relative risk of study drug discontinuation due to adverse events was 5.96 (95% CI 4.12–8.62) for the continuous isoniazid group compared to the 6-month treatment duration group [43].

Only one of the studies included in this meta-analysis reported data on adherence [46]. The proportion of patients who reported or were observed taking more than 90% of their assigned treatment regimen in the allotted time frame were 84% in the 6-month isoniazid arm and 89% in the continuous isoniazid arm. In the continuous arm, the median treatment duration was 3.3 years (IQR 2.1–4.3 years). The proportion of patients who had to permanently discontinue medication was highest for the continuous isoniazid arm (36.5%) compared to the rifapentine-isoniazid arm (1.8%), rifampicin-isoniazid arm (3.8%), and 6-month isoniazid arm (1.9%). Similarly, this was the only study included in the meta-analysis which reported resistance rates for incident TB cases in both the continuous and 6-month isoniazid groups [46]. They found 1 case of isoniazid resistance in the continuous isoniazid arm and no cases in the 6-month arm (RR 5.96, 95% CI 0.24–146). The other two studies reported that the observed rate of INH resistance in incident cases was similar between the continuous and 6-month isoniazid groups, and not different from the expected rate of INH resistance in the population [44, 45].

Data on the cost-effectiveness of continuous isoniazid are limited. One study utilized data from a National Institute for Research in Tuberculosis clinical trial conducted in India in conjunction with a previously published model of TB and HIV co-infection [47]. Compared to no preventive therapy, 6 months of isoniazid increased life expectancy by 0.8 months at a lifetime cost per person of $57,00 US dollars; continuous isoniazid increased life expectancy by 1.0 months at a lifetime per person cost of $5,780 US dollars. The incremental cost-effectiveness ratio (ICER) for continuous isoniazid was $4,290 per year of life saved compared to $1,140 for 6 months of isoniazid. The WHO Commission on Macroeconomics and Health suggests that a regimen is cost-effective if the incremental cost-effectiveness ratio is <3 times the Gross Domestic Product (GDP) per capita ($2,490 US dollars in India in 2009) making the 6-month isoniazid regimen (ICER $1,140) cost-effective and the continuous isoniazid regimen (ICER $4,290) not cost-effective [47]. A second study developed a decision-analytic model utilizing previously published data with health care costs from South Africa [48]. This study also considered varying antiretroviral coverage (55% versus 90%), presence of infection control practices, type of TB diagnostic algorithms (sputum smear and chest radiography versus Xpert MTB/RIF), and presence of intensified case finding in addition to isoniazid duration (6 versus 36 months) in the strategies evaluated. The 36-month strategy was most cost-effective when packaged together with expanded ART provision (90%), infection control, Xpert MTB/RIF, and intensified TB case finding. The incremental cost-effectiveness ratio (ICER) per TB case averted was $28,936 US dollars [48]. A third study, also utilizing a decision-analytic model, used primary data and key results from a clinical trial conducted in Botswana [44]. This study considered TST results in the decision to provide isoniazid and a CD4+ lymphocyte cut-off of 250 and 500 cells/ μL in the decision to provide ART in addition to isoniazid duration (6 versus 36 months). Providing 36 months of isoniazid to only TST-positive PLWH and providing ART only when CD4+ counts were <250 cells/μL was the most cost-effective strategy with respect to ICER per TB case averted ($1,612 US dollars) and ICER per death averted ($2,418) [49].

The WHO GDG noted that consideration be given to the increased resource needs for implementation of continuous isoniazid at the programmatic level, as it is known that global implementation of the 6-month regimen among PLWH is low. The preference for TST before starting therapy was recognized as a possible barrier. Additionally, the committee expressed concerns about lower completion rates and adherence with a longer regimen as well as the development of isoniazid resistance. Continued research on operationalizing continuous isoniazid among PLWH is needed.

Rifamycin-Containing Regimens

There are several rifamycin-containing regimens available as options for treatment of LTBI and endorsed by the WHO and/or CDC as discussed below.

Rifampicin

The WHO guidelines recommend daily rifampicin for 3–4 months as an alternative to isoniazid monotherapy in countries with low TB incidence [26, 27]. The CDC guidelines also recommend daily rifampicin for 4 months as an alternative regimen [28]. Rifampicin-containing regimens must be prescribed with caution to PLWH in high-burden settings. This population has a higher possibility of sub-clinical or undiagnosed TB, which if treated with monotherapy could result in rifampicin resistance.

In addition, there are several drug-drug interactions between rifampicin and ART. Currently, a non-nucleoside reverse transcriptase inhibitor (NNRTI) plus two nucleoside reverse transcriptase inhibitors (NRTIs) is the recommended first-line ART regimen with rifampicin-based therapy. The preferred NRTI backbone is abacavir (ABC), zidovudine (AZT), or tenofovir (TDF) with either lamivudine (3TC) or emtricitabine (FTC). The preferred NNRTI is efavirenz (600 mg) although nevirapine can be considered for patients with an intolerance or contraindication to efavirenz [50–52]. Nevirapine induces its own metabolism and is, therefore, initiated using a lead-in dose of 200 mg/day for 2 weeks before increasing to full doses of 400 mg/day. Rifampicin can also lower serum nevirapine levels in addition to this auto-induction [53]. When evaluated in clinical trials, nevirapine with lead-in dosing has been shown to be associated with low serum nevirapine concentrations, decreased antiviral efficacy, emergence of drug resistance, and death [54–57]. Therefore, experts suggest starting nevirapine at full-dose when it is co-administered with rifampicin [50, 51]. However, when compared to efavirenz, even full-dose nevirapine has been shown to be associated with poor virologic outcomes and higher rates of discontinuation due to toxicity [58]. These data strengthen the preference for efavirenz over nevirapine as the NNRTI of choice during rifampicin co-administration.

Integrase inhibitor-based regimens are now increasingly used as first-line therapy, including in resource-limited settings. Raltegravir and dolutegravir are metabolized mainly by uridine 5′-diphospho-glucuonosyltransferase 1A1, which is induced by rifampicin. Co-administration of these agents with rifampicin reduces serum levels of both drugs. For this reason, rifabutin is preferred over rifampicin by the CDC guidelines. However, there are emerging data that rifampicin may be co-administered with raltegravir if doses are increased to 800 mg twice daily [59–61] or with dolutegravir if doses are increased to 50 mg twice daily [62, 63]. Based on these data and expert opinion, the CDC currently favors the increased dose of raltegravir and dolutegravir when co-administered with rifampicin until further data are available [51]. Additional details on drug-drug interactions between rifampicin and integrase inhibitors can be found in the chapter on pharmacologic considerations during the co-treatment of HIV and TB.

In a network meta-analysis of studies including both HIV-negative and HIV-positive persons, rifampicin monotherapy was effective at preventing active TB when compared to placebo (OR 0.41, 95% CI 0.19–0.85). Rifampicin monotherapy had a lower odds of hepatotoxicity than either 6 or 9-months of isoniazid. The odds ratio for hepatotoxicity with rifampicin monotherapy was 0.03 (95% CI 0.00–0.48) compared to 6-months isoniazid and 0.17 (95% CI 0.06–0.47) compared to 9-months of isoniazid. Stratifying results by HIV-status revealed no statistically significant difference in effect estimates [37].

Since the publication of this meta-analysis, the first randomized controlled trial investigating the effectiveness of 4 months of rifampicin (and compared to 9 months of isoniazid) has been published. Participants were eligible if they were 18 years of age or older, and had a positive TST. Among 6012 participants enrolled, 242 (4.0%) were HIV-positive. There were 3443 participants enrolled in the rifampicin group; 4 developed confirmed active TB and 4 developed clinically diagnosed active TB over 7732 person-years of follow-up. Among 3416 participants enrolled in the isoniazid group, 4 developed confirmed active TB and 5 developed clinically diagnosed active TB over 7652 person-years of follow-up. The rate difference for rifampicin minus isoniazid was <0.01 cases/100 p-y (95% CI −0.14 to 0.16) for confirmed active TB and < 0.01 cases/100 p-y (95% CI −0.23 to 0.22) for confirmed or clinically diagnosed TB. The upper bound of the 95% confidence interval was below the prespecified noninferiority margin of 0.75; thus the 4-month rifampicin regimen was noninferior to 9 months of isoniazid. Treatment completion was better for rifampicin compared to isoniazid (rate difference 15.1%, 95% CI 12.7 to 17.4%). There were fewer grade 3–5 adverse events (rate difference – 1.1%, 95% CI −1.9—0.4%) and fewer grade 3–5 hepatotoxicity events (rate difference −1.2%, 95% CI −1.7 to −0.7%) for rifampicin compared to isoniazid.

Rifabutin

The CDC guidelines note that rifabutin can be substituted for rifampicin among PLWH on ART regimens that are known to interact with rifampicin, such as protease-inhibitor (PI) based or integrase inhibitor based ART [28]. Rifabutin is not recommended in the WHO LTBI treatment guidelines [26, 27]. Although rifabutin is a less potent inducer of CYP isoenzymes, rifabutin itself is metabolized by CYP3A enzymes. Co-administration with PI-based ART, therefore, leads to CYP3A inhibition and increased serum rifabutin levels. Experts suggest decreasing rifabutin doses to 150 mg daily when co-administered with PI-based ART [64]. If rifabutin is not available, double dose lopinavir/ritonavir is the only PI that can be used with rifampicin although these patients must be monitored closely for hepatotoxicity. No dose adjustments are required when rifabutin is co-administered with either raltegravir or dolutegravir.

Rifabutin has been evaluated in a phase II pilot study including 44 participants with LTBI and HIV co-infection. Participants were eligible if they were TST positive adults with confirmed HIV-infection; they were excluded if CD4+ counts were < 200 cells/μL. Participants were randomized to receive either rifabutin 300 mg and isoniazid 750 mg twice weekly for three months (arm 1, n = 16), rifabutin 600 mg and isoniazid 750 mg twice weekly for three months (arm 2, n = 14), or isoniazid 300 mg daily for 6 months (arm 3, n = 14). The study was terminated early prior to reaching enrollment goals by the pharmaceutical sponsor. Three, one, and four subjects did not complete treatment in arms 1, 2, and 3, respectively. Adverse events were reported for four, nine, and seven participants in arms 1, 2, and 3, respectively. During follow-up two cases of active TB were identified (both in the isoniazid monotherapy arm) [65].

Rifampicin and Isoniazid

The WHO guidelines recommend rifampicin plus isoniazid daily for 3 months as an alternative to isoniazid monotherapy in countries with a low TB incidence [26, 27]. ART considerations when using this regimen are the same as those for rifampicin monotherapy as described above.

In a network meta-analysis of studies including both HIV-negative and HIV-positive persons, rifampicin plus isoniazid was effective in preventing active TB when compared to placebo (OR 0.53, 95% CI 0.36–0.78). Furthermore, when using standard meta-analysis methods for direct pairwise comparisons between regimens, rifampicin plus isoniazid did not differ from isoniazid of any duration with respect to prevention of active TB. Rifampicin plus isoniazid also did not differ from isoniazid of either 6 or 9-months in duration with respect to the development of hepatotoxicity. Stratifying results by HIV-status revealed no statistically significant difference in effect estimates [37].

Isoniazid and Rifapentine

The WHO guidelines recommend rifapentine plus isoniazid weekly for 12 weeks (3HP) in countries with a low TB incidence (strong recommendation) and those with a high TB incidence (conditional recommendation) [26, 27]. The CDC guidelines first recommended 3HP among PLWH in December 2011 if they were otherwise healthy and not taking antiretroviral therapy and if 3HP was given by directly-observed therapy [29]. This recommendation was based largely on the results of the PREVENT TB trial which compared 3HP given weekly by directly observed therapy to 9H given daily by self-administered therapy [66]. The primary endpoint was active TB and the non-inferiority margin was 0.75%. There were 403 PLWH enrolled in the United States, Brazil, Spain, Peru, Canada, and Hong Kong; median CD4+ count was 495 (IQR 389–675 cells/μL). Participants receiving or planning to initiate ART during the first 90 days after enrollment were excluded. Cumulative TB rates were 1.01% in the 3HP arm and 3.50% in the 9H arm (rate difference – 2.49, upper 95% CI of the difference 0.60%). Treatment completion was higher in the 3HP arm compared to the 9H arm (89% versus 64%, $p < 0.001$). Discontinuation of study drug due to adverse events was similar (3% in 3HP arm versus 4% in 9H arm, $p = 0.79$) [66].

In June 2018, these recommendations were updated to also include PLWH who were taking efavirenz or raltegravir-based ART. Moreover, self-administered therapy was now an allowed mode of administration in the United States based on local resources and patient characteristics [30]. The inclusion of PLWH taking standard dose efavirenz or raltegravir-based ART were based on pharmacokinetic studies [67, 68]. Co-administration of dolutegravir and rifapentine is not currently recommended, but some pharmacokinetic data on this topic are discussed in the chapter on pharmacologic considerations during co-treatment of HIV and TB. Self-administered 3HP was evaluated in the iAdhere study, which compared that administration method to directly-observed therapy [69]. The primary endpoint was treatment completion and the non-inferiority margin was 15%. Among 1002 participants enrolled in the United States, Spain, Hong Kong, and South Africa, 11 (1.1%) were known to be HIV-positive. Overall, treatment completion was 87% (95% CI 83–91%) in the direct observation group, 74% (95% CI 70–79%) in the self-administration group, and 76% (95% CI 71%–81%) in the self-administration with text message reminders group. In the United States, treatment completion was 85% (95% CI 81–89%), 78% (95% CI 73%–83%) and 77% (95% CI 71–82%), respectively. Therefore, treatment completion of self-administered therapy was non-inferior to directly observed therapy in the United States but not overall or outside of the United States [69].

This short course regimen (3HP) has been evaluated in a cost-effectiveness study in both high- and low-burden settings. Using a cohort of PLWH in a Ugandan HIV clinic, 3HP was estimated to avert 9 cases of TB and 1 death for every 1000 PLWH on ART when compared to isoniazid monotherapy [70]. Incremental cost effectiveness was estimated at $9,402 US dollars per disability-adjusted life year averted. In this model, the cost-effectiveness of 3HP was highly dependent on the cost of rifapentine, completion of treatment, and the prevalence of LTBI. The authors concluded that in comparison to isoniazid monotherapy, that 3HP would only be cost-effective in this high burden setting if the price of rifapentine was reduced and treatment completion rates were high (>85%) [70]. In a US setting, the cost-effectiveness of 3HP was initially evaluated using a cost per dose of 3HP of $12.31 US dollars [71]. In this analysis, 3HP was estimated to avert 5.2 cases of TB for every 1000 people treated. The 3HP regimen would cost $21,525 and $4,294 more per TB case prevented compared to 9H from a health system and societal perspective; it would cost $4,565 and $911 per quality adjusted life year, respectively. In the US, activities costing <$50,000/QALY gained are generally considered to be cost-effective [71]. These estimates were updated after the price of rifapentine was reduced and the cost per dose of 3HP dropped to $6 [72]. In the updated analysis, the cost to the health system per TB case averted by 3HP decreased to $8,816 and the cost to the health system per QALY gained decreased to $1,879. Additionally the authors estimated that switching from 9H to self-administered 3HP would lead to an additional $141 saved per individual treated from a health system perspective and $231 saved per individual treated from a societal perspective. Therefore, in low burden settings the 3HP regimen was deemed to be cost-effective when compared to 9H, with additional savings following the reduction in rifapentine pricing and ability to provide self-administered therapy [72].

The BRIEF-TB trial conducted by the AIDS Clinical Trial Group evaluated an ultra-short 1-month course of daily self-administered rifapentine and isoniazid for treatment of LTBI among PLWH [73]. Participants were eligible if they were at least 13 years old and lived in a high burden TB setting or were TST/IGRA positive. Concomitant ART with an efavirenz or nevirapine based regimen was allowed. Participants were stratified by ART status and CD4+ count and randomized to either 9 months of isoniazid 300 mg daily or one month of weight based-rifapentine plus isoniazid 300 mg daily. The primary outcome was a combined endpoint of active TB, death due to TB, and death due to unknown causes. There were 3000 participants; 50% were on ART and median CD4+ count was 470 cells/ μL (IQR 346–635 cells/μL). Only 634 (21%) had a positive TST/IGRA. A non-inferiority margin of 1.25/100 person-years was set based on an assumed 2.0/100 p-y incidence of the primary endpoint in the 9H arm. The incidence of the primary endpoint was 0.69/100 person-years in the 1HP arm and 0.72/100 p-y in the 9H arm (IR difference − 0.025, upper 95% CI 0.31). Since the upper bound of the 95% CI was below the non-inferiority margin, the 1HP regimen was deemed non-inferior to the 9H regimen. There was no difference in serious adverse events between arms (5.6% 1HP versus 7.1% 9H, $p = 0.1$). There was a higher incidence of targeted safety events in the 9H arm (5.1/100 p-y) compared to the 1HP arm (3.3/100 p-y, $p = 0.03$). Treatment completion was higher in the 1HP arm compared to the 9H arm (97% versus 90%, $p < 0.01$). There was one case of rifampicin-resistance TB in each arm and 1 case of isoniazid-resistant TB in the 9H arm [73]. Additional studies conducted in low-burden TB settings and on the pharmacokinetics of integrase inhibitors when co-administered with daily rifapentine dosing are needed.

Rifapentine was approved by the US Food and Drug Administration for the treatment of latent TB in 2014. Despite the drop in price as discussed above, the drug has not been widely registered outside the US, raising concerns for global access to rifapentine. In 2015, the 20th World Health Organization (WHO) meeting on the selection and use of essential medications recommended the addition of Rifapnetine to the WHO Model List of Essential Medications (EML) [74]. The addition of a medication to the EML often leads to increased demand from country governments and subsequently stimulates drug manufacturers to invest in registration in these countries.

Effect of Antiretroviral Therapy in Preventing the Progression of Latent TB Infection

Many studies have demonstrated the protective effect of ART on the incidence of active TB disease. In 2012, a systematic review and meta-analysis was conducted which included randomized controlled trials, prospective cohort studies, and retrospective cohort studies if they compared TB incidence by ART status among HIV-positive adults for a median of at least 6 months in developing countries [75]. The authors found that ART was strongly associated with a reduction in TB incidence regardless of CD4+ count: HR 0.16 (95% CI 0.07–0.36) for CD4+ <200, HR 0.34 (95% CI 0.19–0.60) for CD4+ 200–350, HR 0.43 (95% CI 0.30–0.63) for CD4+ >350, and HR 0.35 (95% CI 0.28–0.44) for any CD4+ count [75].

Following the publication of this systematic review an additional clinical trial using a 2-by-2 factorial design was conducted to assess the benefits of early ART, 6-months of isoniazid preventive therapy, or both among PLWH in Ivory Coast – the TEMPRANO trial [76]. Participants were included if they had CD4+ counts <800 cells/μL and did not meet criteria for ART initiation according to WHO guidelines. They were randomly assigned to deferred ART (ART initiation per WHO guidelines), deferred ART plus isoniazid, early ART (immediate ART initiation), or early ART plus isoniazid. The primary endpoint was a composite endpoint including AIDS, non-AIDS defining cancer, non-AIDS defining bacterial disease, or death from any cause at 30 months. There were 2056 participants enrolled and followed for 4757 person-years. Tuberculosis accounted for 42% of the 204 primary endpoints observed. Early ART initiation was associated with a decreased incidence of TB (aHR 0.50, 95% CI 0.32–0.79) compared to deferred ART initiation among all patients. When stratified by baseline CD4+ count, early ART initiation was associated with a similar decrease in TB incidence for those with baseline CD4+ count <500 cells/μL (aHR 0.48, 95% CI 0.27–0.87) and a trend toward a decreased incidence for those with a baseline CD4+ count >500 cells/μL (aHR 0.54, 95% CI 0.26–1.09). Isoniazid preventive therapy was associated with a decreased incidence of TB overall (aHR 0.44, 95% CI 0.28–0.69), regardless of baseline CD4+ count (aHR 0.42 [95% CI 0.23–0.76] for CD4 < 500 cells/μL and aHR 0.47 [95% CI 0.23–0.97] for CD4 > 500 cells/μL). The group who received both early ART plus isoniazid had the fewest number of TB events (n = 11), followed by those who received deferred ART plus isoniazid (n = 16), early ART alone (n = 17), and deferred ART alone (n = 41) [76]. Participants who completed this trial follow-up were invited to participate in additional post-trial phase. The primary endopoint of this post-trial phase was death due to any cause from the time of inclusion in the parent trial. The hazard ratio of death was 0.63 (95% CI 0.41–0.97) after adjusting for ART timing and 0.61 (95% CI 0.39–0.94) after adjusting for ART timing, CD4+ count, and other patient characteristics [77]. These data suggest that there is additive benefit of ART and isoniazid preventive therapy in the prevention of TB.

In summary, tuberculosis is likely a spectrum of disease rather than a dichotomous condition of latent infection or active disease; the spectrum is impacted by immunocompromising conditions, particularly HIV infection. Both antiretroviral therapy and anti-mycobacterial therapy are important for the treatment of LTBI. Several anti-mycobacterial regimens are endorsed by the World Health Organization and US Centers for Disease Control and Prevention for treatment of latent TB infection among PLWH, although LTBI treatment options are limited by drug-drug interactions with antiretroviral drugs. Further research is needed to advance our understanding the TB disease spectrum, to develop novel and shorter, well-tolerated anti-mycobacterial treatment regimens, and to better characterize the interactions of anti-mycobacterial and antiretroviral drugs.

Acknowledgements No additional assistance outside the efforts of the authors was contributed for this article. Sources of funding include the National Institutes of Health (P30 AI110527). Both authors report no conflicts of interest with respect to this work.

References

1. Young DB, Gideon HP, Wilkinson RJ (2009) Eliminating latent tuberculosis. Trends Microbiol 17(5):183–188
2. Stead WW (1995) Management of health care workers after inadvertent exposure to tuberculosis: a guide for the use of preventive therapy. Ann Intern Med 122(12):906–912
3. Joshi R, Reingold AL, Menzies D, Pai M (2006) Tuberculosis among health-care workers in low- and middle-income countries: a systematic review. PLoS Med 3(12):e494
4. Morrison J, Pai M, Hopewell PC (2008) Tuberculosis and latent tuberculosis infection in close contacts of people with pulmonary tuberculosis in low-income and middle-income countries: a systematic review and meta-analysis. Lancet Infect Dis 8(6):359–368
5. Stein CM, Nsereko M, Malone LL, Okware B, Kisingo H, Nalukwago S et al (2018) Long-term stability of resistance to latent M. tuberculosis infection in highly exposed TB household contacts in Kampala, Uganda. Clin Infect Dis
6. Santin M, Munoz L, Rigau D (2012) Interferon-gamma release assays for the diagnosis of tuberculosis and tuberculosis infection in HIV-infected adults: a systematic review and meta-analysis. PLoS One 7(3):e32482
7. Chen J, Zhang R, Wang J, Liu L, Zheng Y, Shen Y et al (2011) Interferon-gamma release assays for the diagnosis of active tuberculosis in HIV-infected patients: a systematic review and meta-analysis. PLoS One 6(11):e26827
8. Metcalfe JZ, Everett CK, Steingart KR, Cattamanchi A, Huang L, Hopewell PC et al (2011) Interferon-gamma release assays for active pulmonary tuberculosis diagnosis in adults in low- and middle-income countries: systematic review and meta-analysis. J Infect Dis 204(Suppl 4):S1120–S1129
9. Cattamanchi A, Smith R, Steingart KR, Metcalfe JZ, Date A, Coleman C et al (2011) Interferon-gamma release assays for the diagnosis of latent tuberculosis infection in HIV-infected individuals: a systematic review and meta-analysis. J Acquir Immune Defic Syndr 56(3):230–238
10. Martineau AR, Newton SM, Wilkinson KA, Kampmann B, Hall BM, Nawroly N et al (2007) Neutrophil-mediated innate immune resistance to mycobacteria. J Clin Invest 117(7):1988–1994
11. Clay H, Volkman HE, Ramakrishnan L (2008) Tumor necrosis factor signaling mediates resistance to mycobacteria by inhibiting bacterial growth and macrophage death. Immunity 29(2):283–294
12. van Zyl-Smit RN, Pai M, Peprah K, Meldau R, Kieck J, Juritz J et al (2009) Within-subject variability and boosting of T-cell interferon-gamma responses after tuberculin skin testing. Am J Respir Crit Care Med 180(1):49–58
13. van Zyl-Smit RN, Zwerling A, Dheda K, Pai M (2009) Within-subject variability of interferon-g assay results for tuberculosis and boosting effect of tuberculin skin testing: a systematic review. PLoS One 4(12):e8517
14. Pai M, Joshi R, Dogra S, Zwerling AA, Gajalakshmi D, Goswami K et al (2009) T-cell assay conversions and reversions among household contacts of tuberculosis patients in rural India. Int J Tuberc Lung Dis 13(1):84–92

15. Wayne LG, Lin KY (1982) Glyoxylate metabolism and adaptation of *Mycobacterium tuberculosis* to survival under anaerobic conditions. Infect Immun 37(3):1042–1049
16. Roberts DM, Liao RP, Wisedchaisri G, Hol WG, Sherman DR (2004) Two sensor kinases contribute to the hypoxic response of *Mycobacterium tuberculosis*. J Biol Chem 279(22):23082–23087
17. Sherman DR, Voskuil M, Schnappinger D, Liao R, Harrell MI, Schoolnik GK (2001) Regulation of the *Mycobacterium tuberculosis* hypoxic response gene encoding alpha -crystallin. Proc Natl Acad Sci U S A 98(13):7534–7539
18. McKinney JD, Honer zu Bentrup K, Munoz-Elias EJ, Miczak A, Chen B, Chan WT et al (2000) Persistence of *Mycobacterium tuberculosis* in macrophages and mice requires the glyoxylate shunt enzyme isocitrate lyase. Nature 406(6797):735–738
19. Reed MB, Gagneux S, Deriemer K, Small PM, Barry CE, 3rd. The W-Beijing lineage of *Mycobacterium tuberculosis* overproduces triglycerides and has the DosR dormancy regulon constitutively upregulated. J Bacteriol 2007;189(7):2583–2589
20. Rustad TR, Harrell MI, Liao R, Sherman DR (2008) The enduring hypoxic response of *Mycobacterium tuberculosis*. PLoS One 3(1):e1502
21. Chao MC, Rubin EJ (2010) Letting sleeping dos lie: does dormancy play a role in tuberculosis? Annu Rev Microbiol 64:293–311
22. Lin PL, Rodgers M, Smith L, Bigbee M, Myers A, Bigbee C et al (2009) Quantitative comparison of active and latent tuberculosis in the cynomolgus macaque model. Infect Immun 77(10):4631–4642
23. Mtei L, Matee M, Herfort O, Bakari M, Horsburgh CR, Waddell R et al (2005) High rates of clinical and subclinical tuberculosis among HIV-infected ambulatory subjects in Tanzania. Clin Infect Dis 40(10):1500–1507
24. Sterling TR, Pham PA, Chaisson RE (2010) HIV infection-related tuberculosis: clinical manifestations and treatment. Clin Infect Dis 50(Suppl 3):S223–S230
25. Meintjes G, Lawn SD, Scano F, Maartens G, French MA, Worodria W et al (2008) Tuberculosis-associated immune reconstitution inflammatory syndrome: case definitions for use in resource-limited settings. Lancet Infect Dis 8(8):516–523
26. World Health Organization. Guidelines on the management of latent tuberculosis infection. 2015
27. World Health Organization. Latent tuberculosis infection: Updated and consolidated guidelines for programmatic management 2018
28. American Thoracic Society (2000) Targeted tuberculin testing and treatment of latent tuberculosis infection. MMWR Recomm Rep 49(RR-6):1–51
29. Centers for Disease Control and Prevention (2011) Recommendations for use of an isoniazid-rifapentine regimen with direct observation to treat latent *Mycobacterium tuberculosis* infection. MMWR Morb Mortal Wkly Rep 60(48):1650–1653
30. Borisov AS, Bamrah Morris S, Njie GJ, Winston CA, Burton D, Goldberg S et al (2018) Update of recommendations for use of once-weekly isoniazid-rifapentine regimen to treat latent *Mycobacterium tuberculosis* infection. MMWR Morb Mortal Wkly Rep 67(25):723–726
31. Panel on Opportunistic Infections in HIV-Infected Adults and Adolescents. Guidelines for the prevention and treatment of opportunistic infections in HIV-infected adults and adolescents: recommendations from the Centers for Disease Control and Prevention, the National Institutes of Health, and the HIV Medicine Association of the Infectious Diseases Society of America Available at http://aidsinfo.nih.gov/contentfiles/lvguidelines/adult_oi.pdf. Accessed April 8, 2019
32. Ferebee SH (1970) Controlled chemoprophylaxis trials in tuberculosis. A general review. Bibl Tuberc 26:28–106
33. International Union Against Tuberculosis Committee on Prophylaxis (1982) Efficacy of various durations of isoniazid preventive therapy for tuberculosis: five years of follow-up in the IUAT trial. Bull World Health Organ 60(4):555–564

34. Ferebee SH, Mount FW (1962) Tuberculosis morbidity in a controlled trial of the prophylactic use of isoniazid among household contacts. Am Rev Respir Dis 85:490–510
35. Bucher HC, Griffith LE, Guyatt GH, Sudre P, Naef M, Sendi P et al (1999) Isoniazid prophylaxis for tuberculosis in HIV infection: a meta-analysis of randomized controlled trials. AIDS 13(4):501–507
36. Akolo C, Adetifa I, Shepperd S, Volmink J (2010) Treatment of latent tuberculosis infection in HIV infected persons. Cochrane Database Syst Rev 1:CD000171
37. Zenner D, Beer N, Harris RJ, Lipman MC, Stagg HR, van der Werf MJ (2017) Treatment of latent tuberculosis infection: an updated network meta-analysis. Ann Intern Med 167(4):248–255
38. Snider DE Jr, Caras GJ, Koplan JP (1986) Preventive therapy with isoniazid. Cost-effectiveness of different durations of therapy. JAMA 255(12):1579–1583
39. Comstock GW, Baum C, Snider DE Jr (1979) Isoniazid prophylaxis among Alaskan Eskimos: a final report of the bethel isoniazid studies. Am Rev Respir Dis 119(5):827–830
40. Comstock GW, Ferebee SH (1967) Hammes LM. A controlled trial of community-wide isoniazid prophylaxis in Alaska. Am Rev Respir Dis 95(6):935–943
41. Comstock GW (1999) How much isoniazid is needed for prevention of tuberculosis among immunocompetent adults? Int J Tuberc Lung Dis 3(10):847–850
42. World Healt Organization. Recommendation on 36 months isoniazid preventive therapy to adult and adolescents living with HIV in resource-constrained and high TB- and HIV-prevalence settings. 2015
43. Den Boon S, Matteelli A, Ford N, Getahun H (2016) Continuous isoniazid for the treatment of latent tuberculosis infection in people living with HIV. AIDS 30(5):797–801
44. Samandari T, Agizew TB, Nyirenda S, Tedla Z, Sibanda T, Shang N et al (2011) 6-month versus 36-month isoniazid preventive treatment for tuberculosis in adults with HIV infection in Botswana: a randomised, double-blind, placebo-controlled trial. Lancet 377(9777):1588–1598
45. Swaminathan S, Menon PA, Gopalan N, Perumal V, Santhanakrishnan RK, Ramachandran R et al (2012) Efficacy of a six-month versus a 36-month regimen for prevention of tuberculosis in HIV-infected persons in India: a randomized clinical trial. PLoS One 7(12):e47400
46. Martinson NA, Barnes GL, Moulton LH, Msandiwa R, Hausler H, Ram M et al (2011) New regimens to prevent tuberculosis in adults with HIV infection. N Engl J Med 365(1):11–20
47. Pho MT, Swaminathan S, Kumarasamy N, Losina E, Ponnuraja C, Uhler LM et al (2012) The cost-effectiveness of tuberculosis preventive therapy for HIV-infected individuals in southern India: a trial-based analysis. PLoS One 7(4):e36001
48. Gupta S, Abimbola T, Date A, Suthar AB, Bennett R, Sangrujee N et al (2014) Cost-effectiveness of the Three I's for HIV/TB and ART to prevent TB among people living with HIV. Int J Tuberc Lung Dis 18(10):1159–1165
49. Smith T, Samandari T, Abimbola T, Marston B, Sangrujee N (2015) Implementation and operational research: cost-effectiveness of antiretroviral therapy and isoniazid prophylaxis to reduce tuberculosis and death in people living With HIV in Botswana. J Acquir Immune Defic Syndr 70(3):e84–e93
50. World Health Organization (2009) Treatment of tuberculosis guidelines. In: 4th ed.
51. Nahid P, Dorman SE, Alipanah N, Barry PM, Brozek JL, Cattamanchi A et al (2016) Official American Thoracic Society/Centers for Disease Control and Prevention/Infectious Diseases Society of America Clinical Practice Guidelines: treatment of drug-susceptible tuberculosis. Clin Infect Dis 63(7):e147–ee95
52. Panel on Antiretroviral Guidelines for Adults and Adolescents. Guidelines for the use of antiretroviral agents in adults and adolescents living with HIV. Department of Health and Human Services. Available at http://www.aidsinfo.nih.gov/ContentFiles/AdultandAdolescentGL.pdf. Accessed April 8, 2019
53. Cohen K, van Cutsem G, Boulle A, McIlleron H, Goemaere E, Smith PJ et al (2008) Effect of rifampicin-based antitubercular therapy on nevirapine plasma concentrations in South African adults with HIV-associated tuberculosis. J Antimicrob Chemother 61(2):389–393

54. van Oosterhout JJ, Kumwenda JJ, Beadsworth M, Mateyu G, Longwe T, Burger DM et al (2007) Nevirapine-based antiretroviral therapy started early in the course of tuberculosis treatment in adult Malawians. Antivir Ther 12(4):515–521
55. Boulle A, Van Cutsem G, Cohen K, Hilderbrand K, Mathee S, Abrahams M et al (2008) Outcomes of nevirapine- and efavirenz-based antiretroviral therapy when coadministered with rifampicin-based antitubercular therapy. JAMA 300(5):530–539
56. Avihingsanon A, Manosuthi W, Kantipong P, Chuchotaworn C, Moolphate S, Sakornjun W et al (2008) Pharmacokinetics and 48-week efficacy of nevirapine: 400 mg versus 600 mg per day in HIV-tuberculosis coinfection receiving rifampicin. Antivir Ther 13(4):529–536
57. Swaminathan S, Padmapriyadarsini C, Venkatesan P, Narendran G, Ramesh Kumar S, Iliayas S et al (2011) Efficacy and safety of once-daily nevirapine- or efavirenz-based antiretroviral therapy in HIV-associated tuberculosis: a randomized clinical trial. Clin Infect Dis 53(7):716–724
58. Bonnet M, Bhatt N, Baudin E, Silva C, Michon C, Taburet AM et al (2013) Nevirapine versus efavirenz for patients co-infected with HIV and tuberculosis: a randomised non-inferiority trial. Lancet Infect Dis 13(4):303–312
59. Grinsztejn B, De Castro N, Arnold V, Veloso VG, Morgado M, Pilotto JH et al (2014) Raltegravir for the treatment of patients co-infected with HIV and tuberculosis (ANRS 12 180 Reflate TB): a multicentre, phase 2, non-comparative, open-label, randomised trial. Lancet Infect Dis 14(6):459–467
60. Taburet AM, Sauvageon H, Grinsztejn B, Assuied A, Veloso V, Pilotto JH et al (2015) Pharmacokinetics of raltegravir in HIV-infected patients on rifampicin-based antitubercular therapy. Clin Infect Dis 61(8):1328–1335
61. Wenning LA, Hanley WD, Brainard DM, Petry AS, Ghosh K, Jin B et al (2009) Effect of rifampin, a potent inducer of drug-metabolizing enzymes, on the pharmacokinetics of raltegravir. Antimicrob Agents Chemother 53(7):2852–2856
62. Dooley KE, Sayre P, Borland J, Purdy E, Chen S, Song I et al (2013) Safety, tolerability, and pharmacokinetics of the HIV integrase inhibitor dolutegravir given twice daily with rifampin or once daily with rifabutin: results of a phase 1 study among healthy subjects. J Acquir Immune Defic Syndr 62(1):21–27
63. Dooley KE, Kaplan R, Mwelase N, Grinsztejn B, Ticona E, Lacerda M et al (2019) Dolutegravir-based antiretroviral therapy for patients co-infected with tuberculosis and HIV: a multicenter, noncomparative, open-label, randomized trial. In: Clin Infect Dis
64. Loeliger A, Suthar AB, Ripin D, Glaziou P, O'Brien M, Renaud-Thery F et al (2012) Protease inhibitor-containing antiretroviral treatment and tuberculosis: can rifabutin fill the breach? Int J Tuberc Lung Dis 16(1):6–15
65. Matteelli A, Olliaro P, Signorini L, Cadeo G, Scalzini A, Bonazzi L et al (1999) Tolerability of twice-weekly rifabutin-isoniazid combinations versus daily isoniazid for latent tuberculosis in HIV-infected subjects: a pilot study. Int J Tuberc Lung Dis 3(11):1043–1046
66. Sterling TR, Villarino ME, Borisov AS, Shang N, Gordin F, Bliven-Sizemore E et al (2011) Three months of rifapentine and isoniazid for latent tuberculosis infection. N Engl J Med 365(23):2155–2166
67. Podany AT, Bao Y, Swindells S, Chaisson RE, Andersen JW, Mwelase T et al (2015) Efavirenz pharmacokinetics and pharmacodynamics in HIV-infected persons receiving rifapentine and isoniazid for tuberculosis prevention. Clin Infect Dis 61(8):1322–1327
68. Weiner M, Egelund EF, Engle M, Kiser M, Prihoda TJ, Gelfond JA et al (2014) Pharmacokinetic interaction of rifapentine and raltegravir in healthy volunteers. J Antimicrob Chemother 69(4):1079–1085
69. Belknap R, Holland D, Feng PJ, Millet JP, Cayla JA, Martinson NA et al (2017) Self-administered versus directly observed once-weekly isoniazid and rifapentine treatment of latent tuberculosis infection: a randomized trial. Ann Intern Med 167(10):689–697
70. Johnson KT, Churchyard GJ, Sohn H, Dowdy DW (2018) Cost-effectiveness of preventive therapy for tuberculosis with isoniazid and rifapentine versus isoniazid alone in high-burden settings. Clin Infect Dis 67(7):1072–1078

71. Shepardson D, Marks SM, Chesson H, Kerrigan A, Holland DP, Scott N et al (2013) Cost-effectiveness of a 12-dose regimen for treating latent tuberculous infection in the United States. Int J Tuberc Lung Dis 17(12):1531–1537
72. Shepardson D, MacKenzie WR (2014) Update on cost-effectiveness of a 12-dose regimen for latent tuberculous infection at new rifapentine prices. Int J Tuberc Lung Dis 18(6):751
73. Swindells S, Rramchandani R, Gupta A, Benson CA, Leon-Cruz JT, Mwelase N et al (2019) One month of rifapentine plus isoniazid to prevent HIV-related tuberculosis. N Engl J Med 380(11):1001–1011
74. Organization WH. Report of the WHO Expert Committee, 2015 (including the 19th WHO Model List of Essential Medicines and the 5th WHO Model List of Essential Medicines for Children)
75. Suthar AB, Lawn SD, del Amo J, Getahun H, Dye C, Sculier D et al (2012) Antiretroviral therapy for prevention of tuberculosis in adults with HIV: a systematic review and meta-analysis. PLoS Med 9(7):e1001270
76. TEMPRANO ANRS 12136 Study Group (2015) Danel C, Moh R, Gabillard D, Badje A, Le Carrou J, et al. A trial of early antiretrovirals and isoniazid preventive therapy in Africa. N Engl J Med 373(9):808–822
77. Badje A, Moh R, Gabillard D, Guehi C, Kabran M, Ntakpe JB et al (2017) Effect of isoniazid preventive therapy on risk of death in west African, HIV-infected adults with high CD4 cell counts: long-term follow-up of the Temprano ANRS 12136 trial. Lancet Glob Health 5(11):e1080–e10e9

Treatment of Drug-Sensitive Tuberculosis in Persons with HIV

Alice K. Pau, Safia Kuriakose, Kelly E. Dooley, and Gary Maartens

Abstract Tuberculosis (TB) is the most common opportunistic infection and the leading cause of death in patients with human immunodeficiency virus (HIV) worldwide. Persons with advanced HIV infection and those not on antiretroviral therapy (ART) are at highest risks of morbidity and mortality. Thus, early diagnosis and treatment of both HIV and TB are keys to treatment success. Concurrent treatment of both infections can be challenging due to high pill burden, overlapping toxicities, drug interactions, adherence concerns, and the potential of immune reconstitution inflammatory syndrome. This chapter provides an overview of treatment regimens and duration of therapy for drug-susceptible TB, management of adverse drug reactions, monitoring parameters, adherence interventions, and appropriate timing for ART initiation in this patient population.

Keywords Tuberculosis · HIV · Anti-tuberculosis therapy · Antiretroviral therapy Adherence · Directly observed therapy · Drug induced liver injury · Toxicities · Immune reconstitution inflammatory syndrome · Mortality

---

A. K. Pau (✉)
National Institute of Allergy and Infectious Diseases, National Institutes of Health, Bethesda, MD, USA
e-mail: apau@niaid.nih.gov

S. Kuriakose
Clinical Monitoring Research Program Directorate, Frederick National Laboratory for Cancer Research Sponsored by the National Cancer Institute, Frederick, MD, USA
e-mail: safia.kuriakose@nih.gov

K. E. Dooley
Division of Clinical Pharmacology, Center for Tuberculosis Research, Johns Hopkins University School of Medicine, Baltimore, MD, USA
e-mail: kdooley1@jumi.edu

G. Maartens
Division of Clinical Pharmacology, Department of Medicine, University of Cape Town, Cape Town, South Africa
e-mail: gary.maartens@uct.ac.za

I. Sereti et al. (eds.), *HIV and Tuberculosis*,
https://doi.org/10.1007/978-3-030-29108-2_9

General Principles of Tuberculosis Treatment

Tuberculosis (TB) is the most common opportunistic infection globally and the leading cause of death in persons with HIV, especially in areas with high TB prevalence, such as sub-Saharan Africa [1]. The severity of TB and related complications in persons with HIV correlate well with the degree of immune suppression (as measured by CD4 cell count) at the time of TB diagnosis and whether or not the patient is receiving antiretroviral therapy (ART) [2]. Thus, early diagnosis and treatment of both infections are keys to treatment success and reduction of TB and HIV associated morbidity and mortality.

All patients with documented TB [defined as having a positive smear of acid fast bacilli (AFB) or molecular test or culture diagnosis of *Mycobacterium tuberculosis* (MTB), along with compatible clinical signs and symptoms] or high clinical suspicion of TB (compatible symptoms in a highly endemic area but without microbiologic confirmation) should be initiated on TB treatment immediately [3–5]. The primary goals of therapy are to reduce TB-associated disease progression and death, and to cure the patient of TB. Another important goal, for patients with pulmonary or laryngeal TB, is to reduce the duration of infectivity and to prevent further spread of TB to others. To accomplish these goals, first-line TB therapy consists of an initial "intensive" phase of a four-drug regimen with isoniazid, a rifamycin antibiotic (generally rifampicin or rifabutin), ethambutol, and pyrazinamide, and is used for approximately 2 months. This multiple drug combination rapidly reduces bacillary load and halts replication of bacilli. Furthermore, combination therapy is particularly important while awaiting drug susceptibility testing results, as it should be effective against MTB isolates that may harbor mono-drug resistance, particularly to isoniazid. After the intensive phase, therapy should be continued with 2 drugs (isoniazid and a rifamycin) for approximately 4 months. The purpose of this continuation phase treatment is to eradicate the slower growing bacilli in order to achieve cure. The duration of therapy may be extended if the patient has severe disease at presentation, has slower than expected clinical response, has smear or culture positivity at the end of the 2-month intensive phase treatment, has certain types of extrapulmonary MTB infection (e.g. TB of the spine), has adverse drug reactions necessitating substitution for isoniazid or rifampin with other TB drugs or discontinuation of pyrazinamide, or if the patient has periods of non-adherence to the regimen [3, 5].

Because of the long duration of therapy, non-adherence or loss to follow-up is a major hurdle for successful TB treatment. Failure to complete a treatment course may result in acquired drug resistance, relapse of TB, and further transmission of MTB to others.

Indications for TB Treatment

All patients with documented TB by microbiological (AFB smear and/or culture and/or approved molecular genetic testing) should commence TB therapy promptly. The availability of a more sensitive rapid diagnostic test, such as the Xpert RIF/TB Ultra assay, should reduce the proportion of patients needing empiric TB treatment.

Empiric TB Therapy for Suspected TB

If TB is strongly suspected, therapy should be initiated empirically pending diagnostic confirmation. The World Health Organization (WHO) has issued recommendations for empiric TB therapy in persons with HIV and suspected TB and negative rapid diagnostic tests, while awaiting culture results [6]. Xpert RIF/TB assay is the preferred rapid diagnostic test. However, the AFB smear is still the only available rapid test in many resource-poor settings. Diagnosis of TB should be highly suspected if a patient presents with any one of the following symptoms: cough for more than 2 weeks, weight loss, night sweats, or fever. Danger signs are any one of the following: respiratory rate >30 breaths/min, heart rate >120 beats/min, temperature > 39°C, or inability to walk unaided. In ambulatory patients without danger signs, TB therapy is recommended if the chest radiograph is compatible with TB, or if there are features of extrapulmonary TB. In seriously ill patients with HIV (defined as having danger signs), in addition to empiric TB therapy, broad spectrum antibiotics are recommended and treatment for *Pneumocystis* pneumonia should also be considered, especially in patients with CD4 count <200 cells/mm^3.

Few studies have evaluated the performance of the WHO algorithms for TB diagnosis. Studies of the algorithm report relatively poor diagnostic performance, but high negative predictive values indicate the algorithm has value in ruling out TB [7, 8]. In a South African study evaluating the seriously ill algorithm, none of the classic TB symptoms predicted TB in patients with cough; predictors of TB were cough ≥14 days, inability to walk unaided, temperature >39 °C, low haemoglobin, low white cell count, and chest radiographic features of TB [9].

Treatment of Drug Susceptible Tuberculosis (Also See Table 1)

MTB divides slowly and may exist in either a replicating, metabolically active state or a dormant, slowly or non-replicating state [10]. Successful treatment requires multiple drugs for two main reasons—to kill both actively-multiplying and semi-dormant (so-called "persister") bacilli and to prevent selection of drug resistant mutants.

Table 1 Management of drug susceptible tuberculosis (TB) in persons with HIV

	Recommendations	Comments
Treatment Regimens for Drug Susceptible TB		
Treatment Regimens for Drug Susceptible TB	For uncomplicated pulmonary TB *Intensive phase* – Isoniazid (+ pyridoxine) + rifampicin (or rifabutin) + pyrazinamide + ethambutol (see Table 2 for dosing recommendations) × 2 months, followed by *Continuation phase* – Isoniazid (+ pyridoxine) + rifampicin (or rifabutin) × 4 months	Treatment should be given daily. Intermittent therapy (2–3 times weekly is not recommended for HIV-infected patients). Therapy may be extended if treatment interruption is needed. Use of rifampicin vs. rifabutin depends on ART regimen (see "Co-treatment of Tuberculosis and HIV: Pharmacologic Considerations" chapter for recommendation). Directly observed therapy should be done if feasible. Treatment duration may need to be extended based on clinical response.
Total duration of Therapy Based on Clinical Situations	Pulmonary TB with good clinical response—6 months Pulmonary TB with positive AFB smear or culture at 2 months of TB treatment—9 months Extrapulmonary TB w/CNS involvement—at least 9 months Extrapulmonary TB w/bone or joint involvement—9 months Extrapulmonary TB in other sites—6 months	
Management of Drug-Induced Liver Injury (DILI)		
Asymptomatic and ALT <5× ULN	Continue TB drugs and monitor more frequently	If patient becomes symptomatic or ALT >5× ULN—see recommendation below.
– Symptoms consistent with DILI and ALT >3× ULN, or – Asymptomatic and ALT >5× ULN	Stop TB drugs except ethambutol and any concomitant drugs with DILI potential. Start TB regimen with ethambutol + a fluoroquinolone + linezolid or an aminoglycoside. Evaluate for other causes of liver disease, including viral hepatitis. Obtain additional laboratory tests associated with liver function such as bilirubin, alkaline phosphatase, prothrombin time, and albumin. Close monitoring for resolution of symptoms and laboratory abnormalities.	Initiation of a new regimen of ethambutol + a fluoroquinolone + linezolid or an aminoglycoside will allow for continue TB treatment and avoid development of TB drug resistance.

(continued)

Table 1 (continued)

	Recommendations	Comments
Rechallenging with TB Drugs	Rechallenge can begin after resolution of symptoms and jaundice and ALT decrease to <2.5 times ULN (or close to baseline if ALT is elevated at baseline) See text for detailed discussion regarding drug rechallenge recommendations	Rechallenge with pyrazinamide should be limited to patients with TB meningitis or disseminated TB
When to Start Antiretroviral Therapy (ART) in ART-Naïve Patients or Those Not on Therapy at the Time of TB Diagnosis		
CD4 count <50 cells/mm^3 and without TB meningitis	Within 2 weeks of starting TB therapy	Early ART in patients with CD4 < 50 cells/mm^3 has been associated with reduced mortality Monitor for clinical signs and symptoms of IRIS
CD4 count ≥ 50 cells/mm^3	Within 8 weeks of starting TB therapy	Monitor for clinical signs and symptoms of IRIS Delay ART until after completion of TB treatment is not recommended
TB meningitis	ART may be delayed to after 4–8 weeks of TB treatment with close monitoring for adverse events	Early ART has been associated with higher rate of Grade 4 adverse events in patients with TB meningitis

ALT alanine aminotransferase, *ART* antiretroviral therapy, *CNS* central nervous system, *DILI* drug related liver injury, *HIV* human immunodeficiency virus, *IRIS* immune reconstitution inflammatory syndrome, *TB* tuberculosis, *ULN* upper limit of normal

Receipt of ART is a key factor for TB treatment success. Six-month TB treatment duration in persons with HIV not receiving ART is associated with higher treatment failure and recurrence rates compared to those receiving ≥8 months of treatment [11–13]. However, in patients on ART, rates of TB recurrence are similar for treatment durations of 6 and ≥8 months [11].

Treatment should begin with a combination of rifampicin (or rifabutin), isoniazid (with pyridoxine supplementation), pyrazinamide, and ethambutol for a 2-month intensive phase. Drug resistance testing to at least rifampicin, and when possible isoniazid should be performed on all initial isolates if this is affordable and available [3, 5]. If susceptibility to rifampicin and isoniazid is confirmed, ethambutol may be discontinued before the end of the 2 months.

After the successful completion of the 2 month, 4-drug intensive phase of therapy—defined as resolution or significant improvement of clinical symptoms along with negative or significantly reduced number of AFB seen on sputum smear - patients with drug sensitive TB should remain on isoniazid and rifampicin (or rifabutin). This continuation phase generally lasts for an additional 4 months but may require extension to an additional 3 months. Duration may vary depending on sever-

Table 2 Dosing of anti-tuberculosis medications in adult patients

Medications	Formulations	Once daily dosing	For patients with renal insufficiency	For patients with liver disease
Isoniazid[a,b]	Oral liquid, tablet, intramuscular / intravenous injection	5 mg/kg	No dosage adjustment necessary.	Use with caution in patients with liver disease
Rifampicin[a]	Oral liquid, tablet/capsule, intravenous injection	10 mg/kg	No dosage adjustment necessary.	Use with caution in patients with liver disease
Rifabutin	Capsule	5 mg/kg Usual dose is 300 mg (dosage adjustment may be necessary based on concomitant ARV drug)	CrCl <30 mL/min: give 50% of usual dose Consider monitoring rifabutin concentration, especially if dose adjusted based on drug interaction	Use with caution in patients with liver disease
Pyrazinamide[a]	Oral liquid, tablet, dispersible tablet	40–55 kg: 1000 mg 56–75 kg: 1500 mg 76–90 kg: 2000 mg >90 kg: 2000 mg	CrCl <30 mL/min or hemodialysis: Adjust dose to 25–35 mg/kg three times weekly. For patients on hemodialysis, dose on dialysis day, after dialysis session	Use with caution in patients with liver disease. Contraindicated in patients with severe hepatic impairment
Ethambutol[a]	Oral liquid, tablet	40–55 kg: 800 mg 56–75 kg: 1200 mg 76–90 kg: 1600 mg >90 kg: 1600 mg	CrCl <30 mL/min or hemodialysis: Adjust dose to 20–25 mg/kg three times weekly. For patients on hemodialysis, dose on dialysis day, after dialysis session	No dosage adjustment necessary
Levofloxacin	Tablet, intravenous injection	750–1000 mg	CrCl <30 mL/min: give usual dose three times per week	Limited hepatic metabolism
Moxifloxacin	Tablet, intravenous injection	400 mg	No dosage adjustment necessary.	Use with caution in patients with liver disease

CrCl creatinine clearance, *FDC* fixed dose combination
[a]Available as a component of a variety of brand or generic fixed dose combination (FDC) tablets containing other TB drugs, please refer to product labels of the FDC for dosing information
[b]Isoniazid should be given with pyridoxine 25–50 mg daily to reduce peripheral neuropathy

ity, delayed culture conversion, the site of infection, presence of drug resistance, clinical response, treatment interruptions, or other factors.

Intermittent dosing (e.g. two or three times weekly therapy) is not recommended in persons coinfected with HIV and TB as it is associated with higher rate of treatment failure and acquired drug resistance in both the intensive and continuation phase [11, 12, 14–17]. Therefore, daily (or at least 5 times weekly) dosing is recommended for all persons with HIV and TB.

While on therapy, patients with smear positive pulmonary TB should be monitored with sputum AFB smears (and cultures in resource-rich settings) at months 2 and 5—cultures should be done at these timepoints if AFB smear is positive [4, 5]. A baseline chest radiography (x-ray or CT scan) and follow-up imaging at 2 months should also be performed if resources permit, or sooner if clinically indicated. Patients with extensive disease and delayed culture conversion are at increased risk of TB relapse [18, 19]. In patients with cavitation on initial chest radiograph and whose culture or smear at month 2 remains positive, the continuation phase treatment should be extended by 3 months—for a total treatment duration of 9 months. If sputum cultures remain positive after ≥4 months of therapy or if they revert to positive after initial culture conversion to negative, repeat drug resistance testing should be performed. For more information on treatment of drug resistant TB and susceptibility testing, please see the "Drug Resistant TB and HIV" chapter. If signs and symptoms of TB disease fail to improve or worsen, patients should be evaluated for other potential causes such as another infection or IRIS.

Anti-tuberculosis Medications and Drug Therapy Monitoring

If possible, baseline liver function tests, complete blood count, and serum creatinine should be performed as these results can affect treatment decisions, including drug choice, dosing, and monitoring frequency. Patients should also be monitored regularly during therapy for adverse drug reactions, clinical response to treatment, and adherence. Frequency and type of monitoring depend on several factors including comorbidities (e.g. underlying liver or renal impairment), potential for drug interactions, and side effects. As weight gain is common in patients with successful TB therapy, body weight should also be monitored to determine if weight-based drugs need dosage adjustment. Patients should be educated on potential side effects of their drug regimen and counseled to report these immediately. Parameters for individual drug monitoring are reviewed below. Table 2 provides information on the dosing of the different TB drugs used for patients with drug susceptible TB, including recommendations for dosing in patients with renal and liver diseases. Table 3 lists common and serious adverse effects reported with these TB drugs. For more information on clinical pharmacology and drug-drug interactions, please refer to chapter on "Co-treatment of Tuberculosis and HIV: Pharmacologic Considerations".

Isoniazid

Isoniazid inhibits the synthesis of mycolic acids, which are an essential component of the bacterial cell wall. It has excellent early bactericidal activity against actively replicating MTB, however, it is bacteriostatic against non-replicating MTB. Isoniazid is metabolized via acetylation by N-acetyl transferase 2 (NAT2). Rates of acetylation, and thus drug exposures, differ due to genetic variation in NAT2 alleles, leading to "slow" or "rapid" acetylation. The prevalence of NAT2 alleles varies geographically, with high prevalence of rapid acetylators in Southeast Asia [20]. Slow acetylation results in higher isoniazid plasma concentrations and increases the risk of toxicity [20–22]. Rapid acetylation, which reduces isoniazid exposure, has been associated with treatment failure, acquired drug resistance, and treatment relapse [23]. Isoniazid use has been associated with hepatotoxicity (which may be severe), peripheral neuropathy, drug-induced lupus, and rarely neuropsychiatric effects such as acute psychosis. Pyridoxine (vitamin B6) supplementation (25–50 mg/day) is recommended for all patients with HIV to prevent peripheral neuropathy.

Rifamycin Antibiotics

The rifamycin antibiotics bind to the beta subunit of RNA polymerase to inhibit RNA transcription and protein synthesis. They are bactericidal against MTB and have excellent sterilizing activity. These agents are essential for eliminating persisting organisms and therefore are key to shortened treatment duration to 6-months. Unless an adverse reaction prohibits its use, a rifamycin antibiotic should be maintained as a component of the TB treatment regimen for the entire course of therapy, unless rifamycin resistance is documented.

Three rifamycin antibiotics are currently used to treat TB—rifampicin, rifabutin, and rifapentine. Rifampicin is the most commonly used rifamycin antibiotic and is available in several fixed dose combination (FDC) products. Rifabutin may be used in place of rifampicin when a significant drug-drug interaction prohibits the use of rifampicin or if rifampicin is not tolerated because of a hypersensitivity reaction, as cross-reactivity occurs only in a minority of patients [24]. Intermittently dosed rifapentine, though approved as part of a combination therapy for TB, is not recommended for persons with HIV due to the high rate of treatment failure with acquired rifamycin monoresistance in one study [25]. However, higher doses of rifapentine given once daily have been shown to be more effective than the doses initially used [26, 27], and are being studied in patients with TB and HIV. Higher doses of rifampicin result in earlier sputum culture conversion in phase II studies [28, 29], and are also being evaluated in patients with and without HIV co-infection [30].

All rifamycins may cause red-orange discoloration of body fluids, which patients should be counseled about prior to treatment initiation. Hepatitic or cholestatic

Table 3 Common or serious adverse drug reactions of anti-tuberculosis medications by organ system

Organ System	Isoniazid	Rifampicin	Rifabutin	Pyrazinamide	Ethambutol	Levofloxacin/ Moxifloxacin
Dermatologic (also refer to Hypersensitivity Reactions)	Rash	Rash, flushing, pruritis	Rash	Rash	Rash	Rash, pruritus, photosensitivity
Gastrointestinal[a]	Nausea, vomiting, diarrhea	Nausea, vomiting, anorexia, abdominal pain, diarrhea, dyspepsia	Nausea, vomiting, anorexia, abdominal pain, diarrhea, dyspepsia, taste disturbance	Nausea, vomiting, anorexia	Nausea, vomiting, anorexia, abdominal pain	Nausea, vomiting, abdominal pain, anorexia, diarrhea, taste disturbance
Hematologic	Sideroblastic anaemia	Thrombocytopenia (can be part of hypersensitivity reaction)	Neutropenia (dose related)			
Hepatic[b]	Hepatotoxicity	Hepatotoxicity (cholestatic or hepatitic), transient elevations in liver function tests	Hepatotoxicity	Hepatotoxicity		Hepatotoxicity (less likely than with isoniazid, rifamycin, or pyrazinamide)

(continued)

Table 3 (continued)

Organ System	Isoniazid	Rifampicin	Rifabutin	Pyrazinamide	Ethambutol	Levofloxacin/ Moxifloxacin
Hypersensitivity reactions	Fever, skin eruptions, lymphadenopathy, TEN, DRESS	Hypersensitivity reactions can be associated with fever, flu-like syndrome, hypotension, abnormal LFTs, thrombocytopenia, petechiae, and multi-organ dysfunction—associated with intermittent dosing of rifampicin SJS, TEN, DRESS, and anaphylaxis have also been reported	Hypersensitivity less common than rifampicin, with limited cross-reactivity	Hypersensitivity rare	Hypersensitivity rare	SJS, TEN, anaphylaxis
Musculoskeletal	Arthralgia, myalgia arthritis	Myalgia	Polyarthralgia, myalgia	Arthralgia, myalgia. Gout	Gout	Arthropathy, tendonitis and tendon rupture
Nervous System	Peripheral neuropathy, psychosis				Peripheral neuropathy	Peripheral neuropathy, dizziness, insomnia, headache, anxiety, confusion, depression, convulsions
Ocular	Optic neuritis (concentration dependent)		Uveitis (concentration dependent)		Optic neuropathy (concentration dependent)	

Miscellaneous Adverse reactions	Pancreatitis, hyperglycemia, lupus like syndrome	Red-orange discoloration of body fluids	Red-orange discolouration of body fluids	Hyperuricemia	Hyperuricemia	QTc prolongation, hypo- or hyper-glycemia, *C. difficile*-associated diarrhea; rarely, rupture or tear of aorta
Monitoring recommendations[c]						
General monitoring	Hepatic function (ALT, AST), Renal Function (serum creatinine), Weight					
Drug specific monitoring	Signs and symptoms of DILI, peripheral neuropathy, and other adverse effects	Signs and symptoms of DILI and other adverse effects	CBC—monitor CBC monthly Monitor for signs and symptoms of uveitis and other adverse effects	Signs and symtoms of DILI and other side effects If available, uric acid monitoring may be considered in patients at risk of hyperuricemia/gout	Patients should have visual acuity (Snellen Chart) at baseline and at monthly intervals, and educated to report changes in vision (including colour perception).	Signs and symptoms of adverse effects

ALT alanine aminotransferase, *CBC* Complete Blood Count, *DILI* Drug Induced Liver Injury *DRESS* Drug Reaction with Eosinophilia and Systemic Symptoms, *SJS* Stevens Johnson Syndrome, *TEN* Toxic Epidermal Necrolysis

[a]Gastrointestinal symptoms may be related to hepatotoxicity

[b]Pyrazinamide is the most likely offending agent, followed by isoniazid, and then rifampin. Hepatotoxicity with levofloxacin and moxifloxacin is rare

[c]Frequency and type of laboratory monitoring depend on several factors including pre-existing comorbidities (e.g. liver or renal impairment), drug interactions, and adverse events. Monthly intervals are suggested but may not be feasible in all settings

drug-induced liver injury (DILI) is also a potential adverse effect. Hypersensitivity reactions, including a flu-like syndrome, are more common with intermittent dosing strategies (which are not recommended in HIV/TB coinfected patients). Other immune-mediated adverse drug reactions include rashes, thrombocytopenia, and interstitial nephritis. Uveitis and neutropenia are toxicities that are unique to rifabutin [31, 32].

Rifamycin antibiotics, especially rifampicin and rifapentine, are potent inducers of many cytochrome P450 and Phase 2 metabolic enzymes, as well as drug transporters (e.g. P-glycoprotein), which may lead to significant decrease in systemic exposure of drugs that are substrates of the induced enzymes and/or transporters [33]. Therefore, all concomitant drugs, including antiretroviral drugs, should be carefully evaluated for drug-drug interactions (see "Co-treatment of Tuberculosis and HIV: Pharmacologic Considerations", chapter for detailed discussion).

Pyrazinamide

The exact mechanism of pyrazinamide's activity against MTB has not been well established. Like rifamycin antibiotics, pyrazinamide is a bactericidal agent with excellent MTB sterilizing activity. It is active only in acidic environments. Pyrazinamide is associated with a higher incidence of hepatotoxicity than the other first line TB medications [3]. Other side effects include hyperuricemia, gout, polyarthralgia, and hypersensitivity reactions.

Ethambutol

Ethambutol is a bacteriostatic agent that interferes with mycobacterial cell wall synthesis. It is given at the minimum effective dose, solely to protect against emergence of resistance. Once rifamycin and isoniazid susceptibilities are confirmed, ethambutol may be discontinued during the intensive phase of treatment. Ethambutol is associated with optic neuropathy, manifested by decreased visual acuity, visual field defects, and abnormal color perception. Patients should be evaluated at baseline and at routine intervals while on therapy, with visual acuity testing measured using the Snellen chart. They should also be counseled to report visual changes to their providers. Other causes of visual changes common in persons with advanced HIV such as cytomegalovirus retinitis should be ruled out. If toxicity is detected, ethambutol should be immediately discontinued to avoid permanent damage.

Fluoroquinolones

Levofloxacin and moxifloxacin are fluoroquinolones with rapid bactericidal activity against MTB [34–36]. However, clinical trials in which a fluoroquinolone was substituted for either isoniazid or ethambutol in an effort to shorten the treatment duration to 4-month were not successful [37, 38]. Therefore, this class of drug is generally reserved for use in patients who are unable to tolerate first line drugs (e.g. isoniazid) or who have isoniazid mono-resistant TB [4]. A fluoroquinolone is also a standard part of TB regimens for multiple drug resistant TB [39]. Fluoroquinolones are associated with a variety of adverse drug reactions such as tendonitis and tendon rupture, modest QT prolongation (more marked with moxifloxacin than levofloxacin), peripheral neuropathy, hypo- or hyper-glycemia, and neuropsychiatric effects (e.g. insomnia, acute psychosis, seizures), and rarely rupture of the aorta. They are also susceptible to chelation and decreased drug absorption when given concomitantly with polyvalent cations such as antacids, mineral supplements or multivitamins.

Management of Adverse Drug Reactions

Adverse drug reactions are common when treating both TB and HIV infections, and some severe reactions may require permanent discontinuation of one or more anti-tuberculosis medication. Clinicians should seek expert consultation to construct an effective regimen that includes drug substitutions and to determine if the treatment duration requires extension.

Nausea and vomiting are common early side effects of TB therapy. However, liver function tests should be evaluated if there is new onset nausea and vomiting, which are common symptoms of hepatotoxicity. For more information on drug induced hepatotoxicity, please see the discussion below.

Rash can also occur and if mild, patients can continue TB therapy and be managed symptomatically if appropriate. Patients should be evaluated for features of severe cutaneous adverse drug reactions: mucous membrane involvement, blistering, fever or other systemic symptoms, or evidence of systemic hypersensitivity reactions. It should be noted that hepatotoxicity often presents after the rash, so it is important to repeat liver function tests after a week. A notable exception is ethambutol, which may cause rash but does not cause drug induced liver injury. All drugs should be discontinued immediately if a severe adverse drug reaction is suspected. To identify the causative drug, follow the same approach presented in the "Drug Induced Liver Injury" section to reintroduce each medication.

Management of Drug Induced Liver Injury (DILI) Associated with TB Therapy

DILI is among the most common and most serious complication of TB medications, and can be attributed to isoniazid, rifampicin, or pyrazinamide. HIV infection, older age, female sex, those with underlying liver diseases (such as caused by hepatitis B or C co-infection, chronic alcohol use), and baseline abnormal serum transaminases are all risk factors for DILI [40–45]. DILI may manifest with laboratory abnormalities [best assessed by elevation of serum alanine aminotransferase (ALT) and bilirubin] with or without clinical manifestations such as fever, jaundice, nausea, vomiting, abdominal pain and/or mental status changes. Transient asymptomatic ALT elevation, known as hepatic adaptation, is common.

Managing patients with suspected DILI can be challenging. One needs to balance the need to treat active TB and the risk of life-threatening acute liver injury if therapy is continued. The decision for continuation of TB therapy should be done on a case by case basis, with careful evaluation before stopping drugs, during treatment interruption, and after resumption of therapy. In general, for asymptomatic patients with ALT <5 times upper limit of normal (ULN), many experts recommend continued TB treatment, but with more frequent clinical and laboratory monitoring. In most cases, the laboratory abnormalities resolve without intervention. If the ALT increases to ≥5 times ULN, regardless of symptoms, or if a patient is symptomatic and has an ALT ≥3 times ULN, all TB drugs and any concomitant hepatotoxic drugs should be interrupted [3]. Other causes of liver disease, including viral hepatitis, should be investigated and additional laboratory tests such as bilirubin, alkaline phosphatase, prothrombin time, albumin, and viral hepatitis serology should be evaluated. In patients with active hepatitis B virus (HBV) infection started on ART, immune reconstitution inflammatory syndrome (IRIS) associated with HBV may also be a cause of the liver disease [46]. Additionally, some antiretroviral drugs including nevirapine, efavirenz, maraviroc, protease inhibitors, and dolutegravir may also cause hepatotoxicity.

After discontinuation of isoniazid, rifampicin, and pyrazinamide, TB treatment with ethambutol, a fluoroquinolone (moxifloxacin or levofloxacin), and either linezolid or an aminoglycoside should be commenced as soon as possible and continued throughout the rechallenge period, in order to avoid gaps in treatment and to serve as a background regimen to protect against the development of resistance. Rechallenge is not recommended if the DILI resulted in acute liver failure. In all other cases, TB drug rechallenge can begin once the symptoms and jaundice have resolved, and the ALT has decreased to <2.5 times ULN or, in the case of patients who have pre-treatment ALT elevation, when the ALT declines to close to baseline level. The discontinued TB drugs should be reintroduced one at a time, starting with the drug with the least likelihood of causing liver injury. Among the three key TB drugs - rifampicin, isoniazid, and pyrazinamide, rifampicin has the least propensity for causing DILI and, therefore, should be restarted first. If no increase in ALT and/or symptoms occur in 3–7 days, then isoniazid can be restarted for 3–7 days. The decision to rechallenge with pyrazinamide should be made carefully. Some experts recommend that this rechallenge should be limited to patients with TB meningitis or disseminated TB. If symptoms and/or ALT elevation recur after initiation of any of the drugs, the last drug started should be stopped, and the ALT closely monitored. Once rechallenge has been completed, the composition and duration of the regimen needs to be considered. If rechallenge with isoniazid and rifampicin was successful and the DILI occurred towards the end of the 2-month intensive phase, one may complete the course with isoniazid and rifampicin alone. If the DILI occurred early in the intensive phase, the fluoroquinolone should be continued for the duration of the intensive phase if pyrazinamide is not reintroduced. If isoniazid is not tolerated, the fluoroquinolone should be continued for the duration of therapy. Some experts also recommend continuing ethambutol for the duration of TB treatment. In most cases the course of therapy should be extended from 6 to 9 months [3]. If rifampicin is not tolerated, patients should be treated with a regimen for multiple drug resistant TB, and treatment duration should be prolonged.

Challenges in Treating TB and HIV Coinfection

Successful concurrent treatment of TB and HIV infection is critical for reducing morbidity and mortality associated with both infections. However, co-treatment can be complicated and there are several key challenges as discussed below.

Polypharmacy, Pill Burden, Adherence, and Directly Observed Therapy

HIV treatment generally requires two to three active antiretroviral agents while TB requires an initial four drug regimen. Additionally, many patients may also receive cotrimoxazole, and some may receive other medications to treat comorbidities, other co-infections, or to prevent other OIs. A number of HIV and TB treatment regimens are available as fixed dose combination (FDC) tablets, and are commonly used in many countries. Using FDC tablets for both TB and HIV treatment can help reduce pill burden, which improves adherence, limiting the risk of treatment discontinuation and drug resistance. FDC tablets may also simplify prescription writing and pharmacy inventory control. Although FDC tablets are recommended for the management of patients with drug susceptible TB [4], some patients may require individualized medication regimens or doses that need to be managed with separate formulations—this is often a problem in resource-limited settings where individual drugs are not always available.

A key component of adherence is involving the patient in the care plan and designing it collaboratively to achieve treatment success. Patients (and family or caregivers) should receive education on the general disease course of TB, transmission potential, treatment plan, TB and HIV treatment goals, proper medication administration schedule, side effects, drug resistance, and the importance of strict adherence. Adherence interventions may include consistent communication with patients (such as home visits or phone calls), psychological support, and material support (such as food assistance, financial aid, or transportation) to alleviate indirect costs associated with TB diagnosis and treatment [4].

If the healthcare setting has adequate capacity, TB treatment should be administered through directly observed therapy (DOT) for all patients with HIV and TB. DOT typically requires that the TB drugs be given at least 5 days per week be observed [3]. DOT with documentation of doses administered should be continued as much as possible in the outpatient setting - in clinics, or by community workers or family members. The WHO has reported cure rates >80% with programs incorporating DOT [4]. DOT improves treatment completion rates and treatment responses, and reduces loss to follow-up [47]. It can also facilitate early identification of potential adverse drug reactions, disease progression, and immune reconstitution inflammatory syndrome (IRIS). Community or home-based DOT is preferred and should be administered by trained lay providers or healthcare workers [4]. However, DOT may not be feasible in all settings. If resources are limited, some factors to consider if prioritizing individuals who will most benefit from DOT include severity of disease, likelihood for non-adherence, delayed culture conversion, and transmission risk. In settings where in-person DOT is not possible, alternative strategies such as virtual DOT using video conferencing via smart phone can be a better and sometimes more acceptable option [48, 49]. Virtual DOT together with intensive patient education and weekly home visits [50] have also been successful.

Patients with both HIV and TB infections require collaborative care from providers treating both diseases. Studies have found that patients co-infected with TB and HIV who attend clinics where integrated TB and HIV care is provided achieve better clinical outcomes than those cared for in separate clinics [51, 52]. If this approach is not possible, close communication between providers in the HIV and TB clinic is critical for treatment success.

Drug-Drug Interactions

Polypharmacy in patients with both HIV and TB makes drug interactions a serious concern. Rifamycin antibiotics are an integral part of TB treatment but have significant drug interaction potential due to their ability to induce drug metabolizing enzymes and drug transporters [33, 53]. Many drugs, including some antiretroviral drugs, require dose adjustment or should not be co-administered with rifampicin. Rifabutin is an alternative when rifampicin cannot be used due to drug-drug interactions, as its CYP3A4 induction potential is only ~40% as potent as rifampicin's. But unique among rifamycins, rifabutin is also a CYP3A4 substrate and therefore its metabolism maybe altered in the presence of CYP3A4 inducers or inhibitors. This may result in bi-directional interactions, which requires careful evaluation for rifabutin dose adjustment. Despite their complexities, rifamycin antibiotics are an integral part of TB treatment and unless there is contraindications, they should be used for the full treatment duration. Please refer to the "Co-treatment of Tuberculosis and HIV: Pharmacologic Considerations" chapter for further discussion.

Adverse Drug Reactions

Both anti-tuberculosis and antiretroviral medications can have significant, often overlapping side effects that necessitate regular monitoring and if needed, treatment modifications. Severe reactions may include drug induced hepatotoxicity and hypersensitivity reactions, among others (Table 3). The initiation of seven or more drugs within a short timeframe complicates assessment of causality and patient management when toxicities develop. First line TB treatments are more efficacious and better tolerated than second line alternatives. Therefore, first line agents should not be discontinued without careful evaluation.

Immune Reconstitution Inflammatory Syndrome (IRIS)

IRIS results from an inflammatory reaction that can occur after ART initiation in patients with HIV infection. Patients with HIV not known to have TB may develop unmasking TB IRIS after ART initiation, or patients on TB therapy may develop paradoxical IRIS after ART initiation, manifested as clinical worsening after initial response to TB therapy. Recognition and management of TB IRIS can be complicated, as the clinical signs and symptoms may be similar to an undiagnosed opportunistic infection, inadequate response to TB treatment, or adverse drug reactions to TB or HIV treatment. Depending on the severity of IRIS, drug therapy to control the symptoms may include NSAIDs or in more severe cases, systemic corticosteroids. However, prednisolone, the active metabolite of prednisone, is a CYP3A4 substrate so its metabolism is susceptible to induction by rifampicin. Therefore, recommended doses of prednisone for TB-associated IRIS should take into account the reductions in steroid exposures with rifampicin-based TB treatment. Please refer to chapter on IRIS.

Starting ART in Patients with TB

For persons co-infected with HIV, ongoing HIV replication and profound immunodeficiency are key contributing factors to TB disease progression and death [2, 54]. Thus, initiation of effective ART to suppress viral replication and improve immune function should always be a crucial part of the treatment plan for patients with active TB [55]. When considering treating both HIV infection and TB, a number of factors should be considered: (1) whether the patient is already on ART; (2) potential for drug-drug interactions, with special attention to rifamycins' enzyme induction effect on antiretroviral drugs; (3) overlapping drug toxicities; (4) adherence concerns; and (5) the possibility of IRIS.

Patients Known to Have HIV Infection and Receiving ART

Patients with HIV infection who are receiving ART at the time of TB diagnosis should remain on ART. HIV viral load should be determined to assess for treatment response. Clinicians should carefully review each antiretroviral drug in the regimen for potential of significant pharmacokinetic interactions with rifamycin antibiotics, and use the guidance in the "Co-treatment of Tuberculosis and HIV: Pharmacologic Considerations", chapter to determine whether there is a need for antiretroviral drug change or dosage modification. Before switching one antiretroviral drug to an alternative agent, clinicians should review the patient's past antiretroviral history and responses to therapy, any history of drug intolerance, as well as any available current and historic resistance testing. For example, even though efavirenz has been shown to have no clinically significant interaction with rifampicin, it should not be used in a patient with prior history of virologic failure on efavirenz-based therapy.

Patients Not Receiving ART at the Time of TB Diagnosis

Some patients may have had stopped ART before TB diagnosis. If available, ART history should be assessed as noted earlier, in order to determine the appropriate antiretroviral choice. The timing for when to restart ART in relationship to TB treatment should follow the recommendation as stated below.

Because of the high prevalence of HIV infection in patients with TB, it is recommended that all TB patients should be tested for HIV [3, 4]. As a result, a substantial number of patients with TB receive the diagnosis of HIV infection for the first time after TB diagnosis, and therefore, are not receiving ART when TB treatment is initiated. All persons with active TB should be started on TB treatment promptly without delay. All patients with TB and HIV co-infection should also receive ART. As noted earlier, major concerns about concomitant treatment of both infections include high pill burden, drug-drug interactions, overlapping toxicities, and difficulties with adherence. However, the argument for treating HIV is that untreated HIV infection may lead to additional opportunistic infections, especially in patients with severe immunosuppression.

Several large randomized controlled trials, mostly conducted in countries with high burden of HIV and TB in Asia and Africa, addressed the optimal timing of ART initiation in patients with TB and not on ART [56–61]. The study endpoints for these trials were death, AIDS related events, TB outcomes, and/or IRIS events. As expected, in most of these studies, low CD4 count, anemia, and low body mass index are frequently reported as predictors of poor outcomes, including deaths.

Recommendations for When to Start ART in TB Patients

The results from the randomized controlled trials provided guidance for the optimal time to start ART in patients with TB. These studies showed that patients with CD4 count <50 cells/mm^3 are at highest risk of disease progression and death if initiation of ART is delayed to more than 4 weeks after starting TB treatment [56, 58, 60]. Therefore, patients with CD4 count ≤50 cells/mm^3, ART should be initiated within 2 weeks of TB treatment, or as soon as feasible [5, 62]. Early initiation of ART in these patients may, however, lead to higher incidence of paradoxical TB IRIS—thus patients should be closely monitored after ART initiation for any signs and symptoms associated with IRIS and drug related adverse events. An exception to starting ART early may be made in patients with TB meningitis, where serious adverse events were more frequently observed in patients who received early ART (<7 days of TB treatment) without survival benefit [63]; in these TB meningitis cases, a delay in therapy until up to 4–8 weeks may be warranted, with close follow-up after initiation of ART. For patients with CD4 >50 cells/mm^3, ART should be started within 8 weeks of TB treatment [5, 55, 62]. Waiting until completion of TB treatment before initiation of ART is not recommended.

Conclusion

TB is the most common cause of morbidity and mortality for persons with HIV worldwide, especially in patients with CD4 counts <50 cells/mm^3 at the time of TB diagnosis. Early diagnosis of HIV infection and initiation of ART, may reduce the severity of TB. Successful TB and HIV treatment not only benefits the patient; from a public health standpoint, it reduces transmission of both HIV and MTB to others. Though treatment for both HIV and TB can be complex, both infections are treatable and good clinical outcomes are possible—especially in patients with good medication adherence. Integrated clinics with providers who have expertise in management of both infections can further improve patient outcomes. Despite effective therapy, TB recurrences and drug resistance continue to be challenges, especially in areas with high TB prevalence. Ongoing clinical and laboratory research evaluating new TB treatment strategies—such as dose optimization (especially with older drugs), more potent drugs, more effective drug combinations, and shorter treatment duration—are urgently needed to make further progress in the control of this disease.

References

1. WHO. Global Tuberculosis Report - 2017. 2017 January 1, 2018]; Available from: http://www.who.int/tb/publications/global_report/en/
2. Bisson GP, Zetola N, Collman RG (2015) Persistent high mortality in advanced HIV/TB despite appropriate antiretroviral and antitubercular therapy: an emerging challenge. Curr HIV/AIDS Rep 12(1):107–116
3. Nahid P et al (2016) Official American Thoracic Society/Centers for Disease Control and Prevention/Infectious Diseases Society of America Clinical Practice Guidelines: Treatment of Drug-Susceptible Tuberculosis. Clin Infect Dis 63(7):e147–e195
4. WHO. Guidelines for treatment of drug-susceptible tuberculosis and patient care (2017 update). 2017 January 1, 2018; Available from: http://www.who.int/tb/publications/2017/dstb_guidance_2017/en/
5. Panel on Opportunistic Infections in HIV-Infected Adults and Adolescents. Guidelines for the prevention and treatment of opportunistic infections in HIV-infected adults and adolescents:recommendations from the Centers for Disease Control and Prevention, the National Institutes of Health, and the HIV Medicine Association of the Infectious Diseases Society of America. 2017 December 10, 2017 January 1, 2018]; Available from: http://aidsinfo.nih.gov/contentfiles/lvguidelines/adult_oi.pdf
6. WHO. Improving the diagnosis and treatment smear-negative pulmonary and extrapulmonary tuberculosis among adults and adolescents. Recommendations for HIV-prevalent and resource-constrained settings. . 2007 January 1, 2018; Available from: http://whqlibdoc.who.int/hq/2007/WHO_HTM_TB_2007.379_eng.pdf
7. Padmapriyadarsini C et al (2013) Evaluation of a diagnostic algorithm for sputum smear-negative pulmonary tuberculosis in HIV-infected adults. J Acquir Immune Defic Syndr 63(3):331–338
8. Wilson D et al (2011) Evaluation of the World Health Organization algorithm for the diagnosis of HIV-associated sputum smear-negative tuberculosis. Int J Tuberc Lung Dis 15(7):919–924
9. Griesel R et al (2017) Optimizing tuberculosis diagnosis in HIV-infected inpatients meeting the criteria of seriously ill in the WHO algorithm. Clin Infect Dis
10. Mitnick CD, McGee B, Peloquin CA (2009) Tuberculosis pharmacotherapy: strategies to optimize patient care. Expert Opin Pharmacother 10(3):381–401
11. Khan FA et al (2010) Treatment of active tuberculosis in HIV-coinfected patients: a systematic review and meta-analysis. Clin Infect Dis 50(9):1288–1299
12. Swaminathan S et al (2010) Efficacy of a 6-month versus 9-month intermittent treatment regimen in HIV-infected patients with tuberculosis: a randomized clinical trial. Am J Respir Crit Care Med 181(7):743–751
13. Perriens JH et al (1995) Pulmonary tuberculosis in HIV-infected patients in Zaire. A controlled trial of treatment for either 6 or 12 months. N Engl J Med 332(12):779–784
14. Li J et al (2005) Relapse and acquired rifampin resistance in HIV-infected patients with tuberculosis treated with rifampin- or rifabutin-based regimens in New York City, 1997–2000. Clin Infect Dis 41(1):83–91
15. Nettles RE et al (2004) Risk factors for relapse and acquired rifamycin resistance after directly observed tuberculosis treatment: a comparison by HIV serostatus and rifamycin use. Clin Infect Dis 38(5):731–736
16. Vashishtha R et al (2013) Efficacy and safety of thrice weekly DOTS in tuberculosis patients with and without HIV co-infection: an observational study. BMC Infect Dis 13:468
17. Burman W et al (2006) Acquired rifamycin resistance with twice-weekly treatment of HIV-related tuberculosis. Am J Respir Crit Care Med 173(3):350–356
18. Jo KW et al (2014) Risk factors for 1-year relapse of pulmonary tuberculosis treated with a 6-month daily regimen. Respir Med 108(4):654–659
19. Horne DJ et al (2010) Sputum monitoring during tuberculosis treatment for predicting outcome: systematic review and meta-analysis. Lancet Infect Dis 10(6):387–394

20. Perwitasari DA, Atthobari J, Wilffert B (2015) Pharmacogenetics of isoniazid-induced hepatotoxicity. Drug Metab Rev 47(2):222–228
21. Ohno M et al (2000) Slow N-acetyltransferase 2 genotype affects the incidence of isoniazid and rifampicin-induced hepatotoxicity. Int J Tuberc Lung Dis 4(3):256–261
22. Huang YS et al (2002) Polymorphism of the N-acetyltransferase 2 gene as a susceptibility risk factor for antituberculosis drug-induced hepatitis. Hepatology 35(4):883–889
23. Pasipanodya JG, Srivastava S, Gumbo T (2012) Meta-analysis of clinical studies supports the pharmacokinetic variability hypothesis for acquired drug resistance and failure of antituberculosis therapy. Clin Infect Dis 55(2):169–177
24. Chien JY et al (2014) Safety of rifabutin replacing rifampicin in the treatment of tuberculosis: a single-centre retrospective cohort study. J Antimicrob Chemother 69(3):790–796
25. Vernon A et al (1999) Acquired rifamycin monoresistance in patients with HIV-related tuberculosis treated with once-weekly rifapentine and isoniazid. Tuberculosis Trials Consortium. Lancet 353(9167):1843–1847
26. Savic RM et al (2017) Defining the optimal dose of rifapentine for pulmonary tuberculosis: exposure-response relations from two phase II clinical trials. Clin Pharmacol Ther 102(2):321–331
27. Dorman SE et al (2015) Daily rifapentine for treatment of pulmonary tuberculosis. A randomized, dose-ranging trial. Am J Respir Crit Care Med 191(3):333–343
28. Boeree MJ et al (2017) High-dose rifampicin, moxifloxacin, and SQ109 for treating tuberculosis: a multi-arm, multi-stage randomised controlled trial. Lancet Infect Dis 17(1):39–49
29. Steingart KR et al (2011) Higher-dose rifampin for the treatment of pulmonary tuberculosis: a systematic review. Int J Tuberc Lung Dis 15(3):305–316
30. Velasquez GE et al (2018) Efficacy and safety of high-dose rifampin in pulmonary tuberculosis. A randomized controlled trial. Am J Respir Crit Care Med 198(5):657–666
31. Schimkat M et al (1996) Rifabutin-associated anterior uveitis in patients infected with human immunodeficiency virus. Ger J Ophthalmol 5(4):195–201
32. Jacobs DS et al (1994) Acute uveitis associated with rifabutin use in patients with human immunodeficiency virus infection. Am J Ophthalmol 118(6):716–722
33. Baciewicz AM et al (2013) Update on rifampin, rifabutin, and rifapentine drug interactions. Curr Med Res Opin 29(1):1–12
34. Cremades R et al (2011) Comparison of the bactericidal activity of various fluoroquinolones against Mycobacterium tuberculosis in an in vitro experimental model. J Antimicrob Chemother 66(10):2281–2283
35. Johnson JL et al (2006) Early and extended early bactericidal activity of levofloxacin, gatifloxacin and moxifloxacin in pulmonary tuberculosis. Int J Tuberc Lung Dis 10(6):605–612
36. Pletz MW et al (2004) Early bactericidal activity of moxifloxacin in treatment of pulmonary tuberculosis: a prospective, randomized study. Antimicrob Agents Chemother 48(3):780–782
37. Gillespie SH et al (2014) Four-month moxifloxacin-based regimens for drug-sensitive tuberculosis. N Engl J Med 371(17):1577–1587
38. Jawahar MS et al (2013) Randomized clinical trial of thrice-weekly 4-month moxifloxacin or gatifloxacin containing regimens in the treatment of new sputum positive pulmonary tuberculosis patients. PLoS One 8(7):e67030
39. WHO. WHO Treatment Guidelines for Drug-Resistant Tuberculosis (2016 Update). 2016 January 1, 2018; Available from: http://apps.who.int/iris/bitstream/10665/250125/1/9789241549639-eng.pdf?ua=1
40. Gaude GS, Chaudhury A, Hattiholi J (2015) Drug-induced hepatitis and the risk factors for liver injury in pulmonary tuberculosis patients. J Family Med Prim Care 4(2):238–243
41. Abbara A et al (2017) Drug-induced liver injury from antituberculous treatment: a retrospective study from a large TB centre in the UK. BMC Infect Dis 17(1):231
42. Nader LA et al (2010) Hepatotoxicity due to rifampicin, isoniazid and pyrazinamide in patients with tuberculosis: is anti-HCV a risk factor? Ann Hepatol 9(1):70–74

43. Dossing M et al (1996) Liver injury during antituberculosis treatment: an 11-year study. Tuber Lung Dis 77(4):335–340
44. Lee AM et al (2002) Risk factors for hepatotoxicity associated with rifampin and pyrazinamide for the treatment of latent tuberculosis infection: experience from three public health tuberculosis clinics. Int J Tuberc Lung Dis 6(11):995–1000
45. Shakya R, Rao BS, Shrestha B (2004) Incidence of hepatotoxicity due to antitubercular medicines and assessment of risk factors. Ann Pharmacother 38(6):1074–1079
46. Crane M et al (2009) Immunopathogenesis of hepatic flare in HIV/hepatitis B virus (HBV)-coinfected individuals after the initiation of HBV-active antiretroviral therapy. J Infect Dis 199(7):974–981
47. Reis-Santos B et al (2015) Directly observed therapy of tuberculosis in Brazil: associated determinants and impact on treatment outcome. Int J Tuberc Lung Dis 19(10):1188–1193
48. Nguyen TA et al (2017) Video directly observed therapy to support adherence with treatment for tuberculosis in Vietnam: a prospective cohort study. Int J Infect Dis 65:85–89
49. Macaraig M et al (2017) A national survey on the use of electronic directly observed therapy for treatment of tuberculosis. J Public Health Manag Pract
50. Kaplan R et al (2016) An integrated community TB-HIV adherence model provides an alternative to DOT for tuberculosis patients in Cape Town. Int J Tuberc Lung Dis 20(9):1185–1191
51. Webb Mazinyo E et al (2016) Adherence to concurrent tuberculosis treatment and antiretroviral treatment among co-infected persons in South Africa, 2008–2010. PLoS One 11(7):e0159317
52. Hermans SM et al (2012) Integration of HIV and TB services results in improved TB treatment outcomes and earlier prioritized ART initiation in a large urban HIV clinic in Uganda. J Acquir Immune Defic Syndr 60(2):e29–e35
53. Egelund EF et al (2017) The pharmacological challenges of treating tuberculosis and HIV coinfections. Expert Rev Clin Pharmacol 10(2):213–223
54. Kaplan R et al (2014) Impact of ART on TB case fatality stratified by CD4 count for HIV-positive TB patients in Cape Town, South Africa (2009–2011). J Acquir Immune Defic Syndr 66(5):487–494
55. WHO. Consolidated guidelines on HIV prevention, diagnosis, treatment and care for key populations - 2016 update. 2016 December 10, 2017 January 1, 2018; Available from: http://www.who.int/hiv/pub/arv/arv-2016/en/
56. Havlir DV et al (2011) Timing of antiretroviral therapy for HIV-1 infection and tuberculosis. N Engl J Med 365(16):1482–1491
57. Abdool Karim SS et al (2010) Timing of initiation of antiretroviral drugs during tuberculosis therapy. N Engl J Med 362(8):697–706
58. Blanc FX et al (2011) Earlier versus later start of antiretroviral therapy in HIV-infected adults with tuberculosis. N Engl J Med 365(16):1471–1481
59. Mfinanga SG et al (2014) Early versus delayed initiation of highly active antiretroviral therapy for HIV-positive adults with newly diagnosed pulmonary tuberculosis (TB-HAART): a prospective, international, randomised, placebo-controlled trial. Lancet Infect Dis 14(7):563–571
60. Abdool Karim SS et al (2011) Integration of antiretroviral therapy with tuberculosis treatment. N Engl J Med 365(16):1492–1501
61. Uthman OA et al (2015) Optimal timing of antiretroviral therapy initiation for HIV-infected adults with newly diagnosed pulmonary tuberculosis: a systematic review and meta-analysis. Ann Intern Med 163(1):32–39
62. Panel on Antiretroviral Guidelines for Adults and Adolescents. Guidelines for the Use of Antiretroviral Agents in Adults and Adolescents Living with HIV. Department of Health and Human Services. 2017 December 10, 2017 January 1, 2018]; Available from: https://aidsinfo.nih.gov/guidelines/html/1/adult-and-adolescent-arv/27/tb-hiv
63. Torok ME et al (2011) Timing of initiation of antiretroviral therapy in human immunodeficiency virus (HIV)--associated tuberculous meningitis. Clin Infect Dis 52(11):1374–1383

Drug-Resistant Tuberculosis and HIV

Sara C. Auld, Neel R. Gandhi, and James C. M. Brust

Abstract Multidrug-resistant (MDR) and extensively drug-resistant (XDR) TB are associated with substantially worse outcomes than drug-susceptible TB—especially in the setting of HIV co-infection. Although global TB incidence has decreased over the past decade, drug-resistant TB remains a substantial threat to control of TB worldwide. This chapter reviews the epidemiology of drug-resistant TB, common genetic mutations conferring resistance to first and second-line TB drugs, as well as current diagnostic methods and principles of treatment. Therapy for drug-resistant TB is complicated, associated with frequent side effects, and a field of active research. Several new and repurposed drugs, such as bedaquiline and delamanid, have recently come to market or are in active clinical development. Guidelines change frequently and consultation with a clinician who has expertise in the co-management of drug-resistant TB and HIV is recommended.

Keywords Multidrug-resistant · Extensively drug-resistant · MDR TB · XDR TB · Mono-resistance

S. C. Auld
School of Medicine and Rollins School of Public Health, Emory University, Atlanta, GA, USA
e-mail: sara.auld@emory.edu

N. R. Gandhi
Rollins School of Public Health and School of Medicine, Emory University, Atlanta, GA, USA
e-mail: neel.r.gandhi@emory.edu

J. C. M. Brust (✉)
Albert Einstein College of Medicine, Montefiore Medical Center, Bronx, NY, USA

I. Sereti et al. (eds.), *HIV and Tuberculosis*,
https://doi.org/10.1007/978-3-030-29108-2_10

History and Epidemiology

In 1948, the British Medical Council conducted the first ever randomized, controlled clinical trial to determine the efficacy of streptomycin for the treatment of pulmonary tuberculosis (TB) [1–3]. Over the course of 15 months, this landmark study enrolled 107 patients who were studied for up to 6 months following treatment with either streptomycin monotherapy or bed rest—the standard of care at the time. The patients assigned to streptomycin treatment had marked improvements in their radiographic findings, temperature trends and sedimentation rates, and were less than half as likely to die as those in the control group. However, high-level resistance to streptomycin developed in 85% of cases for whom resistance testing was performed and, in a number of cases, resistance emerged within several days of initiating treatment [1, 4]. In the years that followed, resistance to every new anti-TB drug was observed soon after each drug's introduction, including the four drugs that make up the current first-line regimen: isoniazid, pyrazinamide, ethambutol and rifampin [5–7]. It became clear that *Mycobacterium tuberculosis* (*Mtb*) was a highly adaptive pathogen and that drug-resistance would remain a challenge for patients, healthcare providers and TB control programs.

Although there were scattered outbreaks of drug-resistant TB in the 1960s and 1970s [8], the emergence of HIV, first described in 1981, laid the foundation for a dramatic upsurge in both TB and drug-resistant TB [9, 10]. In the early 1990s in the United States, there were a series of outbreaks of multidrug-resistant (MDR) TB, defined as resistance to at least isoniazid and rifampin, among people with HIV [11–15]. Many of these outbreaks were nosocomial, occurring in hospitals and residential facilities, and included transmission of MDR TB to healthcare workers [14–16]. In New York City, the prevalence of drug resistance increased from 10% to 23% between 1983 and 1991, and drug-resistant disease was associated with significantly higher mortality [11]. Soon after, similar HIV-associated outbreaks of drug-resistant TB were reported in Spain, Italy and Argentina [17–21].

In the wake of these outbreaks, there was increasing alarm about global underdiagnosis and underreporting of MDR TB and the potential for drug-resistant disease to spread further, given the rising rates of HIV in many parts of the world [16, 22]. A series of targeted drug-resistance surveys conducted in the late 1990s demonstrated drug resistance in all countries studied [23, 24]. Rates of MDR TB were particularly high in Eastern Europe; in Estonia, for example, the prevalence of mono-resistant and MDR TB among all TB cases were 36.9% and 14.1%, respectively. The first comprehensive estimates of the global MDR TB burden were published in 2002, with an estimated 273,000 new cases of MDR TB worldwide, representing 3.2% of all new TB cases [25]. There was, however, substantial regional heterogeneity, with MDR TB rates ranging from 0.7% in Western Europe, North America and Asia, to as high as 5.5% in Eastern Europe and 7.9% in the Eastern Mediterranean regions.

By the mid-2000s, the convergence of the MDR TB and HIV epidemics were creating "the perfect storm." [26] Co-infected patients were experiencing delayed diagnosis, inadequate initial treatment with prolonged infectious periods, and unacceptably high mortality. High rates of MDR TB among people with HIV continued to be reported from Europe and the Americas [27–31], but also from less developed countries that had not been recognized in the initial wave of MDR TB outbreaks. With the involvement of each successive global region, the challenges in combating HIV and MDR TB co-infection mounted, as each epidemic carried unique features. In Eastern Europe, rising rates of HIV infection, a growing epidemic of intravenous drug use, and high incarceration rates were fueling the further spread of MDR TB [32–38]. In southeast Asia, the MDR TB and HIV epidemics were driven by social instability from refugees and internal migration as well as intravenous drug use [39–42]. Across sub-Saharan Africa, HIV had become a generalized epidemic and skyrocketing rates were widely cited as driving the drug-resistant TB epidemic [43–46]. While there had not yet been large-scale surveillance in the heavily populated countries of India and China, multiple reports of MDR TB and HIV co-infection underscored the likely magnitude of the syndemic [47–52].

As the global prevalence of MDR TB increased, the first reports of extensively drug-resistant (XDR) TB (i.e., MDR TB with additional resistance to fluoroquinolones and second-line injectable agents) began to emerge [53, 54]. The most dramatic reports initially came from South Africa in 2006, a country already burdened with the world's worst HIV epidemic. At a rural hospital in Tugela Ferry, South Africa, the first cases of XDR TB and HIV co-infection were described. The most notable feature of that report, and a larger follow-up analysis, was that XDR TB/HIV co-infected patients had rapid and high mortality, with a median survival of less than 30 days [55, 56]. From 2002–2009, nearly three-quarters of all XDR TB cases reported to the WHO were from South Africa [57], but over time, additional reports from other countries, including China and India, all pointed to a growing epidemic of XDR TB and exceedingly high mortality among people with HIV [50, 52, 58–63].

In 2016 it was estimated that there were 490,000 new cases of MDR TB and an additional 110,000 new cases with mono-resistance to rifampin [64]. XDR TB has been reported by 123 countries and it is estimated that 6.2% of all MDR TB cases—nearly 50,000 cases worldwide—are, in fact, XDR TB. Global surveillance data indicate that the highest rates of MDR TB (as a proportion of total TB cases) are in the Russian Federation and former Soviet Union, which, together with India and China, comprise 45% of the total global burden of MDR TB [64, 65]. Among 14 countries designated by the WHO as having a high burden of both MDR TB and TB-HIV co-infection, however, eight are in sub-Saharan Africa (Fig. 1), underscoring the threat that drug-resistant TB poses worldwide.

A number of studies have sought to identify an association between HIV and the development of drug-resistant TB, but neither a systematic review in 2009, nor a later analysis of surveillance data from Kazakhstan found an association between MDR TB and HIV [33, 66]. The Kazakhstan study did identify overlapping risk factors for the two infections [33], and a subsequent meta-analysis found a small, but significant, association between HIV and MDR TB, despite a moderate degree of heterogeneity in the included studies [67]. These studies, however, relied upon

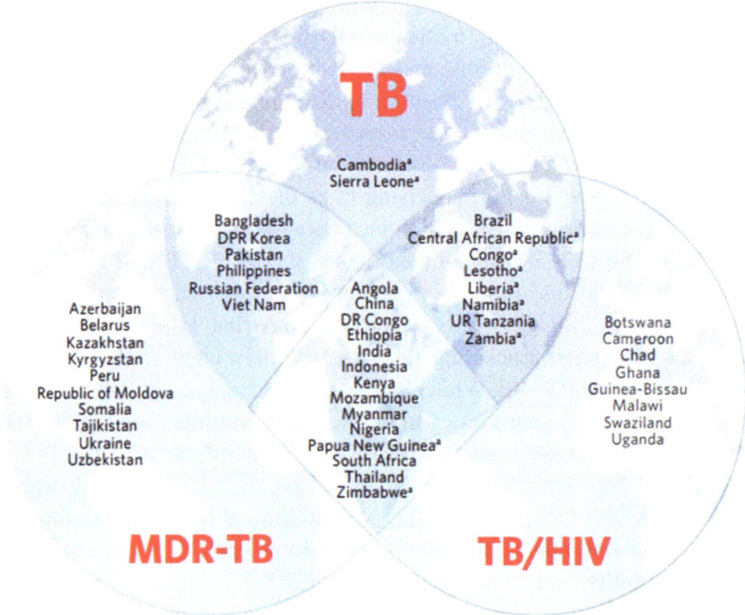

Fig. 1 Countries in the three TB high-burden country lists that will be used by WHO during the period 2016–2020, and their areas of overlap. DPR Korea, Democratic People's Republic of Korea; DR Congo, Democratic Republic of the Congo; HIV, human immunodeficiency virus; MDR, multidrug-resistant; TB, tuberculosis; UR Tanzania, United Republic of Tanzania; WHO, World Health Organization [a]Indicates countries that are included in the list of 30 high-burden countries for TB on the basis of the severity of their TB burden (i.e. TB incidence per 100,000 population), as opposed to the top 20, which are included on the basis of their absolute number of incident cases per year. (From the WHO Global Tuberculosis Report 2017)

the identification of epidemiologic associations, and there has been limited exploration of whether there is a biological relationship whereby HIV infection directly increases the risk of drug-resistant TB. Several potential mechanisms of such a relationship have been posited, including malabsorption of TB drugs, drug interactions and poor adherence to co-treatment [26, 68–70], but there are very limited data supporting these theories. Most studies have found that TB/HIV co-infected patients have high levels of medication adherence, and two recent meta-analyses had conflicting results as to whether people with HIV had different pharmacokinetic exposures to first-line TB drugs as compared to those without HIV [71, 72]. As such, it remains unclear whether HIV increases the risk of MDR TB, or whether shared risk factors contribute to an increased risk of both diseases. Even if risk factors such as poverty, substance use and access to healthcare confound the relationship between drug-resistant TB and HIV, the two epidemics have clearly had a catastrophic convergence in countries from Eastern Europe to sub-Saharan Africa and Southeast Asia.

Development of Drug-Resistance

From the earliest trials of streptomycin, it was clear that *Mtb* could develop drug resistance under selective drug pressure, particularly with monotherapy or inadequate therapy (Fig. 2) [1, 4]. This resistance arises as a result of spontaneous genetic mutations that occur at a predictable rate, and not from horizontal gene transfer of resistant mutations, as is common with many bacterial pathogens [73, 74]. The prevailing belief for many years was that drug-resistant TB was primarily a problem of *acquired resistance*, whereby the majority of MDR TB was created *de novo* by treatment failure, poor medication adherence and physician error [75]. As such, prevention of MDR TB focused primarily on strengthening existing TB control programs and improving treatment adherence among patients with drug-susceptible TB [75–81].

Recently, however, there has been increasing recognition that transmission of *Mtb* strains that are already drug-resistant plays a major role in the development of MDR and XDR TB. Evidence for transmission of drug-resistant TB (i.e. *transmitted, or primary resistance)* includes multiple reports from the 1960s involving pediatric cases of drug-resistant TB, in addition to the HIV-associated outbreaks in the 1990s where MDR TB had been transmitted in nosocomial settings [8, 14, 15,

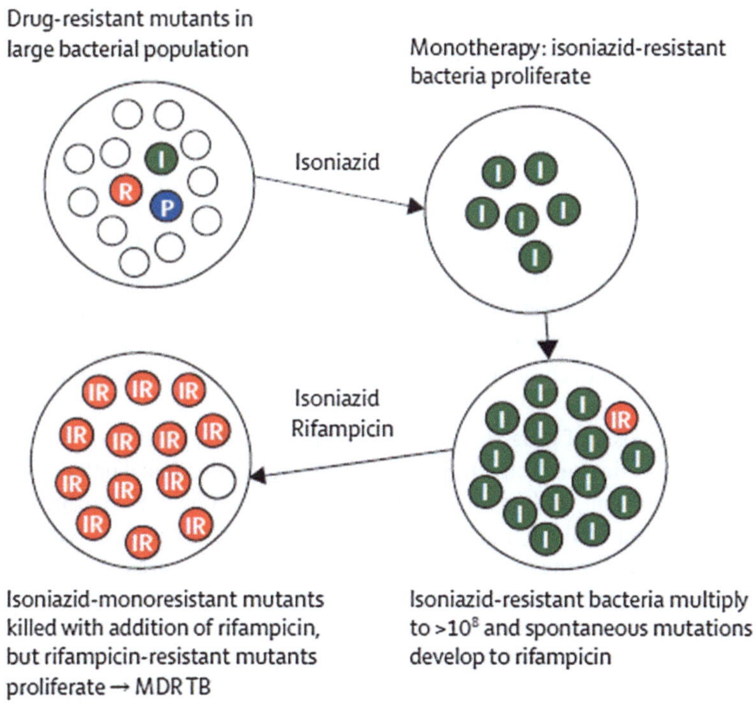

Fig. 2 Acquisition of resistance. I = isoniazid. R = rifampicin. P = pyrazinamide. MDR = multidrug-resistant. TB = tuberculosis. (Adapted from Albino JA, Reichman LB. The treatment of tuberculosis. Respiration 1998; 65: 237–55)

17–19, 82–84]. Large studies of transmission have now been reported in a number of settings, from South Africa to Peru and China, with up to 90% of cases of MDR and XDR TB attributable to transmission in some settings [85–89].

The increasing reports of transmitted resistance likely reflect the natural history of a drug-resistant epidemic. The initial emergence of drug-resistant TB strains will occur as a result of acquired resistance following the introduction of new drugs, but later cases and growth of the epidemic are more likely to be due to direct transmission of resistant strains. Modeling data support this notion of maturation of the drug-resistant TB epidemic and indicate that the relative proportion of MDR TB cases, among all incident cases of TB, is likely to increase in high-burden countries [90, 91]. At this time, it is not known how HIV impacts the likelihood of transmission of MDR or XDR TB, and specifically, whether HIV might accelerate transmission, potentially by increasing the number of individuals vulnerable to infection and disease. Genetic analyses also indicate that HIV may increase *Mtb* strain evolution [92], although these findings have not been replicated and the implications for transmission are thus not clear.

Diagnosis

Drug-resistant TB is indistinguishable from drug-susceptible TB based on its symptomatology, clinical findings, and radiologic patterns. Drug-resistant TB can present as pulmonary or extra-pulmonary disease, similar to drug-susceptible TB. The diagnosis of drug-resistant TB requires specific testing to assess the *Mtb* strain's susceptibility to anti-tuberculous medications, which can be done either phenotypically or genotypically. While several of the available modalities for diagnosing active TB disease (reviewed in chapter "Diagnosis of HIV-Associated Tuberculosis" on the Diagnosis of HIV-associated TB) can also assess susceptibility, many do not. Specifically, among currently available diagnostic assays, smear microscopy, culture (without the addition of phenotypic drug-susceptibility testing [DST]) and urine lipoarabinomannan (LAM) can diagnose active TB disease, but not drug resistance.

Prompt diagnosis of drug resistance is essential. A delay between the onset of symptoms and initiation of effective treatment can result in clinical deterioration for that individual patient, and ongoing transmission to their contacts. Because mortality rates are substantially higher in drug-resistant TB—particularly in HIV-infected patients—and because the majority of drug-resistant TB cases arise due to transmission, minimizing the time to diagnosis is critical to reducing the incidence of drug-resistant TB and its associated morbidity and mortality.

Phenotypic susceptibility testing examines whether an *Mtb* strain can grow even though an antibiotic is included in the growth medium. For example, if a strain grows despite the presence of isoniazid (at a pre-specified critical concentration), it is considered resistant to isoniazid. Phenotypic testing relies upon growth of the mycobacterium either on solid media (e.g., Lowenstein Jensen [LJ] or Middlebrook agar) or in liquid (e.g., Mycobacteria Growth Indicator Tube [MGIT]); *Mtb* typically requires 6–8 weeks to grow on solid media and 2–4 weeks in liquid culture. Critical concentrations of drugs to include in these media have been developed for most first- and second-line TB drugs [93, 94]; however, phenotypic testing is less reproducible for several drugs and thus, DST is not routinely performed for them (e.g., pyrazinamide, ethambutol, ethionamide and para-amino salicylic acid).

Considering that culture and DST is a two-step process of first growing the bacteria in culture and then inoculating them onto drug-containing media, the typical turnaround time for culture and DST on solid media is 6–12 weeks. Efforts have been made to decrease this turnaround time by using indicators that can identify bacterial growth earlier than visualizing colonies with the naked eye. Automated indicators, such as those in the BACTEC-460 or MGIT, can reduce the turnaround time to 4–8 weeks. Other approaches are to visualize microscopic cords that *Mtb* characteristically makes using the Microscopic Observation Drug Susceptibility (MODS) assay, or color changes in the media by reduction of nitrate or an indicator dye in the Nitrate reductase assay (NRA) or Colorimetric redox indicator (CRI) [95]. The MODS assay has a median turnaround time of 7 days to diagnose drug resistance, as compared to 22 days for automated liquid culture and 68–70 days on solid media [96, 97].

Genotypic susceptibility testing relies on the demonstration of a mutation or polymorphism in a mycobacterial gene that is known to confer resistance. Specific resistance-conferring genes have been identified for most anti-tuberculous agents (see Table 1). Polymorphisms in these genes result in a loss of effectiveness of that particular drug, typically by causing a conformational change in a binding site for the drug, disruption of enzymes that are needed to convert a medication into its

Table 1 Genes associated with drug resistance [98–101]

Anti-tuberculous drug	Resistance-conferring genes[a]
Isoniazid	*inhA-mabA, katG (mshA)*
Rifampin	*rpoB*
Pyrazinamide	*pncA (rpsA)*
Fluoroquinolones	*gyrA, gyrB*
Streptomycin	*rpsL, rrs, (gidB)*
Amikacin	*rrs*
Kanamycin	*eis, rrs*
Capreomycin	*rrs, tlyA*
Ethionamide	*inhA (ethA)*
Linezolid	*rrl, rplC*
Ethambutol	*embB (embC)*
Bedaquiline	*(atpE*[b]*, Rv0678, pepQ*[b]*)*

[a]Genes with less frequent prevalence or uncertain correlation with resistance in parentheses
[b]Mutations found in strains created in laboratory or animal models; clinical correlation pending

active form, or expression of an efflux pump which reduces the intracellular concentration of the drug.

Genotypic testing provides advantages over phenotypic assays, but has a few limitations as well (Table 2). The principal advantage to genotypic testing is a faster turnaround time for making a diagnosis of drug resistance. Unlike phenotypic susceptibility testing, genotypic testing utilizes polymerase chain reactions (PCR) to amplify the pertinent resistance conferring regions, rather than relying on the growth of the mycobacteria. Consequently, genotypic testing can identify drug resistance within several hours to days, based on whether samples are tested in isolation or batched. Several commercial assays are now available using cartridge-based or line probe assays, such as the Xpert MTB/RIF (Cepheid) and GenoType MTBDR platforms (Hain Lifesciences), respectively [102, 103]. Genotypic drug resistance can also be diagnosed by carrying out targeted or "Sanger" gene sequencing, although this testing has been primarily utilized in research settings. Recently, the UK and the US have both announced that they will be performing whole genome sequencing on isolates of all active TB cases and will utilize mutation data from the resistance conferring genes to assess drug susceptibility [104].

Table 2 Comparison of phenotypic drug-susceptibility testing, commercially-available genotypic tests and whole-genome sequencing (Adapted from Dheda, Gumbo et al. Lancet Resp Dis. 2017)[101]

	Phenotypic tests	Xpert MTB/RIF	Line probe assays	Whole-genome sequencing
Time to result	Slow (weeks or months)	Less than 2 h	Rapid (hours or days) when done directly from samples	Rapid (hours or days) if done directly from samples
Sensitivity for detecting resistance	High	High for rifampin; no other drugs included	Sensitivity limited by the number of loci incorporated in test; high for rifampin	Dependent on knowledge of polymorphism; high for rifampin
Safety	High risk, requiring sophisticated microbiological protection	Low risk	Moderate microbiological risk when testing clinical samples. High risk if bacterial cultures are used	Moderate risk when testing clinical samples. High risk if bacterial cultures are used
Quality assessment	Quality assurance via WHO and International Union Against Tuberculosis and Lung Disease reference laboratory network	Test-specific quality assurance schemes not widespread	Test-specific quality assurance schemes not widespread	Quality assurance schemes not available
Efficiency	Separate tests for each drug	Detects resistance to one drug only	Two or three drugs per test	Single analysis for all drugs

The principal drawback of genotypic testing is that they test specific resistance conferring genes, and thus, their sensitivity is reliant on drug resistance polymorphisms falling within that interrogated region. For many of the most important TB medications (e.g., isoniazid, rifampin, fluoroquinolones), this is only a minor concern because the majority of resistance conferring mutations occur in only 1–2 genes and within a relatively narrow number of base pairs within those genes. Consequently, tests can focus on a small area for PCR amplification and still have a high sensitivity for diagnosing drug resistance. This is the strategy employed by early generations of the commercially available genotypic platforms focused on identifying drug resistance to rifampin and/or isoniazid (i.e., Xpert MTB/RIF, Hain, INNO-LiPA).

With other drugs, however, mutations conferring resistance either do not localize to a small region of the resistance conferring genes (e.g., pyrazinamide), or may occur in any one of multiple genes (e.g., aminoglycosides, capreomycin), creating logistical challenges for creating commercially available genotypic platforms and resulting in lower sensitivity if assays only employ a select group of loci [105]. Newer generations of the line probe assay utilize multiple genes for the diagnosis of aminoglycoside and capreomycin resistance, and appear to have improved sensitivity [106]. Initiatives to catalog resistance-conferring mutations and their correlation with phenotypic testing will be critical for the continued development of genotypic testing in the future [107, 108].

Over the past decade, efforts have focused on increasing laboratory capacity to diagnose drug resistance in low- and middle-income countries, largely spearheaded by The Stop TB Partnership's Global Lab Initiative [109]. The efforts have entailed creating laboratories which are capable of conducting phenotypic testing, as well as molecular assays for genotypic DST. Scale-up of DST capability has been impacted the most, however, by the ease of use of the automated Xpert MTB/RIF platform—minimizing the infrastructure and training required to carry out genotypic DST. More than 21,500 GeneXpert machines were deployed between 2010 and 2015, and during that time 16 million tests had been performed in 122 countries, to enhance the availability of DST [110]. Future advances in the Xpert platform, including an Xpert MTB/XDR assay capable of detecting second-line drug resistance, as well as other automated systems in development, hold promise for the availability of DST testing in peripheral clinics of low- and middle-income countries in the next decade. (See https://www.finddx.org/tb/pipeline/ for latest pipeline of TB diagnostics.)

Treatment and Outcomes

While the treatment for drug-susceptible TB consists of oral, well-tolerated medications which are dosed daily, the treatment for drug-resistant TB is substantially more complicated. In general, medications for drug-resistant TB are less potent and are associated with side effects which are both more common and more severe than those associated with first-line therapy. Given the lower potency and sterilizing effect of these medications, more drugs are required in combination and for a longer duration of therapy to achieve cure. Treatment outcomes are typically worse for drug-resistant disease than for drug-susceptible disease. Additionally, while uncontrolled HIV infection has been associated with worse outcomes in drug-resistant TB, survival is improved when patients' HIV is well-controlled on antiretroviral therapy (ART) [111, 112].

Early Experience with MDR TB Treatment

To date, there have been very few published clinical trials in drug-resistant TB. As a result, the development of regimens to treat drug-resistant TB has been largely iterative and based on observational studies beginning in the late 1980s. Until that point, drug resistance had been managed on a case-by-case basis, primarily with single drug substitutions. Earlier work by the British Medical Research Council had shown that certain drugs were more important to the regimen than others. For example, replacement of rifampin (due to either intolerance or resistance) required extension of the treatment duration from 6 months to a minimum of 18 months [113]. Given the limited evidence base to guide treatment, however, the earliest outbreaks of MDR TB in the 1990s posed a substantial challenge for treating physicians, who relied on consensus among colleagues and individual expertise.

Retrospective studies of the early outbreaks in the 1980s and 1990s found that receipt of second-line injectable agents such as capreomycin and receipt of a fluoroquinolone were each associated with improved survival [114]. Early guidelines for MDR TB, therefore, recommended inclusion of both a second-line injectable (kanamycin, amikacin or capreomycin) AND a fluoroquinolone, and recommended a treatment duration of 24 months [115]. Treatment guidelines for MDR TB were refined over the years but remained largely unchanged until 2016, when the option for a short-course regimen was added (see below) [116–118].

Development of the Standard 24-Month Regimen

The 2011 WHO guidelines, which were largely reiterated in the 2016 update, recommend a standard regimen of at least 5 medications, including a later-generation fluoroquinolone, a parenteral agent (for an 'intensive phase' in the first 8 months), and pyrazinamide (Table 3) [118, 119]. Later-generation fluoroquinolones (i.e., moxifloxacin or levofloxacin) are preferred over ofloxacin which is, itself, preferred over ciprofloxacin. The later-generation fluoroquinolones have greater *in vitro* activity (i.e., lower minimum inhibitory concentrations) against *Mtb* [120–125], and one study found that the addition of moxifloxacin was associated with earlier culture conversion in ofloxacin-resistant MDR TB patients [126]. Comparing different later-generation fluoroquinolones, an open-label randomized trial in South Korea found no difference in 3-month culture conversion or in final treatment outcomes

Table 3 Example 24-month and Short-course Regimens

24-month Regimen	Short-course Regimen
Intensive phase (6–8 months)	*Intensive phase (4–6 months)*
Moxifloxacin	Moxifloxacin
Kanamycin[a]	Kanamycin[a]
Pyrazinamide	Prothionamide
Ethionamide	Clofazimine
Terizidone	Pyrazinamide
Continuation phase (12–18 months)	High-dose isoniazid
Moxifloxacin	Ethambutol
Pyrazinamide	*Continuation phase (5 months)*
Ethionamide	Moxifloxacin
Terizidone	Clofazimine
	Pyrazinamide
	Ethambutol

[a]Injectable agents

among MDR TB patients treated with either levofloxacin or moxifloxacin [127, 128]. The remainder of the recommended regimen consists of other second-line medications thought to have activity based on the infecting strain or the treatment history of the patient. These other medications include ethionamide (or its analog, prothionamide), cycloserine (or its analog, terizidone), para-aminosalicylic acid (PAS), and clofazimine. Additional medications, such as amoxicillin/clavulanate and carbapenems, are often added in an effort to include a sufficient number of medications in patients with more resistant strains or with a history of extensive prior treatment [129]. The total recommended treatment duration is no less than 20 months, and typically 24 months.

The Short-Course Regimen

In 2010, a group of investigators in Bangladesh reported high MDR TB cure rates in a cohort of patients treated with a 9–12 month regimen [130]. Patients in the initial cohort had a relapse-free cure rate of nearly 90%, and a subsequent study by the same investigators in more than 500 patients showed similar success rates [131]. These early studies were met with some initial skepticism due to their observational, uncontrolled study design, but there have since been multiple, similarly positive reports of successful implementation of the so-called "Bangladesh" regimen [131–133].

Based on these encouraging findings, WHO modified their MDR TB treatment guidelines in 2016 to formally recommend the short-course regimen for MDR TB patients without second-line drug resistance (i.e., without pre-XDR or XDR TB) [119]. The short-course MDR TB regimen consists of moxifloxacin (a third-generation fluoroquinolone), a second-line injectable, pyrazinamide, high-dose isoniazid, ethionamide (or prothionamide), clofazimine, and ethambutol (Table 3).

The injectable medication is given for a minimum of 4 months (extended to 6 months if the sputum remains smear-positive) and the oral medications are continued for a minimum of 5 additional months (i.e., minimum of 9 months total). Patients with resistance to any of the drugs in the short-course regimen, should be treated with a traditional, individualized regimen lasting an average of 24 months.

Following the WHO endorsement of the short-course regimen, nine countries in West and Central Africa participated in an observational study where patients with rifampin-resistant or MDR TB were treated with the regimen [134]. Overall, 82% of patients had treatment success, including 72% of those patients co-infected with HIV. However, there was a higher proportion of deaths among those with HIV co-infection than those without (19% vs. 5%; $p < 0.001$). A randomized controlled trial of a 9-month regimen, the Standardised Treatment Regimen of Anti-TB Drugs for Patients with MDR-TB (STREAM) trial, found that 79% of patients receiving the short-course regimen and 80% of patients receiving a 20–24 month regimen had favorable outcomes [135]. Although the STREAM trial demonstrated non-inferiority of the short-course regimen compared to the traditional 20–24 month regimen, the regimen remains somewhat controversial both for its reliance on injectable medications and because many patients with MDR TB are not eligible to receive the short-course regimen, given the high prevalence of resistance to drugs included in the regimen in many regions [136].

HIV and Drug-Resistant TB

In the pre-ART era, MDR TB treatment outcomes were exceedingly poor among people with HIV. In the earliest nosocomial outbreaks, the overwhelming majority of patients died, often within a month or two of diagnosis [15, 17]. Among patients with HIV and MDR TB in early New York City outbreaks, mortality rates ranged from 60–80%, as compared to 20–30% among those with drug-susceptible TB [11, 13, 137, 138]. In the decade that followed, similarly poor outcomes were reported in nearly every setting where HIV and drug-resistant TB were found [29, 36, 41, 139–141], and a meta-analysis that included studies through the mid-2000s found that MDR TB treatment success was approximately 10% lower with HIV coinfection [142].

Several more recent reports, however, found similar outcomes for those with and without HIV when those with HIV were treated with ART [58, 111, 143, 144]. These findings were echoed in a meta-analysis that included studies conducted through 2010; patients with MDR TB and HIV co-infection had treatment success rates comparable to those without HIV [145]. People with HIV did have higher mortality in this meta-analysis, although this finding was largely driven by exceedingly poor outcomes in early cohorts where ART use was lower. Importantly, however, these retrospective studies were likely limited by indication and survival bias. That is, patients who did not receive ART may not have survived to receive ART, or may have been sicker or perceived as less adherent to medical care than patients who were prescribed ART. One prospective study from South Africa examined survival in co-infected patients treated with ART as compared to HIV-uninfected patients with MDR TB and found that overall survival in the HIV co-infected group was not significantly different from those without HIV. However, participants with a CD4 count persistently less than 100 cells/mm^3 had a significantly higher mortality [146]. In contrast, those with a low CD4 count at baseline who experienced immunological recovery had similar survival to both those with high CD4 counts throughout, as well as HIV-negative participants. These findings thus emphasized the importance of initiating ART in co-infected patients and of close follow-up to ensure continuous virologic suppression.

Current guidelines for treatment of MDR TB/HIV co-infected patients are largely the same as those for patients with MDR TB alone with a few additional concerns [119]. Most importantly, all HIV-infected patients should be initiated on ART. While no clinical trials have been conducted to determine the optimal timing of ART in MDR TB co-infected patients, several randomized trials in co-infected patients with drug-susceptible TB have demonstrated improved survival with early ART, especially when patients have advanced immunosuppression [147–149].

New and Repurposed Drugs for Drug-Resistant TB

After a 40-year gap with no novel anti-tuberculous agents, there are now several new TB medications in clinical development and two that have received at least limited approval by either the US Food and Drug Administration (FDA) or the European Medicines Agency (EMA). Here, we limit our discussion to those which are either approved or in late-stage clinical trials. (Additional information about new TB drugs can be found at: https://www.newtbdrugs.org/pipeline/clinical.)

Bedaquiline

Bedaquiline is a diarylquinoline that inhibits mycobacterial ATP synthase and has a very long terminal half-life of more than 5 months [150]. On the basis of three phase 2 trials, bedaquiline received accelerated FDA approval for use in drug-resistant pulmonary TB in December 2012, EMA approval in December 2013, and was the first TB drug from a new class of medications to be approved since rifampin in 1967 [151–153]. Generalizability of the early trial results to patients co-infected with HIV is limited, as trial patients were largely HIV-negative and, if HIV-positive, were not on ART. In addition, because bedaquiline causes cardiac conduction abnormalities and prolongs the QT interval, the trials generally restricted use of other QT-prolonging drugs (e.g., moxifloxacin, macrolides, clofazimine). Nevertheless, bedaquiline was introduced in South Africa in 2014 and is now included in the country's standard regimen for MDR, pre-XDR, and XDR TB. Despite initial concerns about its safety, increasing observational data from multiple countries [154–157] suggest that the drug is well-tolerated and effective, although a phase 3 trial is still forthcoming. Resistance to bedaquiline has been reported but appears to be infrequent in clinical use [158]. Although polymorphisms of several different genes have been identified, causing both target-modification as well as upregulation of efflux pumps, genotypic tests for bedaquiline resistance are not yet available. Other mechanisms of resistance may yet be identified as use of bedaquiline increases and in less selected patient populations. Bedaquiline has several important drug-drug interactions with ART (see chapter on "Co-treatment of Tuberculosis and HIV: Pharmacologic Considerations").

Delamanid

Delamanid is a nitro-dihydro-imidazooxazole which inhibits mycolic acid biosynthesis and disrupts metabolism of the cell wall. It received approval by the EMA in 2013 based on the results of a phase 2 study [159]. A subsequent phase 3 trial, however, did not show a significant improvement in 6-month culture conversion compared to placebo when added to an optimized background regimen for the treatment of patients with MDR TB [160]. Based on these data, the WHO issued interim guidance that delamanid could be added to a MDR TB regimen only when the regimen could not otherwise be composed according to WHO recommendations [161]. Given its novel mechanism of action and that it may be better tolerated than other second-line TB medications, additional data are needed to determine the value of delamanid as an addition to drug-resistant TB regimens. Like bedaquiline, delamanid also prolongs the QT interval and there has been concern about co-administering delamanid with bedaquiline. A large multinational cohort study, however, found few serious adverse events when the drugs were given together [162], and preliminary data from the DELIBERATE clinical trial found a minimal effect of co-administration on QT-interval prolongation [163].

Pretomanid

Like delamanid, pretomanid (formerly PA-824), is a nitroimidazole and has shown promising results in phase 2 studies. The NIX-TB trial is an ongoing uncontrolled trial testing a novel, 6-month, all-oral regimen consisting of bedaquiline, pretomanid, and high-dose linezolid for patients with XDR TB. Recent preliminary analyses suggest that patients in the trial have had exceptionally good outcomes, suggesting that the regimen warrants further study [164, 165].

Linezolid

Linezolid is an oxazolidinone and has been used for nearly 20 years in the treatment of gram-positive bacterial infections. Given its potent activity against *Mtb*, it has also been used in the treatment of both MDR and XDR TB. Although multiple case series and a clinical trial have shown benefit of including linezolid in treatment regimens, use of the drug is often limited by toxicities [166–169]. Side effects from linezolid are primarily due to inhibition of mitochondrial function and include lactic acidosis, peripheral neuropathy, and optic neuritis. Additional side effects are leukopenia, anemia, and thrombocytopenia. Toxicities are dose-related and generally emerge after several weeks of therapy. Given the high XDR TB treatment success rates seen in the NIX-TB trial, where linezolid was given with pretomanid and bedaquiline [164], there has been an increased interest in the use of linezolid. The standard treatment regimens for MDR and XDR TB in South Africa now contain both bedaquiline and linezolid. Newer oxazolidinones with less mitochondrial toxicity have been developed and are in clinical development. The most advanced of these, sutezolid, had promising results in mice and in a human phase 2a early bactericidal activity study [170, 171].

Clofazimine

Clofazimine is a riminophenazine that was first synthesized as an anti-tuberculous drug in 1954, but for much of the last half-century was used primarily for the treatment of leprosy [172]. The primary mechanism of action is believed to be related to the induction of redox imbalance and membrane destabilization [173]. The drug is lipophilic, which enables it to target transporters in the outer membrane of *Mtb* [174] and it has a long half-life, which may aid in the targeting of slowly replicating bacterial populations [174]. For many years, clofazimine was classified as a so-called "category 5" drug with "questionable efficacy" against *Mtb*, but recently, there has been increasing evidence for its role in the treatment of drug-resistant TB. Two systematic reviews found that 61–65% of drug-resistant TB patients treated with clofazimine had a favorable outcome [173, 175], and the drug is considered a core component of the 9-month MDR TB regimen (see Short-Course Regimen above). Side effects from clofazimine include gastrointestinal intolerance and brownish skin pigmentation [176], and there have been emerging concerns about cross-resistance with bedaquiline [136, 177]. Despite its apparent effectiveness, low global availability and high cost may limit access to clofazimine for many patients [178].

Mono-Resistance and Resistance "Beyond" MDR TB

Isoniazid Mono-Resistance

It is estimated that approximately 8.5% of TB cases worldwide have resistance to isoniazid without concurrent rifampin resistance [64]. With the global rollout of Xpert MTB/RIF, rifampin susceptibility is now routinely available. However, susceptibility to isoniazid, the other cornerstone of drug-susceptible TB regimens, is not routinely tested in many countries, and patients without rifampin resistance are assumed to have fully susceptible isolates. If these patients are treated with the standard first-line regimen, they will receive a weakened regimen in the intensive phase and effective rifampin monotherapy in the continuation phase. A recent meta-analysis found that 15% of patients with isoniazid mono-resistance had treatment failure or relapse when treated with a standard regimen, as compared to 4% of patients with pan-susceptible isolates [179]. Further, 3.6% of mono-resistant patients developed additional acquired drug resistance, compared with 0.6% of pan-susceptible patients. Thus, despite the marked advance of widespread rifampin resistance testing, these findings have raised concerns about the worldwide reliance on rifampin susceptibility testing without concurrent testing for isoniazid [179, 180]. In 2018, the WHO recommended that patients with isoniazid mono-resistance be treated with a 6-month regimen containing rifampin, pyrazinamide, ethambutol, and levofloxacin [181].

Rifampin Mono-Resistance

The detection of rifampin resistance is typically considered a marker for MDR TB as the overwhelming majority of such isolates have concurrent resistance to isoniazid. Nevertheless, it is estimated that 18% of patients with rifampin resistance have mono-resistance [64]. Without the strong sterilizing activity of rifampin, treatment regimens for patients with rifampin mono-resistance must be treated for a minimum of 12–18 months. Many programs use an MDR TB regimen with the addition or substitution of isoniazid, a practice that is supported by WHO guidelines [119]. A UKMRC trial in Hong Kong showed good success with a regimen of isoniazid, pyrazinamide and streptomycin but this regimen has never been tested in patients with HIV [182].

Pre-XDR and XDR TB

Treatment outcomes for MDR TB patients worsen in a stepwise fashion with the addition of second-line drug resistance [56, 183, 184]. The two most widely used resistance categorizations are "pre-XDR TB" (i.e., MDR TB with resistance to a fluoroquinolone OR a second-line injectable, but not both) and "XDR TB" (resistance to *both* a fluoroquinolone AND a second-line injectable). Strains with yet further resistance, have been described in case series and meta-analyses with terms such as "totally drug-resistant," "super extensively drug-resistant," and "drug resistance beyond XDR," [185–187] but their nomenclature remains controversial, particularly given the lack of standardized drug susceptibility testing for many of the third and fourth line agents, as well as the availability of new drugs (e.g., bedaquiline, delamanid). At the moment, there are no formal guidelines for the management of pre-XDR or XDR TB beyond the basic principles underlying the MDR TB guidelines. Because it can be difficult, however, to find 4 or 5 drugs with likely activity against such strains, it is not surprising that outcomes for such patients are extremely poor [184]. With the availability of new drugs, however, countries like South Africa are now recommending regimens containing bedaquiline, linezolid and clofazimine for pre-XDR and XDR TB patients, which hold promise for better treatment outcomes for pre-XDR and XDR TB [164, 188]. Several studies have found comparable outcomes for XDR TB patients with and without HIV, but this is likely related to the generally poor outcomes for XDR TB [55, 58, 144, 189]. There is hope, however, that XDR TB outcomes will improve with the wider availability and use of newer drugs; a recent cohort study from South Africa reported a mortality rate of 13% for patients with XDR TB receiving a bedaquiline-containing regimen, as compared to 25% for those treated with a standard regimen [157].

Adverse Events

Adverse events due to drug toxicity, whether from the toxicities associated with second- and third-line anti-tuberculous agents or from overlapping drug toxicity with ART, pose another challenge to achieving successful treatment outcomes [26, 189]. Side effects associated with second-line TB medications have been well described [190–197] and figure prominently in treatment because they impact regimen choice, medication adherence, and retention in care. Most reports indicate that the overwhelming majority of MDR TB patients will experience at least one side effect during their treatment course. Adverse drug effects can also lead to less effective treatment regimens, if problematic drugs are stopped without the addition or substitution of alternative agents. Although data are limited, two small studies found no difference in frequency or severity of adverse events in patients with MDR-TB/HIV co-infection [197, 198].

In general, all patients initiating drug-resistant TB treatment must be counseled about likely side effects and how best to manage them. Patients who choose to stop MDR TB therapy prematurely often cite medication side effects among their reasons for doing so [199, 200]. Several studies have shown that with extensive patient education, aggressive symptomatic treatment of emergent toxicities, and support from family members/caregivers, most patients can complete treatment successfully with minimal or no changes to their treatment regimen [191, 197]. Table 4 lists the most common side effects associated with each second-line drug. Most reports of medication toxicity have been retrospective with little standardization of severity or detail regarding duration.

One of the most feared side effects is hearing loss from the second-line injectable agents. This typically begins as high-frequency loss and can progress to complete and irreversible deafness [201–204]. Most guidelines recommend screening patients with audiometric testing at baseline and then monthly while receiving an injectable. If evidence of hearing loss is found, either a reduction in dosing frequency (i.e., from daily to three times weekly) or discontinuation of the injectable is advised [205–207]. Whether such dose reduction mitigates the hearing loss or compromises

Table 4 Dosage and toxicities of medications used to treat drug-resistant TB

Medication	Typical dosage	Important side effects	Comments
Levofloxacin	1000 mg PO daily	QT prolongation; tendonitis	
Moxifloxacin	400–800 mg PO daily	QT prolongation; tendonitis	
Amikacin	15 mg/kg/day (max 1000 mg/d)	Hearing loss; tinnitus; nephrotoxicity; hypokalemia; hypomagnesemia	
Kanamycin	15 mg/kg/day	Hearing loss; tinnitus; nephrotoxicity; hypokalemia; hypomagnesemia	
Capreomycin	15 mg/kg/day (max 1000 mg/day)	Hearing loss; tinnitus; nephrotoxicity; hypokalemia; hypomagnesemia; hypocalcemia	
Ethionamide	15–20 mg/kg/day (usually total 500 or 750 mg per day)	Nausea and vomiting; hypothyroidism; taste disturbance; gynecomastia; alopecia	
Prothionamide	15–20 mg/kg/day (usually total 500 or 750 mg per day)	Nausea and vomiting; hypothyroidism; taste disturbance; gynecomastia; alopecia	
Cycloserine	10–15 mg/kg/day (usually total 500 or 750 mg per day)	Psychosis; seizures; depression; difficulty concentrating	Potential for overlapping neuropsychiatric side effects when given with efavirenz.
Terizidone	10–15 mg/kg/day (usually total 500 or 750 mg per day)	Psychosis; seizures; depression; difficulty concentrating.	Potential for overlapping neuropsychiatric side effects when given with efavirenz.
Pyrazinamide	25 mg/kg daily	Gout; arthralgias; hepatotoxicity; rash; photosensitivity; gastrointestinal upset.	
Ethambutol	15–25 mg/kg daily	Optic neuritis	
Clofazimine	100–200 mg daily	Red discoloration of skin, conjunctiva, cornea, and body fluids; gastrointestinal intolerance; photosensitivity.	
Imipenem-cilastatin	1000 mg IV q12h	Diarrhea; nausea; vomiting; seizure	Must be given with clavulanate 125 mg PO q8–12.

(continued)

Table 4 (continued)

Medication	Typical dosage	Important side effects	Comments
Meropenem	1000–2000 mg IV q8-q12[a]	Diarrhea; nausea; vomiting; seizure	Must be given with clavulanate 125 mg PO q8–12.
Rifabutin	5 mg/kg (max 300 mg) PO daily	Uveitis; leukopenia; thrombocytopenia; hepatotoxicity	• Efavirenz decreases rifabutin AUC by 35%. Consider increasing RFB to 450 mg daily; • Most protease inhibitors increase RFB AUC by 250%. Decrease dose of RFB to 150 mg daily. Rifabutin decreases elvitegravir AUC; avoid co-administration.
Linezolid	600 mg PO daily or BID	Peripheral neuropathy; optic neuritis; lactic acidosis; leukopenia; anemia; thrombocytopenia	
Bedaquiline	400 mg PO daily for 2 weeks THEN 200 mg PO TIW to complete 6-month course.	QT-prolongation.	Efavirenz decreases BDQ AUC by 40–50%. Do not co-administer; Lopinavir/ritonavir doubles BDQ AUC. Clinical significance unknown. No interaction with nevirapine. No interaction with integrase strand transfer inhibitors predicted.
Delamanid	100 mg PO BID	QT-prolongation	

BDQ = bedaquiline; AUC = area under the curve; RFB = rifabutin
[a]Dosage based on published reports. Ideal dose not established

treatment outcomes has never been rigorously examined, however. Preventing such hearing loss is one of several motivating forces driving the development, testing, and implementation of novel, injectable-free regimens.

Hypothyroidism can be induced by either ethionamide/prothionamide or PAS and was once thought to be rare. A number of recent studies, however, have shown—with more proactive monitoring—that it is much more common than previously believed [197, 198, 208–210]. In a cohort of patients from South Africa who were treated with ethionamide, 34% of patients required levothyroxine replacement therapy [197], whereas in a cohort from Lesotho who received concurrent ethionamide and PAS, 69% of patients had a thyroid-stimulating hormone level > 10 mIU/L [209], suggesting an additive effect of the two medications. Unlike injectable-related hearing loss, this effect is largely reversible, but requires regular monitoring. Thyroid-stimulating hormone levels should be checked at baseline, and then repeated after 3 months, 6 months, and then every 6 months thereafter until the end of treatment.

As in the treatment of drug-susceptible TB, drug-induced liver injury (DILI) has also been reported in the treatment of MDR TB, with frequencies ranging from

10–17% of patients [211–213]. Multiple second-line drugs can cause DILI, including fluoroquinolones and ethionamide/prothionamide, as well as the first-line drugs isoniazid and pyrazinamide, which are often included in MDR TB regimens. In one study from South Korea, patients with alcoholic hepatitis or co-infection with either hepatitis B or C were at significantly greater risk of incident hepatotoxicity from MDR TB treatment [211]. Although mild elevations in liver function tests can often be followed clinically, treatment interruptions and sequential rechallenge may be necessary if patients experience significant transaminase elevations (i.e., >5x upper limit of normal) or if elevations are accompanied by symptoms.

Lastly, several second-line TB medications can cause treatment-limiting gastrointestinal discomfort, nausea, and vomiting. PAS and ethionamide/prothionamide are the two most common culprits, and tolerability of ethionamide/prothionamide can often be improved by dividing the dose BID.

For a discussion of drug-drug interactions between second-line TB medications and ART, see chapter "Co-treatment of Tuberculosis and HIV: Pharmacologic Considerations."

Other Considerations in MDR TB Treatment

Standardized Versus Individualized Treatment

Treatment for drug-resistant TB has always been hampered by limitations in diagnostics and challenges in obtaining a comprehensive drug susceptibility profile, as outlined above. Because phenotypic resistance testing to many second-line drugs is technically challenging, unreliable and marred by poor reproducibility [93, 214], many programs either do not routinely test for susceptibility, or restrict testing to only a handful of drugs. As a result, the treating clinician often does not know if each medication in a patient's regimen is truly active against that patient's isolate.

In an effort to save money on diagnostic testing and simplify treatment programs while providing patients the best chance at cure, many programs have used surveillance data to analyze broad trends in resistance patterns within a given community. They then use these data to generate a "standard" regimen which can be given to any patient presenting with MDR TB in that community. The appeal of this approach is that it allows programs to plan for consistent drug-utilization, facilitating pharmacy procurement and preventing stock-outs, and allows clinicians with less experience in drug-resistant TB to manage these patients. An important shortcoming, however, is that without precise drug-susceptibility testing for each patient, those with less common resistance patterns may be inadequately treated by the standard regimen. These patients will be at greater risk of treatment failure and amplification of further resistance.

An alternate approach to standardized therapy is "individualized" treatment, in which the clinician creates a regimen specifically developed for a given patient, based on that patient's known exposure to other drug-resistant TB patients and prior treatment history. Typically, the clinician treats with an aggressive regimen at the outset and then removes extra drugs once the precise drug susceptibility pattern is known. This approach has the disadvantage of added costs associated with diagnostics and medications, and requires additional expertise in the treating clinician. In addition, the pill burden and side effects associated with the aggressive regimen may compromise medication adherence. The advantage, however, is that every patient receives the best possible regimen for his or her disease [215, 216]. Most studies have suggested that individualized treatment achieves superior outcomes compared to standardized treatment [115, 142, 217], but in communities with a fairly homogeneous strain epidemiology and/or limited resources for diagnostic testing, standardized therapy can be a reasonable compromise.

Hospital-Based Versus Community-Based Treatment

Because the 24-month MDR TB regimen requires a daily intramuscular injection and is associated with frequent side effects, many TB programs hospitalize patients with MDR TB for the duration of the intensive phase. This both facilitates administration of an injectable agent and provides a theoretical benefit of isolating such patients from spreading their disease in the community. Yet, keeping patients hospitalized for such long periods—often hundreds of kilometers from their homes and families—is potentially counterproductive. In a number of reports, patients frequently became frustrated with what they viewed as a de facto incarceration and left the hospitals against medical advice, thereby discontinuing their TB treatment [140, 218–220]. In addition, countries with very large MDR TB burdens may not have enough hospital beds to accommodate all patients who need them. Patients, then, are often placed on waiting lists, and receive ineffective, first-line therapy while awaiting a hospital bed where they can initiate drug-resistant TB therapy [221]. During this time, patients are often infectious and likely contribute to ongoing transmission of drug-resistant disease in their communities.

In an effort to improve outcomes and adherence, while also broadening access to drug-resistant TB treatment, a number of programs have explored the feasibility and safety of community-based treatment for patients with drug-resistant TB [222]. Keeping patients at home, where they are supported by their family members, makes them less likely to discontinue therapy. Providing care at decentralized clinics or district hospitals makes it easier for patients to attend follow-up appointments. Many models of decentralized, community-based drug-resistant TB care have been implemented in both urban and rural settings, either with patients traveling to a local clinic for their daily injection or injection teams traveling to a patient's home [221–223]. DOT has been provided in such programs by nurses, community health workers, and family members. Two meta-analyses of community-based treatment studies found higher rates of treatment success in decentralized versus centralized MDR TB programs, with no impact of HIV prevalence on treatment outcomes [224, 225]. These data are encouraging both for TB treatment programs and for patients, as community-based treatment is typically less expensive than hospital-based treatment and is vastly preferred by patients [226, 227]. As part of these programs, it is critical to educate the patient and family members about basic infection control practices in order to minimize the risk of transmission within the home.

Community-based treatment has also been used to provide concurrent treatment for drug-resistant TB and HIV. Outcomes from such programs are heterogeneous; when decentralized care was introduced in several sites within the province of KwaZulu-Natal, South Africa, some programs had superior outcomes compared to inpatient care while others had worse outcomes [228]. An analysis of health systems at each of the sites demonstrated the importance of proper staffing, management, training, and oversight if the benefits of decentralized care are to be realized [229].

Surgery for Drug-Resistant TB

Prior to the discovery of anti-tuberculous chemotherapy, treatment for TB often included surgical collapse procedures, such as thoracoplasty, plombage, induced pneumothorax, pneumoperitoneum, and lung resection (partial or total) [230–233]. With the advent of effective medical therapy and high cure rates, surgical management was largely abandoned in most countries. However, the recent global rise in drug-resistant TB and its associated poor outcomes have brought renewed attention to adjuvant surgical therapy [234].

Even with more than a century of global experience with the procedures, there remain very limited data on its effectiveness. To date, there have been no clinical trials of lung resection surgery for drug-resistant TB, and several meta-analyses, including one individual patient data meta-analysis, have attempted to combine the more than two dozen published case series [235–238]. These meta-analyses all found that partial lung resection surgery, when performed along with medical therapy, provided high rates of cure and treatment success. The studies in question, however, were all uncontrolled and suffered from significant indication bias, such that they included patients who had been specifically selected to undergo major surgery and thus may have been healthier than the general drug-resistant TB population. One meta-analysis utilized individual patient-level data in an attempt to address this concern and created a control group of patients from studies in which surgery was not performed [237]. They found that while lobectomy was associated with improved treatment success compared with no surgery, pneumonectomy was not. In addition, treatment success was more likely if surgical resection was performed after sputum culture-conversion, although this likely represents channeling bias as patients who achieved earlier culture-conversion would have been more likely to achieve treatment success regardless, and may have been more likely to undergo surgery after conversion.

Despite concerns about the generalizability of the published data on adjuvant surgical therapy, there is general consensus that surgery likely has an important role in controlling disease in MDR and XDR TB and surgery is recommended by the

WHO as a potential adjunct to appropriate chemotherapy [239]. In general, surgery is advised for patients who have localized disease and adequate pulmonary reserve to tolerate resection (though this is rarely quantified); further, preference is often given to patients with a more favorable resistance profile who have already achieved sputum culture conversion. The resection should be performed by an experienced surgeon and the patient should complete a full course of medical therapy even if the surgery is successful. Further research is needed to develop risk-stratification tools for surgery in the context of drug-resistant TB and to identify patients most likely to benefit from surgical intervention.

Thinking Beyond Drugs for Drug-Resistant TB

Even with the availability of novel drug regimens and surgical intervention, there continue to be patients who are deemed programmatically incurable. In a case series published from the Western Cape province of South Africa, albeit prior to the introduction of bedaquiline and linezolid, 203 (74%) of 273 patients with XDR TB were deemed incurable and 172 (63%) of these patients were discharged home [240]. Over 20% of these patients were still alive one year after hospital discharge, and the investigators identified downstream cases among contacts with nearly identical TB strains as the discharged index cases. For patients with programmatically incurable disease, there is an urgent need to ensure safe discharge plans, such that these individuals do not pose a risk to their communities and have access to palliative care [63].

Finally, the far-reaching impact of drug-resistant TB on patients' lives and their families must be acknowledged. In a US task force convened in 1992 to develop a national action plan to combat MDR TB, the authors recognized the need for social and economic support for patients undergoing MDR TB treatment, to ensure adherence with prolonged treatment regimens, and to prevent secondary cases for patients in unstable living situations [241]. Despite a growing advocacy movement for patient-centered care and respect for patients' autonomy [242], individuals with TB, and particularly drug-resistant TB, often face high levels of stigma in their communities, which can lead to psychological stress [63]. Successful treatment is further challenged by financial stress and loss of productivity, as TB can impact individuals during their adult working years. By enabling patients to remain in their homes, community-based care (as reviewed above) is one element of patient-centered care. Similarly, peer support groups and psychological counseling have been demonstrated to improve adherence and psychosocial well-being in several pilot projects [243–245]. A recent review of these important "non-medical" patient-centered outcomes found a relatively limited evidence base for psychosocial and economic interventions for patients with MDR TB, despite common reports of depression, stigma, discrimination and financial constraints in this population [246].

References

1. Crofton J, Mitchison DA (1948) Streptomycin treatment of pulmonary tuberculosis. Br Med J 2(4582):769–782
2. Crofton J (2006) The MRC randomized trial of streptomycin and its legacy: a view from the clinical front line. J R Soc Med 99(10):531–534
3. Bothwell LE, Podolsky SH (2016) The emergence of the randomized, controlled trial. N Engl J Med 375(6):501–504
4. Crofton J, Mitchison DA (1948) Streptomycin resistance in pulmonary tuberculosis. Br Med J 2(4588):1009–1015
5. Keshavjee S, Farmer PE (2012) Tuberculosis, drug resistance, and the history of modern medicine. N Engl J Med 367(10):931–936
6. Cegielski JP (2010) Extensively drug-resistant tuberculosis: "there must be some kind of way out of here". Clin Infect Dis 50(Suppl 3):S195–S200
7. Manten A, Van Wijngaarden LJ (1969) Development of drug resistance to rifampicin. Chemotherapy 14(2):93–100
8. Steiner M, Chaves AD, Lyons HA, Steiner P, Portugaleza C (1970) Primary drug-resistant tuberculosis. N Engl J Med 283(25):1353–1358
9. Pneumocystis pneumonia--Los Angeles (1981) MMWR Morb Mortal Wkly Rep 30(21):250–252
10. Barre-Sinoussi F, Chermann J, Rey F et al (1983) Isolation of a T-lymphotropic retrovirus from a patient at risk for acquired immune deficiency syndrome (AIDS). Science 220(4599):868–871
11. Frieden TR, Sterling T, Pablos-Mendez A, Kilburn JO, Cauthen GM, Dooley SW (1993) The emergence of drug-resistant tuberculosis in New York City. N Engl J Med 328(8):521–526
12. Frieden TR, Fujiwara PI, Washko RM, Hamburg MA (1995) Tuberculosis in New York City--turning the tide. N Engl J Med 333(4):229–233
13. Bloch AB, Cauthen GM, Onorato IM, Dansbury KG, Kelly GD, Driver CR (1994) Nationwide survey of drug-resistant tuberculosis in the United States. JAMA 271
14. Nosocomial transmission of multidrug-resistant tuberculosis to health-care workers and HIV-infected patients in an urban hospital--Florida (1990) MMWR Morb Mortal Wkly Rep 39(40):718–722
15. Nosocomial transmission of multidrug-resistant tuberculosis among HIV-infected persons--Florida and New York, 1988–1991. MMWR Morb Mortal Wkly Rep 1991; 40(34): 585–91
16. Iseman MD (1994) Evolution of drug-resistant tuberculosis: a tale of two species. Proc Natl Acad Sci U S A 91(7):2428–2429
17. Multidrug-resistant tuberculosis outbreak on an HIV ward--Madrid, Spain, 1991-1995. MMWR Morb Mortal Wkly Rep 1996; 45(16): 330–333
18. Moro ML, Gori A, Errante I et al (1998) An outbreak of multidrug-resistant tuberculosis involving HIV-infected patients of two hospitals in Milan, Italy. Italian Multidrug-Resistant Tuberculosis Outbreak Study Group. AIDS 12(9):1095–1102
19. Ritacco V, Di Lonardo M, Reniero A et al (1997) Nosocomial spread of human immunodeficiency virus-related multidrug-resistant tuberculosis in Buenos Aires. J Infect Dis 176(3):637–642
20. Rullan JV, Herrera D, Cano R et al (1996) Nosocomial transmission of multidrug-resistant *Mycobacterium tuberculosis* in Spain. Emerg Infect Dis 2(2):125–129
21. Angarano G, Carbonara S, Costa D, Gori A (1998) Drug-resistant tuberculosis in human immunodeficiency virus infected persons in Italy. The Italian Drug-Resistant Tuberculosis Study Group. Int J Tuberc Lung Dis 2(4):303–311
22. Crawford JT (1994) Epidemiology of tuberculosis: the impact of HIV and multidrug-resistant strains. Immunobiology 191(4–5):337–343
23. Pablos-Méndez A, Raviglione MC, Laszlo A et al (1998) Global surveillance for antituberculosis-drug resistance, 1994–1997. N Engl J Med 338(23):1641–1649

24. Espinal MA, Laszlo A, Simonsen L et al (2001) Global trends in resistance to antituberculosis drugs. World Health Organization-International Union against Tuberculosis and Lung Disease Working Group on Anti-Tuberculosis Drug Resistance Surveillance. N Engl J Med 344(17):1294–1303
25. Dye C, Espinal MA, Watt CJ, Mbiaga C, Williams BG (2002) Worldwide incidence of multidrug-resistant tuberculosis. J Infect Dis 185(8):1197–1202
26. Wells CD, Cegielski JP, Nelson LJ et al (2007) HIV infection and multidrug-resistant tuberculosis: the perfect storm. J Infect Dis 196(Suppl 1):S86–S107
27. Vanacore P, Koehler B, Carbonara S et al (2004) Drug-resistant tuberculosis in HIV-infected persons: Italy 1999–2000. Infection 32(6):328–332
28. Aguiar F, Vieira MA, Staviack A et al (2009) Prevalence of anti-tuberculosis drug resistance in an HIV/AIDS reference hospital in Rio de Janeiro, Brazil. Int J Tuberc Lung Dis 13(1):54–61
29. Palacios E, Franke M, Munoz M et al (2012) HIV-positive patients treated for multidrug-resistant tuberculosis: clinical outcomes in the HAART era. Int J Tuberc Lung Dis 16(3):348–354
30. Prach LM, Pascopella L, Barry PM et al (2013) Rifampin monoresistant tuberculosis and HIV comorbidity in California, 1993–2008: a retrospective cohort study. AIDS 27(16):2615–2622
31. Ershova JV, Kurbatova EV, Moonan PK, Cegielski JP (2014) Mortality among tuberculosis patients with acquired resistance to second-line antituberculosis drugs--United States, 1993–2008. Clin Infect Dis 59(4):465–472
32. Dubrovina I, Miskinis K, Lyepshina S et al (2008) Drug-resistant tuberculosis and HIV in Ukraine: a threatening convergence of two epidemics? Int J Tuberc Lung Dis 12(7):756–762
33. van den Hof S, Tursynbayeva A, Abildaev T et al (2013) Converging risk factors but no association between HIV infection and multidrug-resistant tuberculosis in Kazakhstan. Int J Tuberc Lung Dis 17(4):526–531
34. Granich R (2008) HIV and MDR-TB in Ukraine: news from a hazardous MDR-TB and HIV site. Int J Tuberc Lung Dis 12(7):701
35. Morozova I, Riekstina V, Sture G, Wells C, Leimane V (2003) Impact of the growing HIV-1 epidemic on multidrug-resistant tuberculosis control in Latvia. Int J Tuberc Lung Dis 7(9):903–906
36. Post FA, Grint D, Werlinrud AM et al (2014) Multi-drug-resistant tuberculosis in HIV positive patients in Eastern Europe. J Infect 68(3):259–263
37. Faustini A, Hall AJ, Perucci CA (2006) Risk factors for multidrug resistant tuberculosis in Europe: a systematic review. Thorax 61(2):158–163
38. Ruddy M, Balabanova Y, Graham C et al (2005) Rates of drug resistance and risk factor analysis in civilian and prison patients with tuberculosis in Samara Region, Russia. Thorax 60(2):130–135
39. Khue PM, Phuc TQ, Hung NV, Jarlier V, Robert J (2008) Drug resistance and HIV co-infection among pulmonary tuberculosis patients in Haiphong City, Vietnam. Int J Tuberc Lung Dis 12(7):763–768
40. Hemhongsa P, Tasaneeyapan T, Swaddiwudhipong W et al (2008) TB, HIV-associated TB and multidrug-resistant TB on Thailand's border with Myanmar, 2006–2007. Trop Med Int Health 13(10):1288–1296
41. Sungkanuparph S, Eampokalap B, Chottanapund S, Thongyen S, Manosuthi W (2007) Impact of drug-resistant tuberculosis on the survival of HIV-infected patients. Int J Tuberc Lung Dis 11(3):325–330
42. Punnotok J, Shaffer N, Naiwatanakul T et al (2000) Human immunodeficiency virus-related tuberculosis and primary drug resistance in Bangkok, Thailand. Int J Tuberc Lung Dis 4(6):537–543
43. Nunes EA, De Capitani EM, Coelho E et al (2005) Patterns of anti-tuberculosis drug resistance among HIV-infected patients in Maputo, Mozambique, 2002–2003. Int J Tuberc Lung Dis 9(5):494–500

44. Nelson LJ, Talbot EA, Mwasekaga MJ et al (2005) Antituberculosis drug resistance and anonymous HIV surveillance in tuberculosis patients in Botswana, 2002. Lancet 366(9484):488–490
45. Sacks LV, Pendle S, Orlovic D, Blumberg L, Constantinou C (1999) A comparison of outbreak- and nonoutbreak-related multidrug-resistant tuberculosis among human immunodeficiency virus-infected patients in a South African hospital. Clin Infect Dis 29(1):96–101
46. Vorkas C, Kayira D, van der Horst C et al (2012) Tuberculosis drug resistance and outcomes among tuberculosis inpatients in Lilongwe, Malawi. Malawi Med J 24(2):21–24
47. Rajasekaran S, Chandrasekar C, Mahilmaran A, Kanakaraj K, Karthikeyan DS, Suriakumar J (2009) HIV coinfection among multidrug resistant and extensively drug resistant tuberculosis patients--a trend. J Indian Med Assoc 107
48. Deivanayagam CN, Rajasekaran S, Venkatesan R, Mahilmaran A, Ahmed PR, Annadurai S (2002) Prevalence of acquired MDR-TB and HIV co-infection. Indian J Chest Dis Allied Sci 44
49. Magee MJ, Blumberg HM, Broz D, Furner SE, Samson L, Singh S (2012) Prevalence of drug resistant tuberculosis among patients at high-risk for HIV attending outpatient clinics in Delhi, India. Southeast Asian J Trop Med Public Health 43
50. Isaakidis P, Das M, Kumar AM et al (2014) Alarming levels of drug-resistant tuberculosis in HIV-infected patients in metropolitan Mumbai, India. PLoS One 9(10):e110461
51. He GX, Wang HY, Borgdorff MW et al (2011) Multidrug-resistant tuberculosis, People's Republic of China, 2007–2009. Emerg Infect Dis 17(10):1831–1838
52. Lan R, Yang C, Lan L et al (2011) Mycobacterium tuberculosis and non-tuberculous mycobacteria isolates from HIV-infected patients in Guangxi, China. Int J Tuberc Lung Dis 15(12):1669–1675
53. Emergence of *Mycobacterium tuberculosis* with extensive resistance to second-line drugs--worldwide, 2000–2004. MMWR Morb Mortal Wkly Rep 2006; 55(11): 301–305
54. Shah NS, Wright A, Bai G et al (2007) Worldwide emergence of extensively drug-resistant Tuberculosis (XDR-TB). Emerg Infect Dis 13(3):380–387
55. Gandhi NR, Moll A, Sturm AW et al (2006) Extensively drug-resistant tuberculosis as a cause of death in patients co-infected with tuberculosis and HIV in a rural area of South Africa. Lancet 368(9547):1575–1580
56. Gandhi NR, Shah NS, Andrews JR et al (2010) HIV coinfection in multidrug- and extensively drug-resistant tuberculosis results in high early mortality. Am J Respir Crit Care Med 181(1):80–86
57. Chaisson RE, Martinson NA (2008) Tuberculosis in Africa—combating an HIV-Driven Crisis. N Engl J Med 358(11):1089–1092
58. Dheda K, Shean K, Zumla A et al (2010) Early treatment outcomes and HIV status of patients with extensively drug-resistant tuberculosis in South Africa: a retrospective cohort study. Lancet 375(9728):1798–1807
59. Cox HS, McDermid C, Azevedo V et al (2010) Epidemic levels of drug resistant tuberculosis (MDR and XDR-TB) in a high HIV prevalence setting in Khayelitsha, South Africa. PLoS One 5(11):e13901
60. O'Donnell MR, Pillay M, Pillay M et al (2015) Primary capreomycin resistance is common and associated with early mortality in patients with extensively drug-resistant tuberculosis in KwaZulu-Natal, South Africa. J Acquir Immune Defic Syndr 69(5):536–543
61. Zhao M, Li X, Xu P et al (2009) Transmission of MDR and XDR tuberculosis in Shanghai, China. PLoS One 4(2):e4370
62. Dheda K, Gumbo T, Gandhi NR et al (2014) Global control of tuberculosis: from extensively drug-resistant to untreatable tuberculosis. Lancet Respir Med 2(4):321–338
63. Dheda K, Gumbo T, Maartens G, et al. (2017) The epidemiology, pathogenesis, transmission, diagnosis, and management of multidrug-resistant, extensively drug-resistant, and incurable tuberculosis. Lancet Respir Med
64. World Health Organization. Global Tuberculosis Report 2017. Document no. WHO/HTM/TB/2017.23, Geneva

65. Zignol M, Dean AS, Falzon D et al (2016) Twenty years of global surveillance of antituberculosis-drug resistance. N Engl J Med 375(11):1081–1089
66. Suchindran S, Brouwer ES, Van Rie A. Is HIV infection a risk factor for multi-drug resistant tuberculosis? A systematic review. PLoS One 2009; 4(5): e5561
67. Mesfin YM, Hailemariam D, Biadgilign S, Kibret KT (2014) Association between HIV/AIDS and multi-drug resistance tuberculosis: a systematic review and meta-analysis. PLoS One 9(1):e82235
68. Tappero JW, Bradford WZ, Agerton TB et al (2005) Serum concentrations of antimycobacterial drugs in patients with pulmonary tuberculosis in Botswana. Clin Infect Dis 41(4):461–469
69. March F, Garriga X, Rodriguez P et al (1997) Acquired drug resistance in *Mycobacterium tuberculosis* isolates recovered from compliant patients with human immunodeficiency virus-associated tuberculosis. Clin Infect Dis 25(5):1044–1047
70. Peloquin CA, Nitta AT, Burman WJ et al (1996) Low antituberculosis drug concentrations in patients with AIDS. Ann Pharmacother 30(9):919–925
71. Stott KE, Pertinez H, Sturkenboom MGG et al (2018) Pharmacokinetics of rifampicin in adult TB patients and healthy volunteers: a systematic review and meta-analysis. J Antimicrob Chemother 73(9):2305–2313
72. Daskapan A, Idrus LR, Postma MJ, et al. (2018) A systematic review on the effect of HIV infection on the pharmacokinetics of first-line tuberculosis drugs. Clin Pharmacokinet
73. David HL (1970) Probability distribution of drug-resistant mutants in unselected populations of *Mycobacterium tuberculosis*. Appl Microbiol 20(5):810–814
74. Rastogi N, David HL (1993) Mode of action of antituberculous drugs and mechanisms of drug resistance in *Mycobacterium tuberculosis*. Res Microbiol 144(2):133–143
75. Ormerod LP (2005) Multidrug-resistant tuberculosis (MDR-TB): epidemiology, prevention and treatment. Br Med Bull 73–74(1):17–24
76. Iseman MD (1999) Management of multidrug-resistant tuberculosis. Chemotherapy 45(Suppl 2):3–11
77. Migliori GB, D'Arcy Richardson M, Sotgiu G, Lange C (2009) Multidrug-resistant and extensively drug-resistant tuberculosis in the West. Europe and United States: epidemiology, surveillance, and control. Clin Chest Med 30(4):637–665. vii
78. Schaaf HS, Gie RP, Kennedy M, Beyers N, Hesseling PB, Donald PR (2002) Evaluation of young children in contact with adult multidrug-resistant pulmonary tuberculosis: a 30-month follow-up. Pediatrics 109(5):765–771
79. The global plan to stop TB 2006–2015. World Health Organization
80. The global MDR-TB and XDR-TB response plan 2007–2008. World Health Organization
81. Raviglione MC (2007) The new stop TB strategy and the global plan to stop TB, 2006–2015. Bull World Health Organ 85
82. Steiner M, Cosio A (1966) Primary Tuberculosis in Children. N Engl J Med 274(14):755–759
83. Tuberculosis outbreak among persons in a residential facility for HIV-infected persons--San Francisco. MMWR Morb Mortal Wkly Rep 1991; 40(38): 649–52
84. Small PM, Shafer RW, Hopewell PC et al (1993) Exogenous reinfection with multidrug-resistant *Mycobacterium tuberculosis* in patients with advanced HIV infection. N Engl J Med 328(16):1137–1144
85. Gandhi NR, Weissman D, Moodley P et al (2013) Nosocomial transmission of extensively drug-resistant tuberculosis in a rural hospital in South Africa. J Infect Dis 207(1):9–17
86. Shah NS, Auld SC, Brust JCM et al (2017) Transmission of extensively drug-resistant tuberculosis in South Africa. N Engl J Med 376(3):243–253
87. Zelner JL, Murray MB, Becerra MC et al (2016) Identifying hotspots of multidrug-resistant tuberculosis transmission using spatial and molecular genetic data. J Infect Dis 213(2):287–294
88. Yang C, Luo T, Shen X, et al. (2016) Transmission of multidrug-resistant *Mycobacterium tuberculosis* in Shanghai, China: a retrospective observational study using whole-genome sequencing and epidemiological investigation. Lancet Infect Dis

89. Yang C, Luo T, Shen X, et al. (2015) Whole-genome sequencing to delineate transmission of multidrug-resistant tuberculosis in Shanghai, China: a cross-sectional study. Union World Conference on Lung Health 2015; Cape Town, South Africa
90. Sharma A, Hill A, Kurbatova E et al (2017) Estimating the future burden of multidrug-resistant and extensively drug-resistant tuberculosis in India, the Philippines, Russia, and South Africa: a mathematical modelling study. Lancet Infect Dis 17(7):707–715
91. Kendall EA, Fofana MO, Dowdy DW (2015) Burden of transmitted multidrug resistance in epidemics of tuberculosis: a transmission modelling analysis. Lancet Respir Med 3(12):963–972
92. Koch AS, Brites D, Stucki D et al (2017) The Influence of HIV on the Evolution of *Mycobacterium tuberculosis*. Mol Biol Evol 34(7):1654–1668
93. Kim SJ (2005) Drug-susceptibility testing in tuberculosis: methods and reliability of results. Eur Respir J 25(3):564–569
94. Bottger EC (2011) The ins and outs of *Mycobacterium tuberculosis* drug susceptibility testing. Clin Microbiol Infect 17(8):1128–1134
95. World Health Organization (2011) Noncommercial culture and drug-susceptibility testing methods for screening patients at risk for multidrug-resistant tuberculosis. World Health Organization, Geneva
96. Moore DA, Evans CA, Gilman RH et al (2006) Microscopic-observation drug-susceptibility assay for the diagnosis of TB. N Engl J Med 355(15):1539–1550
97. Shah NS, Moodley P, Babaria P et al (2011) Rapid diagnosis of tuberculosis and multidrug resistance by the microscopic-observation drug-susceptibility assay. Am J Respir Crit Care Med 183(10):1427–1433
98. Zhang H, Li D, Zhao L et al (2013) Genome sequencing of 161 *Mycobacterium tuberculosis* isolates from China identifies genes and intergenic regions associated with drug resistance. Nat Genet 45(10):1255–1260
99. Zhang Y, Yew WW (2009) Mechanisms of drug resistance in *Mycobacterium tuberculosis*. Int J Tuberc Lung Dis 13(11):1320–1330
100. Nebenzahl-Guimaraes H, Jacobson KR, Farhat MR, Murray MB (2014) Systematic review of allelic exchange experiments aimed at identifying mutations that confer drug resistance in *Mycobacterium tuberculosis*. J Antimicrob Chemother 69(2):331–342
101. Dheda K, Gumbo T, Maartens G et al (2017) The epidemiology, pathogenesis, transmission, diagnosis, and management of multidrug-resistant, extensively drug-resistant, and incurable tuberculosis. Lancet Respir Med 5:269–281
102. Boehme CC, Nabeta P, Hillemann D et al (2010) Rapid molecular detection of tuberculosis and rifampin resistance. N Engl J Med 363(11):1005–1015
103. Hillemann D, Weizenegger M, Kubica T, Richter E, Niemann S (2005) Use of the genotype MTBDR assay for rapid detection of rifampin and isoniazid resistance in *Mycobacterium tuberculosis* complex isolates. J Clin Microbiol 43(8):3699–3703
104. Walker TM, Cruz AL, Peto TE, Smith EG, Esmail H, Crook DW (2017) Tuberculosis is changing. Lancet Infect Dis 17(4):359–361
105. Tukvadze N, Bablishvili N, Apsindzelashvili R, Blumberg HM, Kempker RR (2014) Performance of the MTBDRsl assay in Georgia. Int J Tuberc Lung Dis 18(2):233–239
106. Tagliani E, Cabibbe AM, Miotto P et al (2015) Diagnostic performance of the new version (v2.0) of GenoType MTBDRsl assay for detection of resistance to fluoroquinolones and second-line injectable drugs: a multicenter study. J Clin Microbiol 53(9):2961–2969
107. Miotto P, Tessema B, Tagliani E et al (2017) A standardised method for interpreting the association between mutations and phenotypic drug resistance in *Mycobacterium tuberculosis*. Eur Respir J 50(6)
108. Sandgren A, Strong M, Muthukrishnan P, Weiner BK, Church GM, Murray MB (2009) Tuberculosis drug resistance mutation database. PLoS Med 6(2):e2
109. The Stop TB Partnership. Global Laboratory Initiative—Advancing TB diagnostics. http://www.stoptb.org/wg/gli/ (accessed 15 Jan 2018)

110. Albert H, Nathavitharana RR, Isaacs C, Pai M, Denkinger CM, Boehme CC (2016) Development, roll-out and impact of Xpert MTB/RIF for tuberculosis: what lessons have we learnt and how can we do better? Eur Respir J 48(2):516–525
111. Shin SS, Modongo C, Boyd R et al (2017) High treatment success rates among HIV-infected multidrug-resistant tuberculosis patients after expansion of antiretroviral therapy in Botswana, 2006–2013. J Acquir Immune Defic Syndr 74(1):65–71
112. Brust JCM, Shah NS, Mlisana K et al (2018) Improved survival and cure rates with concurrent treatment for multidrug-resistant tuberculosis-human immunodeficiency virus coinfection in South Africa. Clin Infect Dis 66(8):1246–1253
113. Fox W, Ellard GA, Mitchison DA (1999) Studies on the treatment of tuberculosis undertaken by the British Medical Research Council tuberculosis units, 1946–1986, with relevant subsequent publications. Int J Tuberc Lung Dis 3
114. Frieden TR, Sherman LF, Maw KL et al (1996) A multi-institutional outbreak of highly drug-resistant tuberculosis: epidemiology and clinical outcomes. JAMA 276(15):1229–1235
115. Tierney DB, Franke MF, Becerra MC et al (2014) Time to culture conversion and regimen composition in multidrug-resistant tuberculosis treatment. PLoS One 9(9):e108035
116. American Thoracic Society, CDC, and Infectious Diseases Society of America (2003) Treatment of tuberculosis. MMWR 52(RR-11)
117. World Health Organization (2008) Guidelines for the programmatic management of drug-resistant tuberculosis. In: WHO/HTM/TB/2008, vol 402. Geneva, Switzerland
118. Falzon D, Jaramillo E, Schunemann HJ et al (2011) WHO guidelines for the programmatic management of drug-resistant tuberculosis: 2011 update. Eur Respir J 38(3):516–528
119. World Health Organization. Treatment guidelines for drug-resistant tuberculosis: 2016 update; October 2016 Revision. WHO/HTM/TB/2016.04; Geneva, Switzerland, 2016
120. Clinical and Laboratory Standards Institute TNCfCLS. Susceptibility testing of mycobacteria, nocardiae, and other aerobic actinomycetes: approved Standard. NCCLS Document 2003; M24-A. Wayne, PA, USA: NCCLS, 2011.
121. Tomioka H, Sato K, Akaki T, Kajitani H, Kawahara S, Sakatani M (1999) Comparative in vitro antimicrobial activities of the newly synthesized quinolone HSR-903, sitafloxacin (DU-6859a), gatifloxacin (AM-1155), and levofloxacin against Mycobacterium tuberculosis and *Mycobacterium avium* complex. Antimicrob Agents Chemother 43(12):3001–3004
122. Ruiz-Serrano MJ, Alcala L, Martinez L et al (2000) In vitro activities of six fluoroquinolones against 250 clinical isolates of *Mycobacterium tuberculosis* susceptible or resistant to first-line antituberculosis drugs. Antimicrob Agents Chemother 44(9):2567–2568
123. Rodriguez JC, Ruiz M, Lopez M, Royo G (2002) In vitro activity of moxifloxacin, levofloxacin, gatifloxacin and linezolid against *Mycobacterium tuberculosis*. Int J Antimicrob Agents 20(6):464–467
124. Gillespie SH, Billington O (1999) Activity of moxifloxacin against mycobacteria. J Antimicrob Chemother 44(3):393–395
125. Tortoli E, Dionisio D, Fabbri C (2004) Evaluation of moxifloxacin activity in vitro against *Mycobacterium tuberculosis*, including resistant and multidrug-resistant strains. J Chemother 16(4):334–336
126. Chien JY, Chien ST, Chiu WY, Yu CJ, Hsueh PR (2016) Moxifloxacin improves treatment outcomes in patients with ofloxacin-resistant multidrug-resistant tuberculosis. Antimicrob Agents Chemother 60(8):4708–4716
127. Koh WJ, Lee SH, Kang YA et al (2013) Comparison of levofloxacin versus moxifloxacin for multidrug-resistant tuberculosis. Am J Respir Crit Care Med 188(7):858–864
128. Kang YA, Shim TS, Koh WJ et al (2016) Choice between levofloxacin and moxifloxacin and multidrug-resistant tuberculosis treatment outcomes. Ann Am Thorac Soc 13(3):364–370
129. Davies Forsman L, Giske CG, Bruchfeld J, Schon T, Jureen P, Angeby K (2015) Meropenem-clavulanic acid has high in vitro activity against multidrug-resistant Mycobacterium tuberculosis. Antimicrob Agents Chemother 59(6):3630–3632

130. Van Deun A, Maug AK, Salim MA et al (2010) Short, highly effective, and inexpensive standardized treatment of multidrug-resistant tuberculosis. Am J Respir Crit Care Med 182(5):684–692
131. Aung KJ, Van Deun A, Declercq E et al (2014) Successful '9-month Bangladesh regimen' for multidrug-resistant tuberculosis among over 500 consecutive patients. Int J Tuberc Lung Dis 18(10):1180–1187
132. Piubello A, Harouna SH, Souleymane MB et al (2014) High cure rate with standardised short-course multidrug-resistant tuberculosis treatment in Niger: no relapses. Int J Tuberc Lung Dis 18(10):1188–1194
133. Kuaban C, Noeske J, Rieder HL, Ait-Khaled N, Abena Foe JL, Trebucq A (2015) High effectiveness of a 12-month regimen for MDR-TB patients in Cameroon. Int J Tuberc Lung Dis 19(5):517–524
134. Trebucq A, Schwoebel V, Kashongwe Z, et al. (2017) Treatment outcome with a short multidrug-resistant tuberculosis regimen in nine African countries. Int J Tuberc Lung Dis
135. Nunn AJ, Phillips PPJ, Meredith SK et al (2019) A trial of a shorter regimen for rifampin-resistant tuberculosis. N Engl J Med 380(13):1201–1213
136. Dheda K, Cox H, Esmail A, Wasserman S, Chang KC, Lange C (2018) Recent controversies about MDR and XDR-TB: Global implementation of the WHO shorter MDR-TB regimen and bedaquiline for all with MDR-TB? Respirology 23(1):36–45
137. Mannheimer SB, Sepkowitz KA, Stoeckle M, Friedman CR, Hafner A, Riley LW (1997) Risk factors and outcome of human immunodeficiency virus-infected patients with sporadic multidrug-resistant tuberculosis in New York City. Int J Tuberc Lung Dis 1(4):319–325
138. Park MM, Davis AL, Schluger NW, Cohen H, Rom WN (1996) Outcome of MDR-TB patients, 1983–1993. Prolonged survival with appropriate therapy. Am J Respir Crit Care Med 153(1):317–324
139. van Altena R, de Vries G, Haar CH et al (2015) Highly successful treatment outcome of multidrug-resistant tuberculosis in the Netherlands, 2000–2009. Int J Tuberc Lung Dis 19(4):406–412
140. Brust JC, Gandhi NR, Carrara H, Osburn G, Padayatchi N (2010) High treatment failure and default rates for patients with multidrug-resistant tuberculosis in KwaZulu-Natal, South Africa, 2000–2003. Int J Tuberc Lung Dis 14(4):413–419
141. Farley JE, Ram M, Pan W et al (2011) Outcomes of multi-drug resistant tuberculosis (MDR-TB) among a cohort of South African patients with high HIV prevalence. PLoS One 6(7):e20436
142. Orenstein EW, Basu S, Shah NS, Andrews JR, Friedland GH, Moll AP (2009) Treatment outcomes among patients with multidrug-resistant tuberculosis: systematic review and meta-analysis. Lancet Infect Dis 9
143. Satti H, McLaughlin MM, Hedt-Gauthier B et al (2012) Outcomes of multidrug-resistant tuberculosis treatment with early initiation of antiretroviral therapy for HIV co-infected patients in Lesotho. PLoS One 7(10):e46943
144. O'Donnell MR, Padayatchi N, Kvasnovsky C, Werner L, Master I, Horsburgh CR Jr (2013) Treatment outcomes for extensively drug-resistant tuberculosis and HIV co-infection. Emerg Infect Dis 19(3):416–424
145. Isaakidis P, Casas EC, Das M, Tseretopoulou X, Ntzani EE, Ford N (2015) Treatment outcomes for HIV and MDR-TB co-infected adults and children: systematic review and meta-analysis. Int J Tuberc Lung Dis 19(8):969–978
146. Brust JCM, Shah NS, Mlisana K et al (2017) Improved survival and cure rates with concurrent treatment for MDR-TB/HIV co-infection in South Africa. Clin Infect Dis
147. Abdool Karim SS, Naidoo K, Grobler A et al (2011) Integration of antiretroviral therapy with tuberculosis treatment. N Engl J Med 365(16):1492–1501
148. Havlir DV, Kendall MA, Ive P et al (2011) Timing of antiretroviral therapy for HIV-1 infection and tuberculosis. N Engl J Med 365(16):1482–1491
149. Blanc FX, Sok T, Laureillard D et al (2011) Earlier versus later start of antiretroviral therapy in HIV-infected adults with tuberculosis. N Engl J Med 365(16):1471–1481

150. Dooley KE, Kim PS, Williams SD, Hafner R (2012) TB and HIV Therapeutics: pharmacology research priorities. AIDS Res Treat 2012:874083
151. Diacon AH, Pym A, Grobusch M et al (2009) The diarylquinoline TMC207 for multidrug-resistant tuberculosis. N Engl J Med 360(23):2397–2405
152. Diacon AH, Pym A, Grobusch MP et al (2014) Multidrug-resistant tuberculosis and culture conversion with bedaquiline. N Engl J Med 371(8):723–732
153. Pym AS, Diacon AH, Tang SJ et al (2016) Bedaquiline in the treatment of multidrug- and extensively drug-resistant tuberculosis. Eur Respir J 47(2):564–574
154. Borisov SE, Dheda K, Enwerem M et al (2017) Effectiveness and safety of bedaquiline-containing regimens in the treatment of MDR- and XDR-TB: a multicentre study. Eur Respir J 49(5)
155. Bastard M, Huerga H, Hayrapetyan A, et al. Safety of multidrug-resistant tuberculosis treatment amongst patients receiving bedaquiline in a compassionate use programme in Armenia and Georgia. 48th Union World Conference on Lung Health 11–14 October, 2017; Guadalajara, Mexico (OA-163-13)
156. Skrahina A, Solodovnikova V, Vetushko D, et al. The use of bedaquiline to treat patients with multi- and extensively drug-resistant tuberculosis in Belarus. 48th Union World Conference on Lung Health 11–14 October, 2017; Guadalajara, Mexico (OA-189-13)
157. Schnippel K, Ndjeka N, Maartens G et al (2018) Effect of bedaquiline on mortality in South African patients with drug-resistant tuberculosis: a retrospective cohort study. In: Lancet Respir Med
158. Ismail NA, Omar SV, Joseph L et al (2018) Defining bedaquiline susceptibility, resistance, cross-resistance and associated genetic determinants: a retrospective cohort study. EBioMedicine 28:136–142
159. Gler MT, Skripconoka V, Sanchez-Garavito E et al (2012) Delamanid for multidrug-resistant pulmonary tuberculosis. N Engl J Med 366(23):2151–2160
160. von Groote-Bidlingmaier F, Patientia R, Sanchez E et al (2019) Efficacy and safety of delamanid in combination with an optimised background regimen for treatment of multidrug-resistant tuberculosis: a multicentre, randomised, double-blind, placebo-controlled, parallel group phase 3 trial. Lancet Respir Med 7(3):249–259
161. WHO position statement on the use of delamanid for multidrug-resistant tuberculosis. World Health Organization; 2018
162. Lomtadze N. Adverse events and serious adverse events among patients receiving MDR-TB treatment with bedaquiline and delamanid. 48th Union World Conference on Lung Health 11–14 October, 2017; Guadalajara, Mexico
163. Dooley KE, Rosenkranz SL, Conradie F, et al. QT Effects of Bedaquiline, Delamanid or both in MDR-TB Patients: The Deliberate Trial. In: Conference on Retroviruses and Opportunistic Infections 2019, Seattle, Washington; 2019
164. Conradie F, Diacon AH, Everitt D, et al. The NIX-TB trial of pretomanid, bedaquiline and linezolid to treat XDR-TB. Programs and Abstracts of the Conference on Retroviruses and Opportunistic Infections, Feb 13-17, 2017. Seattle, WA. Abstract #80LB. 2017
165. Conradie F, Diacon A, Howell P, et al. Sustained high rate of successful treatment outcomes: Interim results of 75 patients in the Nix-TB clinical study of pretomanid, bedaquiline and linezolid. . 49th Union World Conference on Lung Health The Hague, Netherlands 24–27 October 2018 Abstract OA03–213-25
166. Cox H, Ford N (2012) Linezolid for the treatment of complicated drug-resistant tuberculosis: a systematic review and meta-analysis. Int J Tuberc Lung Dis 16(4):447–454
167. Sotgiu G, Centis R, D'Ambrosio L et al (2012) Efficacy, safety and tolerability of linezolid containing regimens in treating MDR-TB and XDR-TB: systematic review and meta-analysis. Eur Respir J 40(6):1430–1442
168. Hughes J, Isaakidis P, Andries A et al (2015) Linezolid for multidrug-resistant tuberculosis in HIV-infected and -uninfected patients. Eur Respir J 46(1):271–274
169. Lee M, Lee J, Carroll MW et al (2012) Linezolid for treatment of chronic extensively drug-resistant tuberculosis. N Engl J Med 367(16):1508–1518

170. Wallis RS, Dawson R, Friedrich SO et al (2014) Mycobactericidal activity of sutezolid (PNU-100480) in sputum (EBA) and blood (WBA) of patients with pulmonary tuberculosis. PLoS One 9(4):e94462
171. Tasneen R, Betoudji F, Tyagi S et al (2015) Contribution of oxazolidinones to the efficacy of novel regimens containing bedaquiline and pretomanid in a mouse model of tuberculosis. Antimicrob Agents Chemother 60(1):270–277
172. Barry VC, Belton JG, Conalty ML et al (1957) A new series of phenazines (rimino-compounds) with high antituberculosis activity. Nature 179(4568):1013–1015
173. Gopal M, Padayatchi N, Metcalfe JZ, O'Donnell MR (2013) Systematic review of clofazimine for the treatment of drug-resistant tuberculosis. Int J Tuberc Lung Dis 17(8):1001–1007
174. Cholo MC, Mothiba MT, Fourie B, Anderson R (2017) Mechanisms of action and therapeutic efficacies of the lipophilic antimycobacterial agents clofazimine and bedaquiline. J Antimicrob Chemother 72(2):338–353
175. Dey T, Brigden G, Cox H, Shubber Z, Cooke G, Ford N (2013) Outcomes of clofazimine for the treatment of drug-resistant tuberculosis: a systematic review and meta-analysis. J Antimicrob Chemother 68(2):284–293
176. Moore VJ (1983) A review of side-effects experienced by patients taking clofazimine. Lepr Rev 54(4):327–335
177. Hartkoorn RC, Uplekar S, Cole ST (2014) Cross-resistance between clofazimine and bedaquiline through upregulation of MmpL5 in *Mycobacterium tuberculosis*. Antimicrob Agents Chemother 58(5):2979–2981
178. Hwang TJ, Dotsenko S, Jafarov A et al (2014) Safety and availability of clofazimine in the treatment of multidrug and extensively drug-resistant tuberculosis: analysis of published guidance and meta-analysis of cohort studies. BMJ Open 4(1):e004143
179. Gegia M, Winters N, Benedetti A, van Soolingen D, Menzies D. Treatment of isoniazid-resistant tuberculosis with first-line drugs: a systematic review and meta-analysis. Lancet Infect Dis 2017; 17(2): 223–234
180. Denkinger CM, Pai M, Dowdy DW (2014) Do we need to detect isoniazid resistance in addition to rifampicin resistance in diagnostic tests for tuberculosis? PLoS One 9(1):e84197
181. WHO treatment guidelines for isoniazid-resistant tuberculosis: supplement to the WHO treatment guidelines for drug-resistant tuberculosis. World Health Organization, Geneva, 2018
182. Controlled trial of 6-month and 9-month regimens of daily and intermittent streptomycin plus isoniazid plus pyrazinamide for pulmonary tuberculosis in Hong Kong. The results up to 30 months. Am Rev Respir Dis 1977; 115(5): 727–735
183. Shah NS, Pratt R, Armstrong L, Robison V, Castro KG, Cegielski JP (2008) Extensively drug-resistant tuberculosis in the United States, 1993–2007. JAMA 300(18):2153–2160
184. Falzon D, Gandhi N, Migliori GB et al (2013) Resistance to fluoroquinolones and second-line injectable drugs: impact on multidrug-resistant TB outcomes. Eur Respir J 42(1):156–168
185. Udwadia ZF, Amale RA, Ajbani KK, Rodrigues C (2012) Totally drug-resistant tuberculosis in India. Clin Infect Dis 54(4):579–581
186. Migliori GB, Sotgiu G, Gandhi NR et al (2013) Drug resistance beyond extensively drug-resistant tuberculosis: individual patient data meta-analysis. Eur Respir J 42(1):169–179
187. Velayati AA, Masjedi MR, Farnia P et al (2009) Emergence of new forms of totally drug-resistant tuberculosis bacilli: super extensively drug-resistant tuberculosis or totally drug-resistant strains in iran. Chest 136(2):420–425
188. Ndjeka N, Conradie F, Schnippel K et al (2015) Treatment of drug-resistant tuberculosis with bedaquiline in a high HIV prevalence setting: an interim cohort analysis. Int J Tuberc Lung Dis 19(8):979–985
189. Meintjes G (2014) Management of drug-resistant TB in patients with HIV co-infection. J Int AIDS Soc 17(4 Suppl 3):19508
190. Furin JJ, Mitnick CD, Shin SS et al (2001) Occurrence of serious adverse effects in patients receiving community-based therapy for multidrug-resistant tuberculosis. Int J Tuberc Lung Dis 5(7):648–655

191. Shin SS, Pasechnikov AD, Gelmanova IY et al (2007) Adverse reactions among patients being treated for MDR-TB in Tomsk, Russia. Int J Tuberc Lung Dis 11(12):1314–1320
192. Bloss E, Kuksa L, Holtz TH et al (2010) Adverse events related to multidrug-resistant tuberculosis treatment, Latvia, 2000–2004. Int J Tuberc Lung Dis 14(3):275–281
193. Torun T, Gungor G, Ozmen I et al (2005) Side effects associated with the treatment of multidrug-resistant tuberculosis. Int J Tuberc Lung Dis 9(12):1373–1377
194. Nathanson E, Gupta R, Huamani P et al (2004) Adverse events in the treatment of multidrug-resistant tuberculosis: results from the DOTS-Plus initiative. Int J Tuberc Lung Dis 8(11):1382–1384
195. Baghaei P, Tabarsi P, Dorriz D et al (2011) Adverse effects of multidrug-resistant tuberculosis treatment with a standardized regimen: a report from Iran. Am J Ther 18(2):e29–e34
196. Vega P, Sweetland A, Acha J et al (2004) Psychiatric issues in the management of patients with multidrug-resistant tuberculosis. Int J Tuberc Lung Dis 8(6):749–759
197. Brust JC, Shah NS, van der Merwe TL et al (2013) Adverse events in an integrated home-based treatment program for MDR-TB and HIV in KwaZulu-Natal, South Africa. J Acquir Immune Defic Syndr 62(4):436–440
198. Isaakidis P, Varghese B, Mansoor H et al (2012) Adverse events among HIV/MDR-TB co-infected patients receiving antiretroviral and second line anti-TB treatment in Mumbai, India. PLoS One 7(7):e40781
199. Shringarpure KS, Isaakidis P, Sagili KD, Baxi RK, Das M, Daftary A (2016) When treatment is more challenging than the disease: a qualitative study of MDR-TB patient retention. PLoS One 11(3):e0150849
200. Deshmukh RD, Dhande DJ, Sachdeva KS et al (2015) Patient and provider reported reasons for lost to follow up in MDRTB treatment: a qualitative study from a drug resistant TB Centre in India. PLoS One 10(8):e0135802
201. Jiang M, Karasawa T, Steyger PS (2017) Aminoglycoside-Induced Cochleotoxicity: A Review. Front Cell Neurosci 11:308
202. Schacht J (1999) Biochemistry and pharmacology of aminoglycoside-induced hearing loss. Acta Physiol Pharmacol Ther Latinoam 49(4):251–256
203. Modongo C, Sobota RS, Kesenogile B et al (2014) Successful MDR-TB treatment regimens including amikacin are associated with high rates of hearing loss. BMC Infect Dis 14:542
204. Klis S, Stienstra Y, Phillips RO, Abass KM, Tuah W, van der Werf TS. Long term streptomycin toxicity in the treatment of Buruli Ulcer: follow-up of participants in the BURULICO drug trial. PLoS Negl Trop Dis 2014; 8(3): e2739
205. Modongo C, Pasipanodya JG, Zetola NM, Williams SM, Sirugo G, Gumbo T (2015) Amikacin concentrations predictive of ototoxicity in multidrug-resistant tuberculosis patients. Antimicrob Agents Chemother 59(10):6337–6343
206. van Altena R, Dijkstra JA, van der Meer ME et al (2017) Reduced chance of hearing loss associated with therapeutic drug monitoring of aminoglycosides in the treatment of multidrug-resistant tuberculosis. Antimicrob Agents Chemother:**61**(3)
207. Seddon JA, Godfrey-Faussett P, Jacobs K, Ebrahim A, Hesseling AC, Schaaf HS (2012) Hearing loss in patients on treatment for drug-resistant tuberculosis. Eur Respir J 40(5):1277–1286
208. Dutta BS, Hassan G, Waseem Q, Saheer S, Singh A (2012) Ethionamide-induced hypothyroidism. Int J Tuberc Lung Dis 16(1):141
209. Satti H, Mafukidze A, Jooste PL, MM ML, Farmer PE, Seung KJ (2012) High rate of hypothyroidism among patients treated for multidrug-resistant tuberculosis in Lesotho. Int J Tuberc Lung Dis
210. Andries A, Isaakidis P, Das M et al (2013) High rate of hypothyroidism in multidrug-resistant tuberculosis patients co-infected with HIV in Mumbai, India. PLoS One 8(10):e78313
211. Lee SS, Lee CM, Kim TH et al (2016) Frequency and risk factors of drug-induced liver injury during treatment of multidrug-resistant tuberculosis. Int J Tuberc Lung Dis 20(6):800–805

212. Keshavjee S, Gelmanova IY, Shin SS et al (2012) Hepatotoxicity during treatment for multidrug-resistant tuberculosis: occurrence, management and outcome. Int J Tuberc Lung Dis 16(5):596–603
213. Saukkonen JJ, Cohn DL, Jasmer RM et al (2006) An official ATS statement: hepatotoxicity of antituberculosis therapy. Am J Respir Crit Care Med 174(8):935–952
214. World Health Organization. Policy guidance on drug-susceptibility testing (DST) of second-line antituberculosis drugs. 2008; WHO/HTM/TB/2008.392
215. Franke MF, Appleton SC, Mitnick CD et al (2013) Aggressive regimens for multidrug-resistant tuberculosis reduce recurrence. Clin Infect Dis 56(6):770–776
216. Ahmad Khan F, Gelmanova IY, Franke MF et al (2016) aggressive regimens reduce risk of recurrence after successful treatment of MDR-TB. Clin Infect Dis 63(2):214–220
217. Bastos ML, Hussain H, Weyer K et al (2014) Treatment outcomes of patients with multidrug-resistant and extensively drug-resistant tuberculosis according to drug susceptibility testing to first- and second-line drugs: an individual patient data meta-analysis. Clin Infect Dis 59(10):1364–1374
218. Moyo S, Cox HS, Hughes J et al (2015) Loss from treatment for drug resistant tuberculosis: risk factors and patient outcomes in a community-based program in Khayelitsha, South Africa. PLoS One 10(3):e0118919
219. Elliott E, Draper HR, Baitsiwe P, Claassens MM (2014) Factors affecting treatment outcomes in drug-resistant tuberculosis cases in the Northern Cape, South Africa. Public Health Action 4(3):201–203
220. Toczek A, Cox H, du Cros P, Cooke G, Ford N. Strategies for reducing treatment default in drug-resistant tuberculosis: systematic review and meta-analysis. Int J Tuberc Lung Dis 2013; 17(3): 299–307
221. Brust JC, Shah NS, Scott M et al (2012) Integrated, home-based treatment for MDR-TB and HIV in rural South Africa: an alternate model of care. Int J Tuberc Lung Dis 16(8):998–1004
222. Mitnick C, Bayona J, Palacios E et al (2003) Community-based therapy for multidrug-resistant tuberculosis in Lima, Peru. N Engl J Med 348(2):119–128
223. Heller T, Lessells RJ, Wallrauch CG et al Community-based treatment for multidrug-resistant tuberculosis in rural KwaZulu-Natal, South Africa. Int J Tuberc Lung Dis 14(4):420–426
224. Ho J, Byrne AL, Linh NN, Jaramillo E, Fox GJ (2017) Decentralized care for multidrug-resistant tuberculosis: a systematic review and meta-analysis. Bull World Health Organ 95(8):584–593
225. Weiss P, Chen W, Cook VJ, Johnston JC (2014) Treatment outcomes from community-based drug resistant tuberculosis treatment programs: a systematic review and meta-analysis. BMC Infect Dis 14:333
226. Okello D, Floyd K, Adatu F, Odeke R, Gargioni G (2003) Cost and cost-effectiveness of community-based care for tuberculosis patients in rural Uganda. Int J Tuberc Lung Dis 7(9 Suppl 1):S72–S79
227. Sinanovic E, Floyd K, Dudley L, Azevedo V, Grant R, Maher D (2003) Cost and cost-effectiveness of community-based care for tuberculosis in Cape Town, South Africa. Int J Tuberc Lung Dis 7(9 Suppl 1):S56–S62
228. Loveday M, Wallengren K, Brust J et al (2015) Community-based care vs. centralised hospitalisation for MDR-TB patients, KwaZulu-Natal, South Africa. Int J Tuberc Lung Dis 19(2):163–171
229. Loveday M, Padayatchi N, Wallengren K et al (2014) Association between health systems performance and treatment outcomes in patients co-infected with MDR-TB and HIV in KwaZulu-Natal, South Africa: implications for TB programmes. PLoS One 9(4):e94016
230. Perelman MI, Strelzov VP (1997) Surgery for pulmonary tuberculosis. World J Surg 21(5):457–467
231. Davies HM (1930) The use of surgery in pulmonary tuberculosis. Br Med J 1(3614):687–689
232. Schiffbauer HE (1930) Indications for surgery in pulmonary tuberculosis. Calif Western Med 32(4):245–248

233. Gravesen J, The surgery of pulmonary tuberculosis main principles (1936) Br Med J 2(3944):269–272
234. Kempker RR, Vashakidze S, Solomonia N, Dzidzikashvili N, Blumberg HM (2012) Surgical treatment of drug-resistant tuberculosis. Lancet Infect Dis 12(2):157–166
235. Xu HB, Jiang RH, Li L (2011) Pulmonary resection for patients with multidrug-resistant tuberculosis: systematic review and meta-analysis. J Antimicrob Chemother 66(8):1687–1695
236. Marrone MT, Venkataramanan V, Goodman M, Hill AC, Jereb JA, Mase SR (2013) Surgical interventions for drug-resistant tuberculosis: a systematic review and meta-analysis. Int J Tuberc Lung Dis 17(1):6–16
237. Fox GJ, Mitnick CD, Benedetti A et al (2016) Surgery as an adjunctive treatment for multidrug-resistant tuberculosis: an individual patient data metaanalysis. Clin Infect Dis 62(7):887–895
238. Harris RC, Khan MS, Martin LJ et al (2016) The effect of surgery on the outcome of treatment for multidrug-resistant tuberculosis: a systematic review and meta-analysis. BMC Infect Dis 16:262
239. WHO treatment guidelines for drug-resistant tuberculosis: 2016 update. WHO; 2016
240. Dheda K, Limberis JD, Pietersen E et al (2017) Outcomes, infectiousness, and transmission dynamics of patients with extensively drug-resistant tuberculosis and home-discharged patients with programmatically incurable tuberculosis: a prospective cohort study. Lancet Respir Med 5(4):269–281
241. National action plan to combat multidrug-resistant tuberculosis. MMWR Recomm Rep 1992; 41(Rr-11): 5–48
242. O'Donnell MR, Daftary A, Frick M et al (2016) Re-inventing adherence: toward a patient-centered model of care for drug-resistant tuberculosis and HIV. Int J Tuberc Lung Dis 20(4):430–434
243. Acha J, Sweetland A, Guerra D, Chalco K, Castillo H, Palacios E (2007) Psychosocial support groups for patients with multidrug-resistant tuberculosis: five years of experience. Glob Public Health 2(4):404–417
244. Khanal S, Elsey H, King R, Baral SC, Bhatta BR, Newell JN (2017) Development of a patient-centred, psychosocial support intervention for Multi-Drug-Resistant Tuberculosis (MDR-TB) Care in Nepal. PLoS One 12(1):e0167559
245. Tola HH, Shojaeizadeh D, Tol A et al (2016) Psychological and educational intervention to improve tuberculosis treatment Adherence in Ethiopia based on health belief model: a cluster randomized control trial. PLoS One 11(5):e0155147
246. Thomas BE, Shanmugam P, Malaisamy M et al (2016) Psycho-socio-economic issues challenging multidrug resistant tuberculosis patients: a systematic review. PLoS One 11(1):e0147397

Co-treatment of Tuberculosis and HIV: Pharmacologic Considerations

Ethel D. Weld, Alice K. Pau, Gary Maartens, and Kelly E. Dooley

Abstract Having HIV and TB worsens the impact of both. The treatment of HIV-TB coinfection is beset by challenges, including drug-drug-interactions, coincident toxicities, and the occurrence of the immune reconstitution inflammatory syndrome. These challenges can be overcome with careful attention to evidence-guided practice and clinical pharmacological aspects of co-treatment. There is a clear mortality benefit to treating both infections; the relative timing of initiation of both treatments will be discussed. This chapter will address pharmacologic considerations in the co-treatment of HIV-related latent or active TB of all sensitivity patterns (drug sensitive and multidrug resistant (MDR). The discussion will identify existing gaps in the evidence and include current recommendations for HIV-TB treatment in special populations, including pregnant and lactating women and children.

Keywords HIV · Tuberculosis · Co-management · Drug-drug-interactions · Pharmacology

E. D. Weld · K. E. Dooley (✉)
Divisions of Clinical Pharmacology & Infectious Diseases, Department of Medicine, Johns Hopkins University School of Medicine, Baltimore, MD, USA
e-mail: kdooley1@jhmi.edu

A. K. Pau
National Institute of Allergy and Infectious Diseases, National Institutes of Health, Bethesda, MD, USA

G. Maartens
Division of Clinical Pharmacology, Department of Medicine, University of Cape Town, Cape Town, South Africa

I. Sereti et al. (eds.), *HIV and Tuberculosis*,
https://doi.org/10.1007/978-3-030-29108-2_11

Introduction

Tuberculosis (TB) is currently a leading cause of infectious disease-related death globally, and it is also the #1 killer of persons living with HIV infection (PLWH) [1]. TB disease, though, can be averted in this population by provision of antiretroviral therapy (ART) and appropriate TB preventative therapy (TBPT) [2–4]. When TB does develop, treatment is largely similar among patients with and without HIV co-infection, consisting of four-drug therapy for drug-sensitive TB, namely isoniazid, a rifamycin, pyrazinamide, and ethambutol. Treatment, though, should be administered daily, rather than intermittently [5–8]; vitamin B6 should be administered as standard of care to prevent peripheral neuropathy given that HIV infection itself also confers risk; and ART drugs and doses must be selected with potential drug interactions kept in mind. TB drug exposures—especially rifampicin—may be reduced in patients with advanced HIV, due to malabsorption or weight-based dosing algorithms that under-dose low-weight persons [9, 10]. Therapeutic drug monitoring, where available, or, more importantly, higher dosing, may be of particular benefit in this population [11]. Management of drug-resistant TB, specifically multidrug resistant (MDR) TB (*Mycobacterium tuberculosis* that is resistant to rifampicin and isoniazid) is similar in patients with and without HIV co-infection, though there are notable drug interactions between ART agents and drugs used to treat MDR-TB, as well as toxicity concerns.

Co-treatment of HIV and TB is challenging, owing to high pill burden, overlapping toxicities, and immune reconstitution inflammatory syndrome [12], but there is a clear benefit to giving TB and HIV treatment concurrently [13–16], and so these challenges must and can be addressed. Clearly, coordination of HIV and TB care is of the utmost importance for successful treatment of both infections. In this chapter, we highlight pharmacologic considerations relevant to the co-management of latent or active TB (including drug-sensitive and MDR-TB) and HIV, identify knowledge gaps, and provide treatment recommendations based on the available evidence, with attention to considerations applicable to pregnant women and children [17, 18].

Prevention of HIV-Associated TB

While many opportunistic infections are only a concern when the CD4 count falls below a certain threshold, risk of TB among PLWH is heightened almost immediately following HIV infection, even when CD4 counts remain high [19–21]. TBPT not only reduces risk of TB disease but it also may lower the risk of death in individuals with HIV infection [22]. A recent meta-analysis of Isoniazid Preventive Therapy (IPT) showed no all-cause mortality benefit except among a subset of patients given an extended 12-month course of IPT (in whom RR of mortality = 0.65; 95% CI(0.47, 0.90)) [4]. And an older Cochrane review from 2010 found no evidence that TBPT reduced all-cause mortality as compared to placebo [23].

However, the TEMPRANO study, notably one of the few individually randomized studies of the impact of IPT on mortality in HIV-infected patients *in the era of ART*, showed a substantially reduced risk of death, which persisted for over 5 years, in the arm given IPT compared with no IPT (HR of death was 0·61 (0.39–0.94) after adjustment for baseline CD4 cell count and ART strategy) [22]. TBPT uptake, though, remains poor—the consequence is thousands of preventable deaths each year [24]. It is, thus, imperative that LTBI treatments be accessible, simple to take, and compatible with available ART.

How might regimens for LTBI be different in different regions and for HIV-infected persons on ART versus HIV-uninfected individuals? There are three main options endorsed by the World Health Organization (WHO) for treatment of LTBI in countries with low TB burden (estimated TB incidence <100 per 100,000 population)—isoniazid alone for 6–9 months (IPT), isoniazid plus rifapentine once-weekly for 12 weeks (3HP), or rifampicin alone for 4 months (4R) [25]. In high-burden countries, however, WHO guidelines recommend that people living with HIV (and their child contacts under the age of 5 years) must be offered IPT for a longer course of 36 months. This recommendation is based on evidence from the randomized, placebo-controlled BOTUSA trial of 6 months versus 36 months of IPT, which demonstrated a significant decrease in cases of incident TB in the group given IPT for 36 months as compared to 6 months, chiefly among cases with positive tuberculin skin test (TST) [26].

Isoniazid (9H) when given as IPT is dosed at 300 mg daily. Tolerability of IPT is excellent, though adverse events are modestly more common in HIV-infected than HIV-uninfected patients [27, 28], and WHO, accordingly, recommends pre-treatment liver function testing and on-treatment monthly clinical assessments for HIV-infected patients receiving IPT [25]. For the most part, isoniazid is compatible with antiretrovirals (ARVs). However, daily isoniazid may increase efavirenz exposures in a subset of patients, namely individuals with slow cytochrome P450 2B6 (CYP2B6) metabolizer status. The major metabolic pathway for efavirenz is CYP2B6. CYP2A6 is a minor metabolic pathway for efavirenz, but it is a more important route of clearance for patients with slow CYP2B6 metabolizer genotype. Isoniazid is an inhibitor of CYP2A6 and can, thus, increase efavirenz concentrations in this subpopulation, and the effect is more pronounced in individuals with high isoniazid concentrations (typically seen in patients with the slow NAT2 acetylator genotype) [29, 30]. The frequency, magnitude, and clinical relevance of this drug interaction remain to be quantified. However, being alert to this potential interaction is important, and clinicians should be on the lookout for central nervous system side effects such as insomnia, vivid dreams, paranoia, psychosis, and suicidality in efavirenz-isoniazid co-treated patients, particularly in populations or geographic areas where slow CYP2B6 metabolizer and/or slow NAT2 genotypes are common [31].

With *3HP* for TB prevention, isoniazid and rifapentine are given less frequently (once-weekly), shorter duration (3 months) and at higher doses (900 mg each) than for treatment of TB disease [32]. This regimen is better-tolerated than 9H in HIV-infected patients, and, interestingly, side effects are less common in persons with versus without HIV co-infection [27]. In the Phase 3 trial of 3HP, ART was not permitted, so subsequent drug interaction studies were required to characterize potential drug interactions of once-weekly HP with commonly-used ARVs. This regimen can be used with the following medications: tenofovir disoproxil fumarate (TDF) [33], emtricitabine, efavirenz [33, 34], and raltegravir [35] without dose adjustments. The effects of rifapentine (dosed once-weekly or daily) on tenofovir alefenamide (TAF) have not been assessed. In one small trial involving healthy HIV-uninfected volunteers, dolutegravir given with once-weekly HP was poorly-tolerated [36]. However, further exploration of this combination in a phase I/II study of patients with HIV and LTBI showed that coadministration of 3HP with DTG was well-tolerated and all participants maintained viral suppression despite a mildly reduced DTG trough while on 3HP. The geometric mean (GM) trough concentration of DTG on Day 58 (pre-HP) was 1003 μg/mL (5th–95th %ile: 500–2080), and during HP treatment was 546 (134–1616); of particular note, all trough levels but one were above the DTG IC90 of 64 μg/mL. Given rifapentine's strong induction effects on CYP3A, even once-weekly dosing is expected to reduce exposures of drugs that are CYP3A substrates, so co-administration of 3HP with boosted protease inhibitors or cobicistat-boosted elvitegravir or bictegravir is not recommended.

Moving the field further towards treatment shortening, a recently completed Phase 3 trial (NCT01404312) evaluated an ultra-short course of isoniazid and rifapentine given daily for 1 month (1HP) for the prevention of TB disease in individuals >13 years of age with HIV infection. Participants were not required to have a positive TST skin test or IGRA test. That study showed that 1HP was non-inferior to a 9-month regimen of daily INH (9H), with fewer adverse events and a higher treatment completion rate. Importantly, over a median follow-up period of 3.3 years, the ultra-short course 1HP regimen had no difference in incidence rates of active TB, TB death, or death from an unknown cause, when compared to 9H. Preliminary drug interaction substudies showed that efavirenz could be used with the once-daily regimen [34]. Raltegravir has been studied in combination with daily rifapentine in healthy volunteers, and the coadministration was well-tolerated and did not change the geometric mean of either RAL C_{max} or geometric mean AUC, though it did lower the RAL trough concentration by 41%. Information is still lacking on the combination of TAF and DTG with 1HP, and that combination is being studied by the AIDS Clinical Trials Group. Other drug-drug interaction studies with daily rifapentine are needed; some are on the launch pad.

Rifampicin (4R), is a potent inducer of metabolizing enzymes, and drug interactions for this prophylactic regimen would be expected to be similar to those seen with rifampicin given as part of multidrug therapy for TB disease (see below).

Treatment of Drug-Sensitive TB

Among the four drugs used in first-line treatment for drug-sensitive TB, rifamycins have unique activity against the *M. tuberculosis* bacilli that are slowly replicating or dormant, those so-called "persisters" that must be eradicated to achieve clinical cure [37]. Up to now, no other drugs have comparable, clinically-proven sterilizing activity.

If a rifamycin is not included in the regimen for the full 6 months of therapy, treatment duration must be prolonged significantly [38, 39]. Thus, rifampicin (or one of its rifamycin cousins) is an essential component of first-line regimens for drug-sensitive TB. Rifampicin, however, is a potent and promiscuous inducer of multiple metabolizing enzymes (including phase 1 enzymes, like cytochrome P450 oxidases, and phase 2 enzymes, e.g. transferases such as UDP-glucuronosyltransferases and sulfotransferases) and drug transporters (such as P-glycoprotein, or P-gp) [40]. The use of rifampicin can cause clinically-meaningful drug interactions with companion drugs, including, most importantly, ARVs. There are two main strategies for managing the drug interaction – (1) use standard isoniazid, rifampicin, pyrazinamide, and ethambutol therapy (HRZE) and choose ART agents that either do not have clinically important drug interactions with rifampicin or that have a drug interaction that can be mitigated with adjustment of dose or dosing frequency; or (2) substitute rifabutin, a less potent inducer of metabolizing enzymes and transporters, for rifampicin, to enable use of ARVs that cannot be used with rifampicin (Table 1). Multiple groups publish guidelines regarding HIV-TB co-treatment that are regularly updated and may serve as good references, as this is a rapidly evolving field (https://www.hiv-druginteractions.org/; https://www.bhiva.org/; https://aidsinfo.nih.gov/guidelines; https://www.cdc.gov/hiv/guidelines/index.html).

Table 1 Recommended and alternative therapies for co-treatment of drug-sensitive TB and HIV infections in adults

Antiretroviral medication[a]	Metabolizing enzymes	Rifamycin[b]	Dose adjustments	Comments
Preferred				
Efavirenz	CYP2B6 > CYP2A6	Rifampicin	None	Some patients may experience elevations in efavirenz concentrations; monitor for CNS side effects
Raltegravir	UGT1A1	Rifampicin	Increase raltegravir to 800 mg twice daily	Raltegravir 400 mg twice daily has been studied in 1 RCT and may be sufficient but clinical experience is too limited to recommend that dosing at this time
Dolutegravir	UGT1A1 > CYP3A	Rifampicin	Increase dolutegravir to 50 mg twice daily[c]	Continue twice daily dolutegravir dosing for 10–14 days following completion of TB treatment
Ritonavir-boosted protease inhibitor	CYP3A	Rifabutin	Decrease rifabutin to 150 mg once daily	Monitor closely for exposure-dependent rifabutin toxicities, to include neutropenia and uveitis

(continued)

Table 1 (continued)

Antiretroviral medication[a]	Metabolizing enzymes	Rifamycin[b]	Dose adjustments	Comments
Alternative				
Efavirenz	CYP2B6 > CYP2A6	Rifabutin	Increase rifabutin dose to 450 mg daily	If efavirenz is used, the rifamycin of choice is rifampicin Rifabutin is an alternative in select cases where resistance or tolerability is drug- and not class-specific
Nevirapine	CYP2B6, CYP3A	Rifampicin	Give twice-daily throughout co-treatment (avoid the once daily lead-in phase)	Less effective than efavirenz-based treatment, close monitoring of HIV viral load is required Watch for hepatotoxicity when given with first-line TB drugs
Nevirapine	CYP2B6, CYP3A	Rifabutin	No dose adjustments needed	Less effective than efavirenz-based treatment, close monitoring of HIV viral load is required Watch for hepatotoxicity when given with first-line TB drugs
Raltegravir	UGT1A1	Rifabutin	No dose adjustments needed	Limited clinical experience with this combination
Rilpivirine	CYP3A	Rifabutin	Increase rilpivirine dose to 50 mg once daily	Discordance between Prescribing Information from the EMA (double rilpivirine dose) and US FDA (coadministration contraindictated)
Dolutegravir	UGT1A1 > CYP3A	Rifabutin	No dose adjustments needed	Limited clinical experience with this combination
Ritonavir-boosted lopinavir	CYP3A4	Rifampicin	Double dose (of both lopinavir and ritonavir)	Alternative if rifabutin is not available. Monitor carefully for drug-induced liver injury Double dosing has only been examined for lopinavir/ritonavir

(continued)

Table 1 (continued)

Antiretroviral medication[a]	Metabolizing enzymes	Rifamycin[b]	Dose adjustments	Comments
Etravirine	CYP3A, CYP2C9, CYP2C19; UGT	Rifabutin	No dose adjustments needed	Limited clinical experience with this combination. Do not use in combination with boosted protease inhibitor
Maraviroc	CYP3A4, CYP3A5	Rifabutin	No dose adjustments needed	Limited clinical experience with this combination.
Maraviroc	CYP3A4, CYP3A5	Rifampicin	600 mg orally once daily	Limited clinical experience

[a]Accompanied by two nucleoside or nucleotide reverse transcriptase inhibitors (NRTI) (tenofovir disoproxil fumarate (TAF is not recommended until PK data become available) or abacavir plus emtricitabine or lamivudine)

[b]As part of multidrug treatment for TB including isoniazid, pyrazinamide, and ethambutol

[c]INSTI mutations must be carefully screened for clinically based on ART history, and virologically where possible; in cases of BID DTG dosing because of prior virologic failure on INSTI, the dosing of DTG with RIF has not yet been clearly established

Rifampicin-Based TB Treatment

Nucleoside/nucleotide reverse transcriptase inhibitors (NRTI). In general, these agents, including tenofovir (as tenofovir disoproxil fumarate or TDF), zidovudine, lamivudine, abacavir, and emtricitabine do not have clinically important drug interactions with HRZE and can be used without dose modifications [41–43]. Tenofovir alafanamide, or TAF, is a new, lower-dose, more potent formulation of tenofovir which appears to have lower risk of bone or kidney toxicities than TDF [44]. It is also active against hepatitis B virus (HBV) and, like TDF, is recommended for persons co-infected with HIV and HBV. TAF and TDF are pro-drugs; both are converted to the active compound, tenofovir diphosphate (TFV-DP) in lymphoid cells, but TAF achieves intracellular TFV-DP concentrations approximately fivefold higher than TDF. TAF is a P-gp substrate, and, unlike TDF, is also a minor substrate of CYP3A4, as well as a substrate of drug transporters such as BCRP, ABCG2, and ABCB1. TAF therefore appears to have higher drug interaction liability than TDF, as evidenced by drug interaction studies with cobicistat (inhibitor of CYP3A and P-gp) and carbamazepine (inducer of CYP3A) [45]. Given that rifampicin induces both P-gp and CYP3A4, there is concern that absorption of TAF will be significantly reduced when it is given with HRZE. Data from Gilead (the manufacturer of TAF) demonstrate that, when given in conjunction with RIF 600 mg daily, TAF 25 mg dosed every 12 h instead of daily is well tolerated and yields similar plasma TAF levels and intracellular PBMC TFV-DP levels as TAF 25 mg dosed once daily without RIF [46]. When TAF-FTC is dosed 25 mg–200 mg once daily with 600 mg of RIF, emtricitabine exposures are not affected but TAF exposures are lowered; the geometric mean ratios (90% CI) of plasma TAF with and without RIF are 0.45 (0.42–0.50) and 0.46 (0.40–0.52) for C_{min} and AUC_{0-24}, respectively. However, intracellular TFV-DP concentrations decrease by 36%; the geometric mean ratio (90% CI) of intracellular active metabolite TFV-DP in the setting of TAF-FTC given with as compared to without RIF is 0.64 (0.54–0.75). It is important to note, however, that these decreased intracellular concentrations are still higher than the intracellular concentrations achieved by standard TDF (GMR (90% CI) of intracellular TFV-DP AUC_{0-24} in the setting of TAF-FTC given *with* RIF as compared to TDF dosed normally *without* RIF is 4.21 (2.98–5.95)). It is likely that achieving with standard dosed TAF the (fivefold higher) intracellular levels of TFV-DP comparable to levels achieved clinically with TDF will be adequate for clinical effect, though that has not been proven. Given that TAF is now a first-line ART agent in many settings, further characterization of the pharmacokinetics and pharmacodynamics of TAF when it is given with TB therapy [specifically the effects of HRZE on intracellular TFV-DP and HIV-1 (and HBV, if co-infected) virologic suppression] in patients with HIV-associated TB is a high priority (Table 2).

Non-nucleoside reverse transcriptase inhibitors (NNRTI). Efavirenz is still first-line treatment for HIV infection in many settings globally. Initial small drug interaction studies involving rifampicin and efavirenz were conducted in healthy HIV-uninfected volunteers without concomitant HZE and without waiting the >4 weeks required for EFV to be fully auto-induced. These demonstrated moderate reductions in efavirenz exposures [47], yet both cohort studies and clinical trials among HIV-TB co-infected patients have demonstrated that the standard adult efavirenz dose of 600 mg daily (given with 2 NRTIs) is highly efficacious in patients receiving full TB treatment with HRZE [48–51]. As noted above, efavirenz exposures are known to be highly variable and correlate with CYP2B6 metabolizer genotype (extensive, intermediate, or slow). Among individuals with slow CYP2B6 metabolizer genotype, treatment with HRZE actually reduces efavirenz clearance (isoniazid's concentration-dependent inhibition of the alternative CYP2A6 pathway counterbalances the inductive effects of rifampicin), resulting in higher drug exposures [52, 53]. There is currently a movement to reduce the standard dose of efavirenz to 400 mg daily to lessen the cost of manufacture of this drug and expand access, as well as reducing efavirenz-associated side effects [54]. A randomized controlled trial, ENCORE1, demonstrated that the 400 mg dose was non-inferior to a 600 mg dose in treatment-naïve patients with HIV infection who were not receiving rifampicin for TB treatment [55]. Whether or not this 400 mg dose can be safely given to patients who are extensive metabolizers of efavirenz and require HRZE for TB treatment has been explored in an open-label study among people with HIV without TB. The study found that the coadministration of INH/RIF and 400 mg daily efavirenz led to minimal changes (<25%) in EFV exposures, and efavirenz concentrations in the range observed in the ENCORE-1 trial. Eighteen % of the subjects in the trial were carriers of slow efavirenz CYP2B6 metabolizer alleles 516 T or 983C; 32% were intermediate metabolizers, and 45% were extensive metabolizers. Of the 26 individuals who were enrolled in the trial, 4 discontinued

Table 2 Knowledge gaps in pharmacokinetics and safety of drugs or regimens used for treatment of latent tuberculosis (TB) infection or TB disease, co-administered with antiretroviral therapy (ART), among people living with HIV infection

Research area	Drug(s) for treatment or prevention of TB	Knowledge gaps: TB agent(s) with ART
Co-treatment of latent TB and HIV	Once-weekly isoniazid plus rifapentine (3HP)	• Safety, rifapentine dosing with dolutegravir, boosted PIs, ETR, EVG/c, BIC • Tenofovir alefenamide (TAF) dosing
	Daily high-dose isoniazid plus rifapentine (1HP)	• Safety, rifapentine dosing with dolutegravir, boosted PIs, ETR, EVG/c, BIC • Tenofovir alefenamide (TAF) dosing
	Daily isoniazid (9H)	• Risk of CNS side effects when 9H is given to patients on efavirenz who are CYP2B6 slow metabolizers
Co-treatment of drug-sensitive TB and HIV	Rifampicin	• PK/dosing of TAF • PK and efficacy of efavirenz 400 mg among CYP2B6 extensive metabolizers • PK of efavirenz with regimens including high-dose rifampicin (e.g. 35 mg/kg) • Population-level efficacy of standard-dose raltegravir (400 mg twice daily) • Dolutegravir dosing with standard or high-dose rifampicin • PK and efficacy of maraviroc given at increased dose of 600 mg twice daily • Potential use (through dose adjustment) of cobicistat- or ritonavir-boosted once-daily protease inhibitors
	Rifabutin	• PK/dosing of TAF • Toxicodynamics of rifabutin and its metabolite for bone marrow suppression and uveitis, when rifabutin is given at 150 mg once daily with a boosted protease inhibitor
	Rifapentine	• PK/dosing of TAF, efavirenz, raltegravir, dolutegravir, boosted PI, EVG/c, BIC with regimens including high-dose rifapentine (e.g. 1200 mg)
Co-treatment of drug-resistant TB and HIV	Moxifloxacin	• Effects of efavirenz on moxifloxacin pharmacokinetics and efficacy
	Ethionamide	• Effects on efavirenz PK among slow CYP2B6 metabolizers
	High-dose isoniazid	• Effects on efavirenz concentrations among slow CYP2B6 metabolizers
	Bedaquiline	• QT risk when co-administered with cobicistat- and ritonavir-boosted protease inhibitors • PK and safety with dose-adjusted efavirenz

(continued)

Table 2 (continued)

Research area	Drug(s) for treatment or prevention of TB	Knowledge gaps: TB agent(s) with ART
TB/HIV co-treatment in pregnant women, special considerations	Rifampicin	• DDI with ART in pregnancy
	Pyrazinamide	• PK and safety in pregnancy
	Rifapentine + INH	• PK/ dosing of both RPT and ART (e.g. TAF, efavirenz, raltegravir, dolutegravir); best timing for and safety of this regimen compared with alternatives
	Moxifloxacin/ levofloxacin	• PK and safety in pregnancy
	Clofazamine	• Role in treatment-shortening regimen in pregnancy • PK and safety in pregnancy
TB/HIV co-treatment in children, special considerations	Rifapentine	• PK, dosing (especially in very young children), acceptability and safety with TAF, raltegravir, dolutegravir, etravirine, bictegravir
	Rifampin, higher dose	• PK, dosing, acceptability and safety with LPV/RTV, EFV, TAF, raltegravir, dolutegravir, etravirine,
	Rifabutin	• PK and dosing for children, dose-exposure-toxicity relationships (which may differ from adults) • Pediatric-friendly palatable formulation

participation because of low efavirenz C_{24} levels below the cutoff of 800 ng/mL at >3 consecutive visits (of those four, 3 were extensive metabolizers and 1 was an intermediate metabolizer). These findings remain to be confirmed in individuals with both HIV and TB.

Theoretically, the additive inductive effects of rifampicin when combined with efavirenz (which induces its own metabolism in a concentration- and time-dependent fashion) could be most discernible in patients with the extensive metabolizer genotype, as these patients generally have the lowest efavirenz concentrations. However, it is worth noting that rifampicin does not significantly add to EFV auto-induction when the standard 600 mg daily dose of efavirenz is given. The target trough concentration for efavirenz remains ill-defined but is likely below the commonly-cited 1 mcg/mL [56], and the efficacy of the 400 mg dose in TB-HIV co-infected patients cannot be deduced by pharmacokinetic (PK) studies alone [57, 58]. Thus, for individuals taking the 400 mg daily dose at the time of their TB diagnosis and the initiation of a rifampicin-based regimen, the efavirenz dose should either be increased to 600 mg daily or there should be close monitoring of virologic outcomes. For those who start EFV-based cART while taking RHZE, a starting dose of 600 mg should be considered. Efavirenz pharmacokinetics in people with HIV-associated TB disease receiving short-course high-dose (1200 mg) daily rifapentine are not altered in a clinically significant way; a study of this combination found efavirenz concentrations above the currently recommended trough of 1 mcg/mL in 91% of participants at week 8. When efavirenz (600 mg or 800 mg) was studied in combination with high-dose RIF (20 mg/kg) in HIV-TB coinfected individuals in the ANRS 12292 Rifavirenz phase 2 trial, there was a trend towards lower efavirenz concentrations than with the same dose of efavirenz alone, but concentrations were still within the therapeutic window, and participants remained virologically suppressed.

DDI studies reveal that concomitant rifampicin reduces nevirapine levels by anywhere from 30–55% [59–61]. Nevirapine is less effective than efavirenz in combination ART regimens, including in TB-HIV co-treatment [62–65]; thus, its use should be avoided in patients being treated concurrently for TB and HIV. If options are few and nevirapine must be used, given that rifampicin further induces the enzymes that are auto-induced by and metabolize nevirapine (CYP2B6 and CYP3A4), the lead-in phase of 200 mg once daily should be dropped, and nevirapine should be given twice daily throughout co-treatment [66, 67]. Rilpivirine's area under the concentration-time curve (AUC) and trough concentrations are reduced by 80% and 89%, respectively, by rifampicin, so the two drugs should not be co-administered [68]. Though the effects of rifampicin on etravirine pharmacokinetics have not been tested, rifampin is predicted to reduce etravirine exposures significantly, so these drugs should not be used together [69]. Doravirine's AUC is reduced by 88% when administered with rifampin, so co-administration is not recommended.

Integrase strand transfer inhibitors (INSTI). Raltegravir, the first-in-class INSTI, is metabolized by UDP-glucuronosyltransferase 1A1 (UGT1A1), a phase 2 enzyme that is induced by rifampicin. In healthy volunteers, rifampicin reduced raltegravir trough concentrations 60% [70], and in patients with TB receiving HRZE, raltegravir trough concentrations were diminished by about 40% [71, 72]. In REFLATE, a three-arm phase 2 non-comparative clinical trial that randomized participants to receive 2 NRTIs plus standard-dose raltegravir (400 mg twice daily), double-dose raltegravir (800 mg twice daily), or efavirenz, virologic suppression was achieved in 76%, 78%, and 63% of participants, respectively, by 24 weeks [73]. While these results are encouraging and suggest that raltegravir could be used with HRZE at standard doses without dose adjustment, particularly in light of early HIV raltegravir monotherapy trials demonstrating that 200 mg twice daily and 400 mg twice daily had similar virologic activity [74], the sample size in the REFLATE trial was small, and experience with the 400 mg twice daily dose in patients receiving rifampicin is limited, so it is still not clear if the 400 mg twice daily dose will be adequate on a population level. A follow-up Phase 3 trial with adequate power to detect inferiority is planned. Dolutegravir, like raltegravir, is mainly metabolized by UGT1A1, but CYP3A is also a minor metabolic pathway (unlike raltegravir). Evidence from a drug interaction study conducted among healthy HIV-uninfected volunteers suggests that dolutegravir dosing should be increased from 50 mg once daily to 50 mg twice daily when given with rifampicin-containing TB treatment [75]; results from a PK study of RIF in combination with dolutegravir 50 mg daily versus 100 mg daily are also forthcoming. In addition, a trial of dolutegravir-based ART among ART-naïve patients with HIV-associated TB in which dolutegravir is dosed at 50 mg twice daily during and for two weeks after completing standard first-line TB treatment among patients with HIV-TB co-infection is completed. Results at 24 weeks were promising, with PK targets reached and rapid virologic response; further, there were no IRIS or drug toxicity events necessitating discontinuation, no deaths, and no acquired resistance. Final 48-week results are as yet unpublished but similarly encouraging (NCT02178592), with 75% (95% CI 65–86) and 82% (95% CI 70–93) viral suppression at 48 weeks in the BID DTG and EFV arms, respectively, and the lower response rate in the DTG arm driven by non-treatment-related snapshot failures, such as loss to follow-up and protocol deviations. Elvitegravir is a primarily a CYP3A substrate, and its concentrations are expected to be markedly reduced when it is given with rifampicin, even in the presence of cobicistat as a boosting agent [76], thus, is not recommended in patients receiving a rifamycin-based regimen. Similarly, bictegravir AUC was lowered 75% by concomitant RIF administration [77]. In a study among HIV-uninfected healthy volunteers, giving bictegravir twice daily (as Biktarvy, which contains 50 mg of bictegravir, plus TAF and emtricitabine) with rifampicin failed to mitigate the drug interaction, as trough concentrations were still reduced by 80% compared to Biktarvy given alone. Bictegravir is not recommended with rifampicin-containing TB treatment at this time.

Protease inhibitors (PI). Protease inhibitors, including those currently in clinical use—darunavir, atazanavir, and lopinavir—are all administered with pharmacoenhancing agents that reduce drug clearance, thus boosting drug concentrations, increasing serum half-life, and allowing for less-frequent dosing. Cobicistat is licensed for use with darunavir, atazanavir, and elvitegravir. Ritonavir and cobicistat are both suicide inhibitors of CYP3A, but they differ in some important ways [78]. While both inhibit CYP3A, P-gp and CYP2D6, ritonavir has more off-target activity than cobicistat; specifically, it induces CYP1A2, CYP2C9, CYP2C19, CYP2B6 and UGT enzymes and is a mixed inducer/inhibitor of CYP3A. Ritonavir also inhibits CYP2C19, CYP2C8, and CYP2C9. With regards to use of boosted protease inhibitors together with first-line TB treatment, all clinical assessments have been conducted among individuals taking ritonavir as the PI boosting agent; whether or not dose adjustment can safety mitigate drug interactions between rifampicin and cobicistat-boosted PIs remains to be tested clinically, though *in vitro* data from a human hepatocyte model suggests that both RTV or COBI (but RTV more potently than COBI) can overcome RIF-induced elevations in the clearance of DRV [79].

Rifampicin severely diminishes concentrations of standard-dose ritonavir-boosted lopinavir (by 80–90%), reductions significant enough to compromise treatment efficacy [80–82]. Doubling the boosted lopinavir dose increases exposures, but this strategy increases pill burden and may also increase risk of liver injury and poor GI tolerance, and the safety of this strategy has not been established in large cohorts [83–86]. A trial to assess the safety and PK of double-dose atazanavir was stopped early due to undue toxicity; double-dose boosted darunavir has not yet been studied. Giving a double-dose boosted PI with HRZE is only recommended when there are no other options (particularly rifabutin), and close monitoring for hepatotoxicity is required. Super-boosting (quadrupling the ritonavir component of the regimen to achieve 1:1 PI/ritonavir doses, e.g. lopinavir/ritonavir 400/400 twice daily) has been associated with high risk of symptomatic hepatitis, gastrointestinal upset, and treatment discontinuation in adults with HIV-TB [87], and it is not recommended in this population.

Other ART agents. Maraviroc is a CYP3A substrate. In a Phase 1 clinical trial conducted among healthy HIV-uninfected subjects, rifampicin reduced maraviroc concentrations by approximately 70%, but increasing the dose from 300 mg twice daily to 600 mg twice daily effectively compensated for this drug interaction [88]; this dose adjustment has not been tested in patients being co-treated for HIV and TB. Rifampicin does not appear to impact enfuvirtide or ibalizumab pharmacokinetics, so these drugs can be used with first-line TB treatment if one is really in a 'pinch' [89].

High-dose rifamycins. Rifampicin is typically given at a dose of 10 mg/kg (or 600 mg) daily, but higher doses hold promise for shortening the TB treatment duration, as demonstrated in a recent Phase 2 trial that evaluated doses as high as 35 mg/kg [90]. The incremental increase in induction of metabolizing enzymes and transporters that may result with higher doses of rifampicin has not been measured in clinical studies, though some drug interaction studies are underway (e.g. RIFAVIRENZ, NCT01986543). Results from this study revealed a trend towards lower EFV levels with the 20 mg/kg RIF dosing, but concentrations that remained in the therapeutic window, and comparable TB culture conversion rates and virologic control between arms. Rifapentine is a rifamycin antibiotic that has a longer half-life and lower minimum inhibitory concentration (MIC) against *M. tuberculosis* than rifampicin [91], and it is being tested in shortened (4-month) regimens for TB in a Phase 3 trial (NCT02410772). The right dosing of ARVs (namely efavirenz, TAF, raltegravir, dolutegravir) with high-dose rifapentine, a potent inducer of the same metabolizing enzymes and transporters as rifampicin [92], is still being established.

Rifabutin-Based TB Treatment

Rifabutin is a much less potent inducer of metabolizing enzymes than rifampicin or rifapentine [93]. Its efficacy for treatment of TB is thought to be comparable to that of rifampicin, though there is insufficient evidence to support this assertion, especially among HIV-infected patients [94]. It is often substituted for rifampicin in the TB treatment regimen to avoid drug-drug interactions in patients who require companion drugs that are simply incompatible with rifampicin. Rifabutin, though, is not available in fixed dose combinations with other first-line TB drugs, and access in some settings is limited by cost and availability. Rifabutin is more lipid-soluble than rifampicin; it has a large volume of distribution with low plasma concentrations and high tissue-to-plasma ratios. Uniquely among rifamycins, rifabutin is a CYP3A substrate, so it can be both the victim and the perpetrator of drug-drug interactions when it is given with CYP3A inducers or inhibitors [95]. Compared with other rifamycins, its therapeutic margin is narrow; it has a distinct side effect of uveitis [96], and significant neutropenia occurs more commonly than with rifampicin.

Protease inhibitors. Rifabutin has no effect on concentrations of ritonavir-boosted protease inhibitors. It is likely that rifabutin's impact on the pharmacokinetics of cobicistat-boosted protease inhibitors is similarly negligible, though this has not been studied expressly. Boosted protease inhibitors, via their inhibition of CYP3A, increase concentrations of rifabutin and its metabolites markedly, and, consequently, dose adjustment of rifabutin is required to avoid concentration-dependent toxicities, specifically uveitis and bone marrow suppression [97]. The standard dose of rifabutin is 300 mg once daily; there is some debate regarding the appropriate dose when this drug is co-administered with a boosted protease inhibitor. Giving the drug at 150 mg thrice-weekly was previously common practice, but this dose produces subtherapeutic concentrations in many patients and has been associated with relapse and acquired rifamycin drug resistance [7, 98–102]. Small PK and safety studies in Vietnam and South Africa have demonstrated that dose of 150 mg daily with a boosted PI produces rifabutin exposures similar to 300 mg daily given alone [103, 104], so daily rather than thrice-weekly dosing is now recommended in most guidelines. However, plasma concentrations of 25-O-desacetyl rifabutin (which has some antimycobacterial effect and contributes to toxicity) are increased markedly (5- to 15-fold) when rifabutin is given at a dose of 150 mg once daily with a boosted PI versus rifabutin 300 mg given alone [103, 104]; the implications for risk of adverse events are not yet clear. Experience, thus, remains limited with the currently-recommended dose of 150 mg once daily, and close monitoring for neutropenia or uveitis is required. A trial of double-dose lopinavir-ritonavir with rifampicin-based TB treatment versus standard-dose lopinavir-ritonavir with (daily) rifabutin-based TB treatment is ongoing with results expected in 2018 (NCT01601626).

Non-Nucleoside Reverse Transcriptase Inhibitors (NNRTI)

Efavirenz reduces rifabutin concentrations by inducing CYP3A; the rifabutin dose must be increased to 450 mg daily if these drugs are used together; it is better to use rifampicin than rifabutin with efavirenz. Nevirapine increases rifabutin and des-rifabutin concentrations very modestly (around 30%), but the magnitude of the drug interaction is highly variable among individuals, and so caution and close monitoring are necessary when co-administering these medications (Viramune package insert). The rifabutin dose that should be used with NVP is 300 mg daily [105]. Rifabutin reduces rilpivirine drug concentrations; a dose increase to 50 mg daily might overcome this interaction, but this adjustment overshoots in some patients and may have QT risk, particularly when the patient is taking other QT prolonging drugs [106]. The reductions (30–40%) in etravirine AUC and trough concentrations caused by rifabutin are not considered to be clinically significant [107]. Rifabutin reduces doravirine drug concentrations by 50%, when co-administered, doravirine dose should be increased from 100 mg once daily to 100 mg twice daily.

Integrase Strand Transfer Inhibitors (INSTIs)

Rifabutin can be given with raltegravir or dolutegravir without dose adjustments [75, 108]. When elvitegravir/cobicistat was given with rifabutin in one study, the elvitegravir trough was lowered by 67% (GMR 32.9; 90% CI 26.9–40.1), rifabutin levels were unaffected, and levels of the rifabutin metabolite 25-O-desacetylrifabutin were increased up to sixfold. Thrice-weekly rifabutin dosing with elvitegravir may be an option, but it has not been studied in patients with TB-HIV co-infection, and there is likely still to be a substantial decrease in elvitegravir concentrations [109]. As previously mentioned, rifabutin, as a modest CYP3A4/Pgp inducer, modestly reduces bictegravir AUC (38%) but the clinical significance of this remains to be explored [77].

Other ARVs

Results from an industry-sponsored maraviroc-rifabutin drug interaction study in healthy HIV-uninfected subjects are pending (RIFAMARA; NCT01894776). NRTIs can be given with rifabutin without dose adjustments with one exception—rifabutin induces P-gp, and the effects of this induction on the pharmacokinetics and pharmacodynamics of TAF have not been explored; currently, this combination is contraindicated (Descovy package insert).

Other Considerations

With HIV-TB co-treatment, pill fatigue, adverse events, and complications of immune reconstitution are all risks. Overlapping toxicities between HIV and first-line TB medications are a concern and include, but are not limited to, hepatotoxicity, gastrointestinal upset, rash, and peripheral neuropathy. Coordination of HIV-TB care is a must, to ensure that medications are dosed correctly, side effects are properly attributed and managed, and medication changes are communicated across care teams, particularly when dose adjustments to either the ART or the TB regimen are in place to compensate for drug interactions.

Treatment of Drug-Resistant TB

MDR-TB is treated with multiple drugs, and therapy is given for a prolonged duration. Standard treatment for MDR-TB is 18–24 months and consists of a fluoroquinolone (moxifloxacin or levofloxacin), an injectable agent (amikacin, kanamycin, or capreomycin), two core second-line agents (among ethionamide, cycloserine, linezolid, and clofazimine), and add-on drugs, typically pyrazinamide with or without ethambutol. A new 'short-course' MDR regimen was introduced in WHO recommendations in 2016 and consists of an intensive phase (4–6 months) of high-dose isoniazid, ethionamide, injectable, moxifloxacin, ethambutol, pyrazinamide, and clofazimine followed by a continuation phase (5 months) of moxifloxacin, ethambutol, pyrazinamide, and clofazimine [110]. Two new drugs have been registered for MDR-TB treatment in recent years: bedaquiline or delamanid can be added to standard background therapy if needed to complete a fully-active regimen or if side effects preclude use of standard second-line TB drugs. Because MDR-TB treatment regimens do not include rifampicin, drug interaction liability is much lower with MDR-TB regimens than with first-line TB treatment. MDR-TB has high mortality, especially among patients with advanced HIV, so rapid initiation of ART may be especially important in this group [111, 112]. Overlapping toxicities, though, present a challenge as does the sizeable pill burden.

Second-line drugs—Metabolic drug interactions between ARVs and second-line drugs that are clinically relevant are few, and most are not well characterized. They include the following: efavirenz may reduce moxifloxacin concentrations (mechanism unknown) [113], isoniazid may boost efavirenz concentrations, particularly in slow CYP2B6 metabolizers (see above), and ethionamide may cause additive hepatotoxicity when combined with efavirenz and nevirapine [114] . Injectables can cause renal dysfunction which can, in turn, affect the pharmacokinetics of renally cleared drugs such as tenofovir disoproxil fumarate (TDF). Lamivudine and emtricitabine are also renally cleared, but the effects of slightly higher levels do not tend to include any significant toxicities. Shared and potentially additive toxicities include the following: peripheral neuropathy (high-dose isoniazid, linezolid, stavudine, HIV disease); central nervous system (CNS) side effects (efavirenz, dolutegravir, high-dose isoniazid, cycloserine, ethionamide, fluoroquinolones); QT prolongation (fluoroquinolones, clofazimine, efavirenz, rilpivirine, bedaquiline, and delamanid (see below)), bone marrow suppression (linezolid, zidovudine, trimethoprim-sulfamethoxazole); hepatotoxicity (efavirenz, nevirapine, high-dose isoniazid, pyrazinamide), nausea and vomiting (ritonavir-boosted PIs, ethionamide, PAS), and nephrotoxicity (injectables and TDF or TAF).

New drugs—Bedaquiline is a first-in-class anti-TB antibiotic that works by inhibiting mycobacterial ATP synthase. Bedaquiline is a CYP3A substrate, and within the range of clinically-achievable exposures, its activity is exposure-dependent [115]. Bedaquiline is highly protein-bound, and its terminal half-life is exceedingly long, approximately 5 months [116]. Overall, bedaquiline is well-tolerated. Its main toxicity is QT prolongation. Drug interactions of concern for HIV-TB co-treatment are mediated via the CYP3A metabolic pathway and include the following: boosted protease inhibitors increase bedaquiline exposures (ECG monitoring is prudent if these drugs must be given together) [117–119], and efavirenz reduces bedaquiline exposures by about 50% (bedaquiline dose adjustment may mitigate this interaction but this has not been tested clinically); therefore, the co-administration of BDQ and EFV should be avoided [120, 121]. Based on knowledge of metabolic pathways, bedaquiline can be given with dolutegravir or raltegravir. Nevirapine causes only a modest reduction in bedaquiline exposures, so the two can be given together without dose adjustments [117–119]. Delamanid is a nitro-imidazole antibiotic that was authorized by the European Medicines Agency for treatment of MDR-TB in 2014. Delamanid is metabolized by albumin. It has poor bioavailability, so its dosing must be separated in time from dosing of other drugs (including ARVs), as its absorption may be impaired. When given with ritonavir-boosted lopinavir, delamanid exposures are very modestly increased; efavirenz has no impact on delamanid pharmacokinetics [122]. It is very well-tolerated, its main toxicity is modest QT prolongation, thought to be mediated by its DM-6705 metabolite, which tends to accumulate with multiple dosing; caution (and monitoring) is warranted when delamanid is given together with other drugs that prolong QT interval.

Pregnant Women with TB-HIV Co-Infection

There is a high burden of TB disease and LTBI among pregnant women globally [123], and TB is a leading cause of death in women of childbearing age [124], particularly among the HIV infected. In settings with high burden of HIV, facility-based studies have shown that 15–34% of the indirect causes of obstetric mortality are attributable to TB. LTBI may be more likely to progress to active TB disease during pregnancy than at other times; there is also an increased risk of TB in the postpartum period, which may reflect a progression to active disease during pregnancy, with diagnostic delays [17, 125]. Therefore, pregnancy and the postpartum period may be critical windows in which to treat LTBI, though the composition and safety of LTBI regimens in pregnant women is not established. IMPAACT 2001 (DAIDS ID 12026) is a study of the PK and safety of a shortened LTBI regimen, once-weekly RPT and INH (3HP), in HIV-1-infected and HIV-1-uninfected pregnant and postpartum women with LTBI; results are anticipated in the coming years. The TB APPRISE trial was the first randomized, controlled trial to evaluate the safety of immediate (antepartum-initiated) versus deferred (postpartum-initiated) IPT among pregnant women with HIV living in settings with high TB incidence. The majority of women in the trial were taking efavirenz for HIV treatment, and TST skin test or IGRA positivity was not required for participation. Individuals who were household contacts of active TB cases were excluded from participation and referred for IPT. The trial showed that there was no difference in either maternal or infant TB rates between the immediate and deferred IPT arms and that adverse events were common but not different between the arms. However, among women given antepartum-initiated IPT, there appeared to be a higher risk of adverse pregnancy outcomes; the reasons for this are still under investigation. The findings of this trial should not sway clinicians from timely initiation of IPT post-partum in women at high risk of developing TB disease. Furthermore, pregnancy is a unique case where the soonest feasible initiation of ART must be prioritized, to prevent maternal-to-child transmission of HIV, so for women with TB disease, ART should be started soon after TB treatment initiation for the good of both the mother and infant [126]. Further strengthening the argument for soonest possible initiation of ART in pregnancies complicated by HIV-TB coinfection, there is some evidence that pregnant women with HIV have increased prenatal transmission of HIV to their infants if they also have active TB [127].

Pregnancy itself confers dramatic physiologic changes that substantially affect the manner in which the body absorbs, metabolizes, and clears drugs [128]. With expansion of plasma volume and the creation of a placental-fetal compartment, there is increased cardiac output and more blood flow to organs of elimination, such as liver and kidney. Decreased peak steady state concentrations of many drugs during pregnancy have long been observed, mainly due to increased renal elimination and increased volume of distribution. Various hepatic metabolizing enzymes are variably induced or inhibited at various stages of pregnancy. Gastrointestinal emptying and transit time are slowed, and gastric pH drops, which can affect drug absorption. In addition, there is an increase in total body water, and decreases in serum albumin. ART and TB drugs are both affected by these changes. Because of this, and the highly variable PK of many ARVs, pregnant women should have frequent monitoring of HIV RNA and TB treatment efficacy throughout pregnancy, with dose and regimen adjustments as warranted. In addition, therapeutic drug monitoring (TDM) to ensure therapeutic levels may play an important role in monitoring pregnant women with TB-HIV coinfection.

There are numerous ART for which appropriate dosing has not been established in late pregnancy, including newer agents such as DTG and TAF. Some ART (like PIs) were initially thought to require dose adjustments in late pregnancy to achieve therapeutic exposures [129], given the changes in physiology that accompany late pregnancy; however, when this was done, some studies showed increased toxicity with higher doses, likely secondary to decreased protein binding and similar unbound drug levels as in non-pregnant individuals [130]. Guidelines for active TB treatment in pregnant women differ for low-burden and high-burden countries in one important way: in high-burden countries, it is felt that the benefits of pyrazinamide (PZA) to pregnant women outweigh its risks, so PZA is recommended (though its safety in pregnancy has not been clearly established). There are also uncertainties surrounding the rifamycin-related DDI in pregnancy. MDR-TB in pregnancy presents unique challenges among women with HIV. Newer shortened, 9–12 month regimens for MDR-TB have been recommended by the WHO since 2016 for adults with MDR-TB, with the exception of pregnant women, largely because aminoglycosides, a key component of these shortened regimens, are teratogenic and should be excluded from MDR-TB regimens in pregnancy. Because treatment shortening improves adherence and treatment completion, treatment shortening (with injectable-sparing regimens) should also be a goal for pregnant women. Dosing, PK, and safety in pregnancy for most of the MDR-TB drugs is unknown and richly deserves further study. Overall, uncertainties about TB drugs in pregnant women abound, including questions about both efficacy and toxicity to both the mother and the fetus, plus the magnitude of rifampicin-related drug interactions is mostly unexplored. Thus, pregnant women are a high priority for further research in this area.

Pediatric Considerations

There is a high prevalence of HIV coinfection among children with tuberculosis, especially in countries disproportionately affected by the HIV epidemic. Children with coinfection, in line with the adult syndemic, have worsened outcomes from both diseases [131]. Children under the age of 2 years are more likely than other age groups to have severe and disseminated TB [132], and also have more rapidly progressive HIV disease with high morbidity and mortality, with a 50% mortality rate in their first 2 years of life if left untreated for HIV [133]. Given these sobering statistics, it is very important to optimize and prescribe effective therapeutic regimens for both HIV and TB in children. Palatability of available formulations is a key variable which can clinch a regimen's success or failure. Fortunately, there have been recent advances in improving taste and palatability of TB drugs and creating child-friendly formulations, e.g. scored dispersible tablets in fruit flavors. Clofazimine and bedaquiline are notable exceptions; no pediatric formulations of these are currently available, though bioavailability testing of a dispersible formulation of bedaquiline has been completed. Pediatric fixed-dose combinations for children with DS-TB that are easily dispersible in liquid and dramatically lower pill burden have recently been granted WHO pre-qualification in 2017--Rifampicin 75 mg + Isoniazid 50 mg + Pyrazinamide 150 mg for the intensive treatment phase, and Rifampicin 75 mg + Isoniazid 50 mg for the continuation phase [134].

An additional challenge for pediatric HIV-TB co-management is that few Drug-Drug Interaction (DDI) studies are carried out in children, despite the dramatic changes in metabolism and clearance of drugs over the course of a child's lifetime. Many of the antituberculosis drugs provoke different toxicities in children than in adults (rifabutin, for example, causes marrow suppression more commonly in children than in adults at the same exposures) [135]. Additionally, ART options in children are more limited. Many antiretrovirals have not been studied in children and thus have no dosing guidance for pediatric populations, particularly under age 3, when renal development and thus elimination as well as the ontogeny of metabolizing enzymes are in dynamic flux. For this reason, enriching trials recruitment in this age group, particularly in children age 0–6 months old, is needed in order to understand fully DDIs among infants and young children.

The preferred ART regimen currently recommended by the WHO for the treatment of DS-TB in children with HIV who do not have a prior history of virologic failure is ABC/AZT + 3TC + LPV/r + RTV (super-boosted LPV) for those under 3 years of age, and ABC/AZT + 3TC + EFV for those over the age of 3 years. Triple nucleoside regimens (such as AZT + 3TC + ABC) should not be used in children with HIV-associated TB who have previously failed ART, as they were shown in the ACTG 5095 trial to be virologically inferior to regimens containing EFV and 2 or more nucleosides.

PIs: Because PI-based regimens are often used in perinatally infected children with HIV under age 3 years, and levels of PIs are decreased by the co-administration of rifampin [136], innovative dosing strategies can be employed to optimize levels of both lopinavir and rifampin. "Super-boosting", or the administration of additional ritonavir to coformulated lopinavir-ritonavir in a ratio of ritonavir:lopinavir of 1:1 achieves therapeutic lopinavir levels when co-administered with rifampin, but is difficult to implement as ritonavir is often unavailable and the syrup formulation has a short expiry time. Double-dosed co-formulated lopinavir-ritonavir does not achieve therapeutic lopinavir concentrations in young children on rifampin-based TB therapy [137]. A population PK model suggested that 8 hourly lopinavir-ritonavir might overcome the induction by rifampicin [138]. The higher rifampin doses recently recommended by the WHO in children have not yet been studied with adjusted doses of lopinavir-ritonavir.

NNRTIs: Efavirenz cannot be used in children under the age of 3 years, because appropriate doses have not yet been established for that age group. Levels of EFV among children above age 3 receiving weight-based standard EFV dosing with and without rifampicin-based TB therapy have been observed to be subtherapeutic [139]; it is unclear whether this is clinically detrimental, however, as >88% of children on NNRTI-based ART, both with and without TB cotreatment, have been observed to attain viral suppression [140]. Levels of nevirapine are diminished by the co-administration of rifampicin in children [141], and low nevirapine trough has been correlated with treatment failure in adults. The clinical significance of this in children is unknown, though one trial showed fairly similar clinical and immunological outcomes for children receiving NVP/3TC/D4T with and without rifampicin-based TB treatment [142]. Studies of higher dose nevirapine with rifampicin have not been done, but some experts advocate increasing the dose by 50%; even in high-income, low burden settings, this would often be unavailable [143]. Because of the lack of evidence, nevirapine-based regimens are not at present recommended for children with TB who require rifampicin.

INSTIs: There is a paucity of data on the use of integrase-strand transfer inhibitors with rifampicin in children. IMPAACT trial 1101 is currently investigating the safety and pharmacokinetics of raltegravir in conjunction with rifampicin-based TB regimens. No PK or safety studies of dolutegravir among children with coexisting TB have yet been performed, though some data may emerge from the ODYSSEY trial, run by the Pediatric European Network for Treatment of AIDS (PENTA), in which dolutegravir-based ART is being assessed in children with HIV infection (NCT02259127). A subset of children with HIV-TB are expected to be enrolled in that trial.

Summary

It is inadvisable to treat only TB, or only HIV, in an individual who suffers from both illnesses, as each worsens the impact of the other. The cotreatment of HIV and TB is imperative, and presents many challenges, particularly in an era with no great alternatives to the rifamycins for drug sensitive TB, and with many lingering dosing and toxicity questions surrounding the second line agents for drug resistant TB, particularly in special populations such as children and pregnant women. The role that TDM can play in individualized monitoring of the exposures of TB drugs and ART, in cases of doubt, should also be considered. Many questions and research gaps remain, and progress demands that they be addressed. However, careful attention to the existing knowledge on the clinical pharmacology of both ART and anti-TB drugs will enable the safe and effective treatment of both of these intertwined and deadly diseases, and the shattering of a formidable microbial alliance that has long burdened humankind.

References

1. World Health Organization (2017) Global Tuberculosis Report WHO/HTM/TB/2017.23
2. Golub JE, Pronyk P, Mohapi L, Thsabangu N, Moshabela M, Struthers H et al (2009) Isoniazid preventive therapy, HAART and tuberculosis risk in HIV-infected adults in South Africa: a prospective cohort. AIDS 23(5):631–6. https://doi.org/10.1097/QAD.0b013e328327964f
3. Briggs MA, Emerson C, Modi S, Taylor NK, Date A (2015) Use of isoniazid preventive therapy for tuberculosis prophylaxis among people living with HIV/AIDS: a review of the literature. J Acquir Immune Defic Syndr 68(Suppl 3):297
4. Ayele HT, Mourik MS, Debray TP, Bonten MJ (2015) Isoniazid prophylactic therapy for the prevention of tuberculosis in HIV infected adults: a systematic review and meta-analysis of randomized trials. PLoS One 10(11):e0142290
5. Nettles RE, Mazo D, Alwood K, Gachuhi R, Maltas G, Wendel K et al (2004) Risk factors for relapse and acquired rifamycin resistance after directly observed tuberculosis treatment: a comparison by HIV serostatus and rifamycin use. Clin Infect Dis 38(5):731–736
6. Li J, Munsiff SS, Driver CR, Sackoff J (2005) Relapse and acquired rifampin resistance in HIV-infected patients with tuberculosis treated with rifampin- or rifabutin-based regimens in New York City, 1997–2000. Clin Infect Dis 41(1):83–91
7. Burman W, Benator D, Vernon A, Khan A, Jones B, Silva C et al (2006) Acquired rifamycin resistance with twice-weekly treatment of HIV-related tuberculosis. Am J Respir Crit Care Med 173(3):350–356
8. Vernon A, Burman W, Benator D, Khan A, Bozeman L (1999) Acquired rifamycin monoresistance in patients with HIV-related tuberculosis treated with once-weekly rifapentine and isoniazid. Tuberculosis Trials Consortium. Lancet 353(9167):1843–1847
9. Chirehwa MT, Rustomjee R, Mthiyane T, Onyebujoh P, Smith P, McIlleron H et al (2015) Model-based evaluation of higher doses of rifampin using a semimechanistic model incorporating autoinduction and saturation of hepatic extraction. Antimicrob Agents Chemother 60(1):487–494
10. McIlleron H, Rustomjee R, Vahedi M, Mthiyane T, Denti P, Connolly C et al (2012) Reduced antituberculosis drug concentrations in HIV-infected patients who are men or have low weight: implications for international dosing guidelines. Antimicrob Agents Chemother 56(6):3232–3238
11. Merle CS, Floyd S, Ndiaye A, Galperine T, Furco A, De Jong BC, et al. (2016) High-dose rifampicin tuberculosis treatment regimen to reduce 12-month mortality of TB/HIV co-infected patients: the RAFA trial results. AIDS 2016 Durban South Africa
12. McIlleron H, Meintjes G, Burman WJ, Maartens G (2007) Complications of antiretroviral therapy in patients with tuberculosis: drug interactions, toxicity, and immune reconstitution inflammatory syndrome. J Infect Dis 196(Suppl 1):63
13. Abdool Karim SS, Naidoo K, Grobler A, Padayatchi N, Baxter C, Gray A et al (2010) Timing of initiation of antiretroviral drugs during tuberculosis therapy. N Engl J Med 362(8):697–706
14. Abdool Karim SS, Naidoo K, Grobler A, Padayatchi N, Baxter C, Gray AL et al (2011) Integration of antiretroviral therapy with tuberculosis. N Engl J Med 365:1492–1501

15. Havlir DV, Kendall MA, Ive P, Kumwenda J, Swindells S, Qasba SS et al (2011) Timing of antiretroviral therapy for HIV-1 infection and tuberculosis. N Engl J Med 365:1482–1491
16. Blanc F, Sok T, Laureillard D, Borand L, Rekacewicz C, Nerrienet E et al (2011) Earlier versus later start of antiretroviral therapy in HIV-infected adults with tuberculosis. N Engl J Med 365:1471–1481
17. Gupta A, Mathad JS, Abdel-Rahman SM, Albano JD, Botgros R, Brown V et al (2016) Toward earlier inclusion of pregnant and postpartum women in tuberculosis drug trials: Consensus Statements from an International Expert Panel. Clin Infect Dis 62(6):761–769
18. Nachman S, Ahmed A, Amanullah F, Becerra MC, Botgros R, Brigden G et al (2015) Towards early inclusion of children in tuberculosis drugs trials: a consensus statement. Lancet Infect Dis 15(6):711–720
19. Sonnenberg P, Glynn JR, Fielding K, Murray J, Godfrey-Faussett P, Shearer S (2005) How soon after infection with HIV does the risk of tuberculosis start to increase? A retrospective cohort study in South African gold miners. J Infect Dis 191(2):150–158
20. Selwyn PA, Hartel D, Lewis VA, Schoenbaum EE, Vermund SH, Klein RS et al (1989) A prospective study of the risk of tuberculosis among intravenous drug users with human immunodeficiency virus infection. N Engl J Med 320(9):545–550
21. Wood R, Maartens G, Lombard CJ (2000) Risk factors for developing tuberculosis in HIV-1-infected adults from communities with a low or very high incidence of tuberculosis. J Acquir Immune Defic Syndr 23(1):75–80
22. Badje A, Moh R, Gabillard D, Guehi C, Kabran M, Ntakpe JB et al (2017) Effect of isoniazid preventive therapy on risk of death in west African, HIV-infected adults with high CD4 cell counts: long-term follow-up of the Temprano ANRS 12136 trial. Lancet Glob Health 5(11):e1089
23. Akolo C, Adetifa I, Shepperd S, Volmink J (2010) Treatment of latent tuberculosis infection in HIV infected persons. Cochrane Database Syst Rev (1):CD000171
24. Chaisson RE, Golub JE (2017) Preventing tuberculosis in people with HIV-no more excuses. Lancet Glob Health 5(11):e1049
25. World Health Organization (2015) Guidelines on the managmeent of latent tuberculosis infection. WHO/HTM/TB/2015.01
26. Samandari T, Agizew TB, Nyirenda S, Tedla Z, Sibanda T, Shang N et al (2011) 6-month versus 36-month isoniazid preventive treatment for tuberculosis in adults with HIV infection in Botswana: a randomised, double-blind, placebo-controlled trial. Lancet 377(9777):1588–1598
27. Sterling TR, Scott NA, Miro JM, Calvet G, La Rosa A, Infante R et al (2016) Three months of weekly rifapentine and isoniazid for treatment of *Mycobacterium tuberculosis* infection in HIV-coinfected persons. AIDS 30(10):1607–1615
28. Mueller Y, Mpala Q, Kerschberger B, Rusch B, Mchunu G, Mazibuko S et al (2017) Adherence, tolerability, and outcome after 36 months of isoniazid-preventive therapy in 2 rural clinics of Swaziland: a prospective observational feasibility study. Medicine (Baltimore) 96(35):e7740
29. Luetkemeyer AF, Rosenkranz SL, Lu D, Grinsztejn B, Sanchez J, Ssemmanda M et al (2015) Combined effect of CYP2B6 and NAT2 genotype on plasma efavirenz exposure during rifampin-based antituberculosis therapy in the STRIDE study. Clin Infect Dis 60(12):1860–1863
30. Dooley KE, Denti P, Martinson N, Cohn S, Mashabela F, Hoffmann J et al (2015) Pharmacokinetics of efavirenz and treatment of HIV-1 among pregnant women with and without tuberculosis coinfection. J Infect Dis 211(2):197–205
31. Leger P, Chirwa S, Turner M, Richardson DM, Baker P, Leonard M et al (2016) Pharmacogenetics of efavirenz discontinuation for reported central nervous system symptoms appears to differ by race. Pharmacogenet Genomics 26(10):473–480
32. Sterling TR, Villarino ME, Borisov AS, Shang N, Gordin F, Bliven-Sizemore E et al (2011) Three months of rifapentine and isoniazid for latent tuberculosis infection. N Engl J Med 365(23):2155–2166

33. Sanofi (2015) An open-label, non-randomized, single sequence, two periods, four-treatment, three parallel groups pharmacokinetic interaction study of repeated oral doses (daily or weekly regimen) of rifapentine on ATRIPLA™ (fixed dose combination of efavirenz, emtricitabine, and tenofovir disoproxil fumarate) given to HIV+ patients
34. Podany AT, Bao Y, Swindells S, Chaisson RE, Andersen JW, Mwelase T et al (2015) Efavirenz pharmacokinetics and pharmacodynamics in HIV-infected persons receiving Rifapentine and isoniazid for tuberculosis prevention. Clin Infect Dis. 61(8):1322–1327
35. Weiner M, Egelund EF, Engle M, Kiser M, Prihoda TJ, Gelfond JA et al (2014) Pharmacokinetic interaction of rifapentine and raltegravir in healthy volunteers. J Antimicrob Chemother 69(4):1079–1085
36. Brooks KM, Pau AK, George JM, Alfaro R, Kellogg A, McLaughlin M et al (2016) Early termination of a PK study between dolutegravir and weekly isoniazid/rifapentine. CROI
37. Dickinson JM, Mitchison DA (1981) Experimental models to explain the high sterilizing activity of rifampin in the chemotherapy of tuberculosis. Am Rev Respir Dis 123(4 Pt 1):367–371
38. Jindani A, Nunn AJ, Enarson DA (2004) Two 8-month regimens of chemotherapy for treatment of newly diagnosed pulmonary tuberculosis: international multicentre randomised trial. Lancet 364(9441):1244–1251
39. Okwera A, Whalen C, Byekwaso F, Vjecha M, Johnson J, Huebner R et al (1994) Randomised trial of thiacetazone and rifampicin-containing regimens for pulmonary tuberculosis in HIV-infected Ugandans. The Makerere University-Case Western University Research Collaboration. Lancet 344(8933):1323–1328
40. Dooley KE, Flexner C, Andrade AS (2008) Drug interactions involving combination antiretroviral therapy and other anti-infective agents: repercussions for resource-limited countries. J Infect Dis 198(7):948–961
41. Burger DM, Meenhorst PL, Koks CH, Beijnen JH (1993) Pharmacokinetic interaction between rifampin and zidovudine. Antimicrob Agents Chemother 37(7):1426–1431
42. Gallicano KD, Sahai J, Shukla VK, Seguin I, Pakuts A, Kwok D et al (1999) Induction of zidovudine glucuronidation and amination pathways by rifampicin in HIV-infected patients. Br J Clin Pharmacol. 48(2):168–179
43. Droste JA, Verweij-van Wissen CP, Kearney BP, Buffels R, Vanhorssen PJ, Hekster YA et al (2005) Pharmacokinetic study of tenofovir disoproxil fumarate combined with rifampin in healthy volunteers. Antimicrob Agents Chemother 49(2):680–684
44. Mills A, Arribas JR, Andrade-Villanueva J, DiPerri G, Van Lunzen J, Koenig E et al (2016) Switching from tenofovir disoproxil fumarate to tenofovir alafenamide in antiretroviral regimens for virologically suppressed adults with HIV-1 infection: a randomised, active-controlled, multicentre, open-label, phase 3, non-inferiority study. Lancet Infect Dis 16(1):43–52
45. Maartens G, Boffito M, Flexner CW (2017) Compatibility of next-generation first-line antiretrovirals with rifampicin-based antituberculosis therapy in resource limited settings. Curr Opin HIV AIDS 12(4):355–358
46. Custodio JM, West SK, Lutz J, Vu A, Xiao D, Collins S, et al. Twice daily administration of Tenofovir Alafenamide In combination with Rifampin: potential for Tenofovir Alafenamide use in HIV-TB coinfection. 2017
47. Lopez-Cortes LF, Ruiz-Valderas R, Viciana P, Alarcon-Gonzalez A, Gomez-Mateos J, Leon-Jimenez E et al (2002) Pharmacokinetic interactions between Efavirenz and rifampicin in HIV-infected patients with tuberculosis. Clin Pharmacokinet 41(9):681–690
48. Manosuthi W, Sungkanuparph S, Thakkinstian A, Vibhagool A, Kiertiburanakul S, Rattanasiri S et al (2005) Efavirenz levels and 24-week efficacy in HIV-infected patients with tuberculosis receiving highly active antiretroviral therapy and rifampicin. AIDS 19(14):1481–1486
49. Friedland G, Khoo S, Jack C, Lalloo U (2006) Administration of Efavirenz (600 mg/day) with rifampicin results in highly variable levels but excellent clinical outcomes in patients treated for tuberculosis and HIV. J Antimicrob Chemother 58(6):1299–1302

50. Pedral-Sampaio DB, Alves CR, Netto EM, Brites C, Oliveira AS, Badaro R (2004) Efficacy and safety of Efavirenz in HIV patients on Rifampin for tuberculosis. Braz J Infect Dis 8(3):211–216
51. Patel A, Patel K, Patel J, Shah N, Patel B, Rani S (2004) Safety and antiretroviral effectiveness of concomitant use of rifampicin and efavirenz for antiretroviral-naive patients in India who are coinfected with tuberculosis and HIV-1. J Acquir Immune Defic Syndr 37(1):1166–1169
52. Bertrand J, Verstuyft C, Chou M, Borand L, Chea P, Nay KH et al (2014) Dependence of Efavirenz- and rifampicin-isoniazid-based antituberculosis treatment drug-drug interaction on CYP2B6 and NAT2 genetic polymorphisms: ANRS 12154 Study in Cambodia. J Infect Dis 209(3):399–408
53. HM MI, Schomaker M, Ren Y, Sinxadi P, Nuttall JJ, Gous H et al (2013) Effects of rifampin-based antituberculosis therapy on plasma efavirenz concentrations in children vary by CYP2B6 genotype. AIDS 27(12):1933–1940
54. Crawford KW, Ripin DH, Levin AD, Campbell JR, Flexner C (2012 July 01) Participants of conference on antiretroviral drug optimization. Optimising the manufacture, formulation, and dose of antiretroviral drugs for more cost-efficient delivery in resource-limited settings: a consensus statement. Lancet Infect Dis 12(7):550–560
55. ENCORE1 Study Group, Carey D, Puls R, Amin J, Losso M, Phanupak P et al (2015) Efficacy and safety of efavirenz 400 mg daily versus 600 mg daily: 96-week data from the randomised, double-blind, placebo-controlled, non-inferiority ENCORE1 study. Lancet Infect Dis. 15(7):793–802
56. Marzolini C, Telenti A, Decosterd LA, Greub G, Biollaz J, Buclin T (2001) Efavirenz plasma levels can predict treatment failure and central nervous system side effects in HIV-1-infected patients. AIDS 15(1):71–75
57. Dickinson L, Amin J, Else L, Boffito M, Egan D, Owen A et al (2016) Comprehensive pharmacokinetic, pharmacodynamic and pharmacogenetic evaluation of once-daily Efavirenz 400 and 600 mg in treatment-naive HIV-infected patients at 96 weeks: results of the ENCORE1 Study. Clin Pharmacokinet 55(7):861–873
58. Dickinson L, Amin J, Else L, Boffito M, Egan D, Owen A et al (2015) Pharmacokinetic and pharmacodynamic comparison of once-daily efavirenz (400 mg vs. 600 mg) in treatment-naive HIV-infected patients: results of the ENCORE1 Study. Clin Pharmacol Ther 98(4):406–416
59. Autar RS, Wit FW, Sankote J, Mahanontharit A, Anekthananon T, Mootsikapun P et al (2005) Nevirapine plasma concentrations and concomitant use of rifampin in patients coinfected with HIV-1 and tuberculosis. Antivir Ther 10(8):937–943
60. Ribera E, Pou L, Lopez RM, Crespo M, Falco V, Ocana I et al (2001) Pharmacokinetic interaction between nevirapine and rifampicin in HIV-infected patients with tuberculosis. J Acquir Immune Defic Syndr 28(5):450–453
61. Nafrialdi NAW (2012) Yunihastuti E, Wiria MS. Influence of rifampicin on nevirapine plasma concentration in HIV-TB coinfected patients. Acta Med Indones 44(2):135–139
62. Swaminathan S, Padmapriyadarsini C, Venkatesan P, Narendran G, Ramesh Kumar S, Iliayas S et al (2011) Efficacy and safety of once-daily nevirapine- or efavirenz-based antiretroviral therapy in HIV-associated tuberculosis: a randomized clinical trial. Clin Infect Dis 53(7):716–724
63. Bhatt NB, Baudin E, Meggi B, da Silva C, Barrail-Tran A, Furlan V et al (2015 Jan) Nevirapine or efavirenz for tuberculosis and HIV coinfected patients: exposure and virological failure relationship. J Antimicrob Chemother 70(1):225–232
64. Shipton LK, Wester CW, Stock S, Ndwapi N, Gaolathe T, Thior I et al (2009) Safety and efficacy of nevirapine- and efavirenz-based antiretroviral treatment in adults treated for TB-HIV co-infection in Botswana. Int J Tuberc Lung Dis 13(3):360–366
65. Bonnet M, Bhatt N, Baudin E, Silva C, Michon C, Taburet AM et al (2013) Nevirapine versus efavirenz for patients co-infected with HIV and tuberculosis: a randomised non-inferiority trial. Lancet Infect Dis 13(4):303–312
66. Boulle A, Van Cutsem G, Cohen K, Hilderbrand K, Mathee S, Abrahams M et al (2008) Outcomes of nevirapine- and efavirenz-based antiretroviral therapy when coadministered with rifampicin-based antitubercular therapy. JAMA 300(5):530–539

67. Ramachandran G, Hemanthkumar AK, Rajasekaran S, Padmapriyadarsini C, Narendran G, Sukumar B et al (2006) Increasing nevirapine dose can overcome reduced bioavailability due to rifampicin coadministration. J Acquir Immune Defic Syndr 42(1):36–41
68. Edurant package insert (2011) Tibotec Therapeutics
69. Kakuda TN, Scholler-Gyure M, Hoetelmans RM (2011) Pharmacokinetic interactions between etravirine and non-antiretroviral drugs. Clin Pharmacokinet 50(1):25–39
70. Wenning LA, Hanley WD, Brainard DM, Petry AS, Ghosh K, Jin B et al (2009) Effect of rifampin, a potent inducer of drug-metabolizing enzymes, on the pharmacokinetics of raltegravir. Antimicrob Agents Chemother 53(7):2852–2856
71. Taburet AM, Sauvageon H, Grinsztejn B, Assuied A, Veloso V, Pilotto JH et al (2015) Pharmacokinetics of Raltegravir in HIV-infected patients on rifampicin-based antitubercular therapy. Clin Infect Dis 61(8):1328–1335
72. Reynolds HE, Chrdle A, Egan D, Chaponda M, Else L, Chiong J et al (2015) Effect of intermittent rifampicin on the pharmacokinetics and safety of raltegravir. J Antimicrob Chemother 70(2):550–554
73. Grinsztejn B, De Castro N, Arnold V, Veloso VG, Morgado M, Pilotto JH et al (2014) Raltegravir for the treatment of patients co-infected with HIV and tuberculosis (ANRS 12 180 Reflate TB): a multicentre, phase 2, non-comparative, open-label, randomised trial. Lancet Infect Dis 14(6):459–467
74. Markowitz M, Nguyen BY, Gotuzzo E, Mendo F, Ratanasuwan W, Kovacs C et al (2009) Sustained antiretroviral effect of raltegravir after 96 weeks of combination therapy in treatment-naive patients with HIV-1 infection. J Acquir Immune Defic Syndr 52(3):350–356
75. Dooley KE, Sayre P, Borland J, Purdy E, Chen S, Song I et al (2012) Safety, tolerability, and pharmacokinetics of the HIV integrase inhibitor dolutegravir given twice daily with rifampin or once daily with rifabutin: results of a phase 1 study among healthy subjects. J Acquir Immune Defic Syndr:15
76. Lee JS, Calmy A, Andrieux-Meyer I, Ford N (2012) Review of the safety, efficacy, and pharmacokinetics of elvitegravir with an emphasis on resource-limited settings. HIV AIDS (Auckl) 4:5–15
77. Zhang H, Custodio JM, Wei X, Wang H, Vu A, Ling J, et al. Clinical pharmacology of the HIV integrase strand transfer inhibitor bictegravir. 2017
78. Tseng A, Hughes CA, Wu J, Seet J, Phillips EJ (2017) Cobicistat versus ritonavir: similar pharmacokinetic enhancers but some important differences. Ann Pharmacother 51(11):1008–1022
79. Roberts O, Khoo S, Owen A, Siccardi M (2017) Interaction of Rifampin and Darunavir-Ritonavir or Darunavir-Cobicistat In Vitro. Antimicrob Agents Chemother 61(5):16
80. Acosta EP, Kendall MA, Gerber JG, Alston-Smith B, Koletar SL, Zolopa AR et al (2007) Effect of concomitantly administered rifampin on the pharmacokinetics and safety of atazanavir administered twice daily. Antimicrob Agents Chemother 51(9):3104–3110
81. Burger DM, Agarwala S, Child M, Been-Tiktak A, Wang Y, Bertz R (2006) Effect of rifampin on steady-state pharmacokinetics of atazanavir with ritonavir in healthy volunteers. Antimicrob Agents Chemother 50(10):3336–3342
82. LaPorte C, Colbers E, Bertz R, Vonchek D, Wikstrom K, Boeree M et al (2004) Pharmacokinetics of adjusted-dose lopinavir-ritonavir combined with rifampin in healthy volunteers. Antimicrob Agents Chemother 48(5):1553–1560
83. Decloedt EH, McIlleron H, Smith P, Merry C, Orrell C, Maartens G (2011) Pharmacokinetics of lopinavir in HIV-infected adults receiving rifampin with adjusted doses of lopinavir-ritonavir tablets. Antimicrob Agents Chemother 55(7):3195–3200
84. Decloedt EH, Maartens G, Smith P, Merry C, Bango F, McIlleron H (2012) The safety, effectiveness and concentrations of adjusted lopinavir/ritonavir in HIV-infected adults on rifampicin-based antitubercular therapy. PLoS One 7(3):e32173
85. L'homme RF, Nijland HM, Gras L, Aarnoutse RE, van Crevel R, Boeree M et al (2009) Clinical experience with the combined use of lopinavir/ritonavir and rifampicin. AIDS 27(7):863–865

86. Sunpath H, Winternheimer P, Cohen S, Tennant I, Chelin N, Gandhi RT et al (2014) Double-dose lopinavir-ritonavir in combination with rifampicin-based anti-tuberculosis treatment in South Africa. Int J Tuberc Lung Dis 18(6):689–693
87. Murphy RA, Marconi VC, Gandhi RT, Kuritzkes DR, Sunpath H (2012) Coadministration of lopinavir/ritonavir and rifampicin in HIV and tuberculosis co-infected adults in South Africa. PLoS One 7(9):e44793
88. Abel S, Jenkins TM, Whitlock LA, Ridgway CE, Muirhead GJ (2008) Effects of CYP3A4 inducers with and without CYP3A4 inhibitors on the pharmacokinetics of maraviroc in healthy volunteers. Br J Clin Pharmacol 65(Suppl 1):38–46
89. Boyd MA, Zhang X, Dorr A, Ruxrungtham K, Kolis S, Nieforth K et al (2003) Lack of enzyme-inducing effect of rifampicin on the pharmacokinetics of enfuvirtide. J Clin Pharmacol 43(12):1382–1391
90. Boeree MJ, Heinrich N, Aarnoutse R, Diacon AH, Dawson R, Rehal S et al (2017) High-dose rifampicin, moxifloxacin, and SQ109 for treating tuberculosis: a multi-arm, multi-stage randomised controlled trial. Lancet Infect Dis 17(1):39–49
91. Heifets L (1999) Microbiological aspects of rifapentine. Drugs Today 35(Suppl. D):7
92. Dooley KE, Bliven-Sizemore EE, Weiner M, Lu Y, Nuermberger EL, Hubbard WC et al (2012) Safety and Pharmacokinetics of Escalating Daily Doses of the Antituberculosis Drug Rifapentine in Healthy Volunteers. Clin Pharmacol Ther 91(5). https://doi.org/10.1038/clpt.2011.323
93. Burman W, Dooley KE, Nuermberger E (2011) The rifamycins: renewed interest in an old drug class. In: Donald P, van Helden P (eds) Antituberculosis chemotherapy, vol 40. Karger AG—Medical and Scientific Publishers, Basel
94. Davies G, Cerri S, Richeldi L (2007) Rifabutin for treating pulmonary tuberculosis. Cochrane Database Syst Rev (4). CD005159
95. Blaschke TF, Skinner MH (1996) The clinical pharmacokinetics of rifabutin. Clin Infect Dis 22(Suppl 1):2
96. Tseng AL, Walmsley SL (1995) Rifabutin-associated uveitis. Ann Pharmacother 29(11):1149–1155
97. Griffith DE, Brown BA (1996) Wallace RJ,Jr. Varying dosages of rifabutin affect white blood cell and platelet counts in human immunodeficiency virus--negative patients who are receiving multidrug regimens for pulmonary *Mycobacterium avium* complex disease. Clin Infect Dis 23(6):1321–1322
98. Khachi H, O'Connell R, Ladenheim D, Orkin C (2009) Pharmacokinetic interactions between rifabutin and lopinavir/ritonavir in HIV-infected patients with mycobacterial co-infection. J Antimicrob Chemother 64(4):871–873
99. Jenny-Avital ER, Joseph K (2009) Rifamycin-resistant *Mycobacterium tuberculosis* in the highly active antiretroviral therapy era: a report of 3 relapses with acquired rifampin resistance following alternate-day rifabutin and boosted protease inhibitor therapy. Clin Infect Dis 48(10):1471–1474
100. Boulanger C, Hollender E, Farrell K, Stambaugh JJ, Maasen D, Ashkin D et al (2009) Pharmacokinetic evaluation of rifabutin in combination with lopinavir-ritonavir in patients with HIV infection and active tuberculosis. Clin Infect Dis 49(9):1305–1311
101. Ramachandran G, Bhavani PK, Hemanth Kumar AK, Srinivasan R, Raja K, Sudha V et al (2013) Pharmacokinetics of rifabutin during atazanavir/ritonavir co-administration in HIV-infected TB patients in India. Int J Tuberc Lung Dis 17(12):1564–1568
102. Jenks JD, Kumarasamy N, Ezhilarasi C, Poongulali S, Ambrose P, Yepthomi T et al (2016) Improved tuberculosis outcomes with daily vs. intermittent rifabutin in HIV-TB coinfected patients in India. Int J Tuberc Lung Dis 20(9):1181–1184
103. Lan NT, Thu NT, Barrail-Tran A, Duc NH, Lan NN, Laureillard D et al (2014) Randomised pharmacokinetic trial of rifabutin with lopinavir/ritonavir-antiretroviral therapy in patients with HIV-associated tuberculosis in Vietnam. PLoS One 9(1):e84866

104. Naiker S, Connolly C, Wiesner L, Kellerman T, Reddy T, Harries A et al (2014) Randomized pharmacokinetic evaluation of different rifabutin doses in African HIV- infected tuberculosis patients on lopinavir/ritonavir-based antiretroviral therapy. BMC Pharmacol Toxicol 15:61
105. Yapa HM, Boffito M, Pozniak A (2016) Critical review: what dose of rifabutin is recommended with antiretroviral therapy? J Acquir Immune Defic Syndr 72(2):138–152
106. Crauwels H, van Heeswijk RP, Stevens M, Buelens A, Vanveggel S, Boven K et al (2013) Clinical perspective on drug-drug interactions with the non-nucleoside reverse transcriptase inhibitor rilpivirine. AIDS Rev 15(2):87–101
107. Kakuda TN, Woodfall B, De Marez T, Peeters M, Vandermeulen K, Aharchi F et al (2014) Pharmacokinetic evaluation of the interaction between etravirine and rifabutin or clarithromycin in HIV-negative, healthy volunteers: results from two Phase 1 studies. J Antimicrob Chemother 69(3):728–734
108. Brainard DM, Kassahun K, Wenning LA, Petry AS, Liu C, Lunceford J et al (2011) Lack of a clinically meaningful pharmacokinetic effect of rifabutin on raltegravir: in vitro/in vivo correlation. J Clin Pharmacol 51(6):943–950
109. Ramanathan S, Mathias AA, German P, Kearney BP (2011) Clinical pharmacokinetic and pharmacodynamic profile of the HIV integrase inhibitor elvitegravir. Clin Pharmacokinet 50(4):229–244
110. World Health Organization (2016) WHO treatment guidelines for drug-resistant tuberculosis: 2016 update
111. Brust JCM, Shah NS, Mlisana K, Moodley P, Allana S, Campbell A et al (2017) Improved survival and cure rates with concurrent treatment for MDR-TB/HIV co-infection in South Africa. Clin Infect Dis:26
112. Satti H, McLaughlin MM, Hedt-Gauthier B, Atwood SS, Omotayo DB, Ntlamelle L et al (2012) Outcomes of multidrug-resistant tuberculosis treatment with early initiation of antiretroviral therapy for HIV co-infected patients in Lesotho. PLoS One 7(10):e46943
113. Naidoo A, Chirehwa M, McIlleron H, Naidoo K, Essack S, Yende-Zuma N et al (2017) Effect of rifampicin and efavirenz on moxifloxacin concentrations when co-administered in patients with drug-susceptible TB. J Antimicrob Chemother 72(5):1441–1449
114. Coyne KM, Pozniak AL, Lamorde M, Boffito M (2009) Pharmacology of second-line antituberculosis drugs and potential for interactions with antiretroviral agents. AIDS 23(4):437–446
115. Svensson EM, Karlsson MO (2017) Modelling of mycobacterial load reveals bedaquiline's exposure-response relationship in patients with drug-resistant TB. J Antimicrob Chemother 72(12):3398–3405
116. van Heeswijk RP, Dannemann B, Hoetelmans RM (2014) Bedaquiline: a review of human pharmacokinetics and drug-drug interactions. J Antimicrob Chemother 69(9):2310–2318
117. Svensson EM, Dooley KE, Karlsson MO (2014) Impact of lopinavir-ritonavir or nevirapine on bedaquiline exposures and potential implications for patients with tuberculosis-HIV coinfection. Antimicrob Agents Chemother 58(11):6406–6412
118. Pandie M, Wiesner L, McIlleron H, Hughes J, Siwendu S, Conradie F et al (2016) Drug-drug interactions between bedaquiline and the antiretrovirals lopinavir/ritonavir and nevirapine in HIV-infected patients with drug-resistant TB. J Antimicrob Chemother 71(4):1037–1040
119. Brill MJ, Svensson EM, Pandie M, Maartens G, Karlsson MO (2017) Confirming model-predicted pharmacokinetic interactions between bedaquiline and lopinavir/ritonavir or nevirapine in patients with HIV and drug-resistant tuberculosis. Int J Antimicrob Agents 49(2):212–217
120. Dooley KE, Park JG, Swindells S, Allen R, Haas DW, Cramer Y et al (2012) Safety, tolerability, and pharmacokinetic interactions of the antituberculous agent TMC207 (Bedaquiline) with efavirenz in healthy volunteers: AIDS Clinical Trials Group Study A5267. J Acquir Immune Defic Syndr 59(5):455–462
121. Svensson EM, Aweeka F, Park JG, Marzan F, Dooley KE, Karlsson MO (2013) Model-based estimates of the effects of efavirenz on bedaquiline pharmacokinetics and suggested dose adjustments for patients coinfected with HIV and tuberculosis. Antimicrob Agents Chemother 57(6):2780–2787

122. Mallikaarjun S, Wells C, Petersen C, Paccaly A, Shoaf SE, Patil S et al (2016) Delamanid coadministered with antiretroviral drugs or antituberculosis drugs shows no clinically relevant drug-drug interactions in healthy subjects. Antimicrob Agents Chemother 60(10):5976–5985
123. World Health Organization (2016) Tuberculosis fact sheet: tuberculosis in women
124. Say L, Chou D, Gemmill A, Tuncalp O, Moller AB, Daniels J et al (2014) Global causes of maternal death: a WHO systematic analysis. Lancet Glob Health 2(6):323
125. Zenner D, Kruijshaar ME, Andrews N, Abubakar I (2012) Risk of tuberculosis in pregnancy: a national, primary care-based cohort and self-controlled case series study. Am J Respir Crit Care Med 185(7):779–784
126. Panel on Antiretroviral Guidelines for Adults and Adolescents. Guidelines for the use of antiretroviral agents in HIV-1-infected adults and adolescents. [Internet]. [cited April 10 2009]. Available from: http://www.aidsinfo.nih.gov/ContentFiles/AdultandAdolescentGL.pdf
127. Pillay T, Khan M, Moodley J, Adhikari M, Coovadia H (2004) Perinatal tuberculosis and HIV-1: considerations for resource-limited settings. Lancet Infect Dis 4(3):155–165
128. Loebstein R, Lalkin A, Koren G (1997) Pharmacokinetic changes during pregnancy and their clinical relevance. Clin Pharmacokinet 33(5):328–343
129. Mirochnick M, Best BM, Stek AM, Capparelli E, Hu C, Burchett SK et al (2008) Lopinavir exposure with an increased dose during pregnancy. J Acquir Immune Defic Syndr 49(5):485–491
130. Bonafe SM, Costa DA, Vaz MJ, Senise JF, Pott-Junior H, Machado RH et al (2013) A randomized controlled trial to assess safety, tolerability, and antepartum viral load with increased lopinavir/ritonavir dosage in pregnancy. AIDS Patient Care STDs 27(11):589–595
131. Hesseling AC, Westra AE, Werschkull H, Donald PR, Beyers N, Hussey GD et al (2005) Outcome of HIV infected children with culture confirmed tuberculosis. Arch Dis Child 90(11):1171–1174
132. Marais BJ, Schaaf HS (2014) Tuberculosis in children. Cold Spring Harb Perspect Med 4(9):a017855
133. Newell ML, Coovadia H, Cortina-Borja M, Rollins N, Gaillard P, Dabis F et al (2008) Mortality of infected and uninfected infants born to HIV-infected mothers in Africa: a pooled analysis. Lancet 364(9441):1236–1243
134. Essential medicines and health products: finished pharmaceutical products [Internet]. Available from: https://extranet.who.int/prequal/content/prequalified-lists/medicines?label=&field_medicine_applicant=&field_medicine_fpp_site_value=&search_api_aggregation_1=&field_medicine_pq_date%5Bdate%5D=&field_medicine_pq_date_1%5Bdate%5D=&field_therapeutic_area=23&field_medicine_status=&field_basis_of_listing=43&field_single_fixed_dose_list%5B%5D=2&field_single_fixed_dose_list%5B%5D=3&field_single_fixed_dose_list%5B%5D=4&field_single_fixed_dose_list%5B%5D=&field_co_packed_list%5B%5D=2
135. Moultrie H, McIlleron H, Sawry S, Kellermann T, Wiesner L, Kindra G et al (2015) Pharmacokinetics and safety of rifabutin in young HIV-infected children receiving rifabutin and lopinavir/ritonavir. J Antimicrob Chemother 70(2):543–549
136. Ren Y, Nuttall JJ, Egbers C, Eley BS, Meyers TM, Smith PJ et al (2008) Effect of rifampicin on lopinavir pharmacokinetics in HIV-infected children with tuberculosis. J Acquir Immune Defic Syndr 47(5):566–569
137. McIlleron H, Ren Y, Nuttall J, Fairlie L, Rabie H, Cotton M et al (2011) Lopinavir exposure is insufficient in children given double doses of lopinavir/ritonavir during rifampicin-based treatment for tuberculosis. Antivir Ther 16(3):417–421
138. Zhang C, McIlleron H, Ren Y, van der Walt JS, Karlsson MO, Simonsson US et al (2012) Population pharmacokinetics of lopinavir and ritonavir in combination with rifampicin-based antitubercular treatment in HIV-infected children. Antivir Ther 17(1):25–33
139. Ren Y, Nuttall JJ, Eley BS, Meyers TM, Smith PJ, Maartens G et al (2009) Effect of rifampicin on efavirenz pharmacokinetics in HIV-infected children with tuberculosis. J Acquir Immune Defic Syndr 50(5):439–443

140. Zanoni BC, Phungula T, Zanoni HM, France H, Feeney ME (2011) Impact of tuberculosis cotreatment on viral suppression rates among HIV-positive children initiating HAART. AIDS 25(1):49–55
141. Oudijk JM, McIlleron H, Mulenga V, Chintu C, Merry C, Walker AS et al (2012) Pharmacokinetics of nevirapine in HIV-infected children under 3 years on rifampicin-based antituberculosis treatment. AIDS 26(12):1523–1528
142. Kamateeka MML, Mudiope P, Mubiru M, Ajuna P, Lutajumwa M, Musoke P (2009) Immunological and virological response to fixed-dose nevirapine based highly active antiretroviral therapy (HAART) in HIV-infected Ugandan children with concurrent active tuberculosis infection on rifampicin-based anti-TB treatment. IAS, Cape Town
143. Kwara A, Ramachandran G, Swaminathan S (2010 Jan) Dose adjustment of the non-nucleoside reverse transcriptase inhibitors during concurrent rifampicin-containing tuberculosis therapy: one size does not fit all. Expert Opin Drug Metab Toxicol 6(1):55–68

HIV and Tuberculosis in Children

Tonya Arscott-Mills, Ben Marais, and Andrew Steenhoff

Abstract TB/HIV co-infection poses a formidable challenge. The added complications of age related immune differences affect the presentation and evaluation in children. Young immune immature children have a naturally increased risk of progression to active TB after primary *M. tuberculosis* infection, which is compounded by HIV co-infection. The pauci-bacillary nature of their disease and difficulties in obtaining good specimens from young children complicates diagnostic evaluations. A TB diagnosis in an HIV-infected child relies on a detailed exposure history, careful clinical assessment and use of all available evidence indicating TB infection and/or disease. However, once diagnosed children usually respond well to therapy, although drug-drug interactions, pill burdens, social circumstances and personal age-appropriate support must be considered in the care of these children.

Keywords Pediatric TB/HIV · Childhood TB/HIV · *Mycobacterium tuberculosis* · Tuberculosis · Epidemiology · Diagnosis · Treatment · Prevention · Extrapulmonary TB · Pediatrics · HIV infection · Anti-tuberulous therapy · Isoniazid · Rifampin · Ethambutol · Pyrazinamide · Anti-retroviral therapy · Point-of-care · Directly observed therapy · Drug-resistance

---

T. Arscott-Mills (✉) · A. Steenhoff
Department of Pediatrics, Perelman School of Medicine at the University of Pennsylvania, Philadelphia, PA, USA

Children's Hospital of Philadelphia Global Health Center, Philadelphia, PA, USA

B. Marais
The Children's Hospital at Westmead Clinical School, University of Sydney, Sydney, Australia

I. Sereti et al. (eds.), *HIV and Tuberculosis*,
https://doi.org/10.1007/978-3-030-29108-2_12

Introduction

In 2017 it was estimated that there were 1.8 million [1.3 million–2.4 million] children living with HIV globally [1]. The greatest burden, 88%, of children live in sub-Saharan Africa [2]. Since 2000, scaling up of prevention of mother to child transmission has greatly reduced the number of new pediatric diagnoses. UNICEF estimates that there were two million new infections among children (0–14 years old) averted since 2000 [2]. In addition, between 2000 and 2015 there was a decline of 60% in the number of AIDS-related deaths among children under 15 years of age [3]. This is largely due to increasing antiretroviral therapy (ART) coverage and earlier ART initiation. However, ART coverage in children lags substantially behind that of adults. Globally, in 2017 an estimated 52% of children with HIV were on ART compared to 59% in adults [1].

HIV-induced immunodeficiency is compounded in young children with immature immune systems. Infants born with HIV have a 2 year survival of 47.5% without ART [4]. In addition, the total CD4 count is not an accurate correlate of disease severity or immune compromise, although there is a trend that children with lower CD4 count *percentages* have a higher likelihood of opportunistic infections [5]. HIV causes a functional and numeric decline in CD4 cells, as well as secondary humeral defects [6, 7]. With ART, the CD4 cells often rise to a normal number, but functional defects may remain as seen with the continued increased susceptibility to certain infections such as TB and diminished vaccine responses [8–10]. This is an important reason to initiate ART as soon as possible after diagnosis and not to wait for immune decline before offering ART to children. Without ART there may also be a depletion of CD8 cells, with poor thymic production, lysis of CD4 cells and apoptosis of bystander T-cells all contributing to Acquired Immunodeficiency Syndrome (AIDS) progression [11, 12].

Increased antenatal HIV screening and prevention of mother to child transmission not only reduces the number of children infected with HIV, but also facilitates the diagnosis of HIV in early infancy providing an opportunity for early initiation of ART [2]. However, in many places, programs are not fully functional and clinicians needs to be alert to the myriad different manifestations with which a HIV-infected child may present. Table 1 summarizes World Health Organization (WHO) clinical staging criteria and Table 2 provides an overview of common syndromes with which HIV-infected children may present.

TB/HIV co-Infection in Children

In 2017 there were an estimated ten million new cases of TB worldwide, including one million children under 15 years of age [13]. With the exception of Asia, the same regions hardest hit by the HIV epidemic also have the highest rates of TB in adults and children, reflecting uncontrolled transmission of *Mycobacterium*

Table 1 World Health Organization clinical HIV staging for children <15 years

Stage 1	Asymptomatic; generalised lymphadenopathy
Stage 2	Unexplained persistent hepatosplenomegaly
	Papular pruritic eruptions or extensive wart virus infection or molluscum contagiosum
	Fungal nail infection
	Angular cheilitis; lineal gingival erythema; recurrent oral ulcerations
	Unexplained persistent parotid enlargement
	Herpes zoster
	Recurrent or chronic URTI; otitis media, otorrhea, sinusitis or tonsillitis
Stage 3	Unexplained moderate malnutrition[a] or wasting
	Unexplained persistent diarrhea (14 days or more)
	Unexplained persistent fever (intermittent or constant, for longer than 1 month)
	Persistent oral candidiasis (after first 6–8 weeks of life)
	Oral hairy leukoplakia; acute necrotising ulcerative gingivitis or periodontitis
	Lymph node or pulmonary tuberculosis
	Severe recurrent bacterial pneumonia or chronic lung disease
	Symptomatic lymphoid interstitial pneumonitis
	Unexplained anaemia, neutropenia and/or chronic thrombocytopenia
Stage 4	Unexplained severe wasting[b], stunting or severe malnutrition
	Pneumocystis pneumonia
	Recurrent severe bacterial infections
	Chronic herpes simplex infection
	Esophageal or airway/lung candidiasis
	Extrapulmonary tuberculosis
	Kaposi's sarcoma
	Cytomegalovirus infection with onset at age older than 1 month
	Central nervous system toxoplasmosis (after 1 month of life)
	Extrapulmonary cryptococcosis (including meningitis)
	HIV encephalopathy
	Disseminated endemic mycosis or non-tuberculous mycobacterial infection
	Chronic cryptosporidiosis or isosporiasis (with diarrhea)
	Cerebral or B-cell non-Hodgkin's lymphoma
	Progressive multifocal leukoencephalopathy
	Symptomatic HIV-associated nephropathy or HIV-associated cardiomyopathy
	HIV-associated rectovaginal fistula (African children)
	Reactivation of American trypanosomiasis (south American children)

(continued)

Table 1 (continued)

Source: Adapted from: WHO case definitions of HIV for surveillance and revised clinical staging and immunological classification of HIV-related disease in adults and children. Geneva: World Health Organization; 2007

[a]For children younger than 5 years, moderate malnutrition is defined as weight-for-height <–2 *z*-score or mid-upper arm circumference ≥115 mm to <125 mm

[b]For children younger than 5 years of age, severe wasting is defined as weight-for-height <–3 *z*-score; stunting is defined as length-for-age/height-for-age <–2 *z*-score; and severe acute malnutrition is either weight for height <–3 *z*-score or mid-upper arm circumference <115 mm or the presence of edema

tuberculosis (MTB) within these communities. See Figs. 1 and 2. In total, 9% of the ten million new cases were estimated to be persons with HIV [13], but in parts of Southern Africa the co-infection rate often exceeds 50% [13]. In high TB/HIV prevalence countries, the peak TB incidence in TB/HIV co-infected adults occurs between 25–54 years [14]. These are men and women of reproductive age who are more likely to have young children within their household. In a prospective isoniazid prophylaxis study done in Cape Town, South Africa, 10% of HIV-exposed infants already had documented TB exposure at the time of screening for enrolment (before 4 months of age) [15]. Household contact is one of the major risk factors for TB in children [16]. There is also evidence that maternal TB during pregnancy may not only worsen HIV disease control in the mother, but also increase the risk of HIV transmission to the infant [17].

TB and HIV co-infection represents a vicious cycle where HIV increases susceptibility to TB and TB worsens HIV disease progression. In HIV uninfected persons the incidence of TB disease in those infected with TB is roughly estimated as 10% over a lifetime, although this varies greatly in different age and vulnerability groups, but TB disease risk is often reported as 10% per year in someone with HIV who is not on ART [18]. HIV targets the CD4 cells, which are key for the body's defence against TB. The risk of TB disease increases as the CD4 count falls, but low CD4 counts alone do not account for the increased TB risk in HIV-infected individuals, as studies have shown that even after immune reconstitution with ART an HIV-infected individual remains at increased risk of TB [19].

The mechanisms of how HIV and TB interact are complex. HIV decreases the body's ability to recognize foreign antigens by inhibiting interleukin 2 production and function, down regulating CD4+ receptors on T-cells and blocking HLA II expression. These all lead to the failure and expansion of CD4+ specific cells to fight TB infection. In addition, mononuclear cells in HIV lose the ability to secrete the T_1 helper cell stimulating cytokines, shifting responses from a pro T_1 helper pathway with IFN –gamma and interleukin (IL) 12 to a pro T_2 helper pathway with increased IL4 and IL10. HIV also decreases CD40 ligand leading to decreased macrophage activation with reduced IL12 production [20]. These dysfunctional responses lead to poorly organized granuloma formation with poor MTB containment.

TB also exacerbates HIV disease progression by providing microenvironments that enhance HIV replication, such as activating macrophages and other mononuclear cells that HIV uses for replication. This leads to increased HIV replication, with

Table 2 Overview of microorganisms and clinical disease syndromes observed in HIV-infected children[a]

Organism	Clinical syndrome
Bacteria	
Haemophilus influenza	U/LRTI, meningitis, bacteremia
Streptococcus pneumoniae	U/LRTI, meningitis, bacteremia
Klebsiella pneumonia	U/LRTI, UTI, meningitis, bacteremia
Salmonella spp.	LRTI, gastroenteritis, meningitis, bacteremia
Escherichia coli	LRTI, UTI, meningitis, bacteremia
Staphylococcus aureus	LRTI, meningitis, bacteremia, bone skin and joint
Virus	
Common respiratory viruses	U/LRTI
Cytomegalovirus (CMV)	Congenital infection, pneumonia, retinitis, esophagitis, colitis
Herpes simplex	Skin and soft tissue, ulcers, keratitis, encephalitis
EBV	Cancers
Varicella zoster	Chickenpox and zoster, keratitis, retinitis encephalitis
Influenza	U/LRTI
Molluscum contagiosum	Skin infections
Human papilloma virus	Skin, mouth and upper airway infections, genital warts, cancer risk
JC and BK virus	Progressive multifocal leukoencephalopathy (rarely reported in children)
Measles	Increased risk of severe disease
Mycobacteria	
Mycobacterium tuberculosis	All forms of tuberculosis
M. bovis BCG	Young infants not on ART are at risk of BCG dissemination and children starting ART of BCG IRIS
M.avium-intracellulare (MAC)	Chronic lung infection
Fungi	
Cryptococcus spp.	CNS, lung, skin infection
Candida spp.	Mucosal, nail and skin infection
Pneumocystis jiroveci (PCP)	Chronic lung infection
Protozoa	
Crypto and microsporidium	Gastrointestinal and bile duct infection
Toxoplasma gondii	Congenital infection and CNS disease

HIV human immunodeficiency virus, *URTI* upper respiratory tract infection, *LRTI* lower respiratory tract infection, *UTI* urinary tract infection, *PLE* progressive multifocal leukoencephalopathy, *ART* antiretroviral therapy, *BCG* Bacille Calmette-Guerin, *IRIS* Immune Reconstitution Inflammatory Syndrome, *CNS* central nervous system

Source: Viral Infections in Children, Volume I 2017, Springer Nature ISBN 978–3–319-54,032-0 Chief Editor: Green RJ Chap. 4: Rabie H, Marais BJ. Tuberculosis and other opportunistic infections in HIV-infected children. pp1–26

[a]Especially in children with poor disease control and significant immune compromise

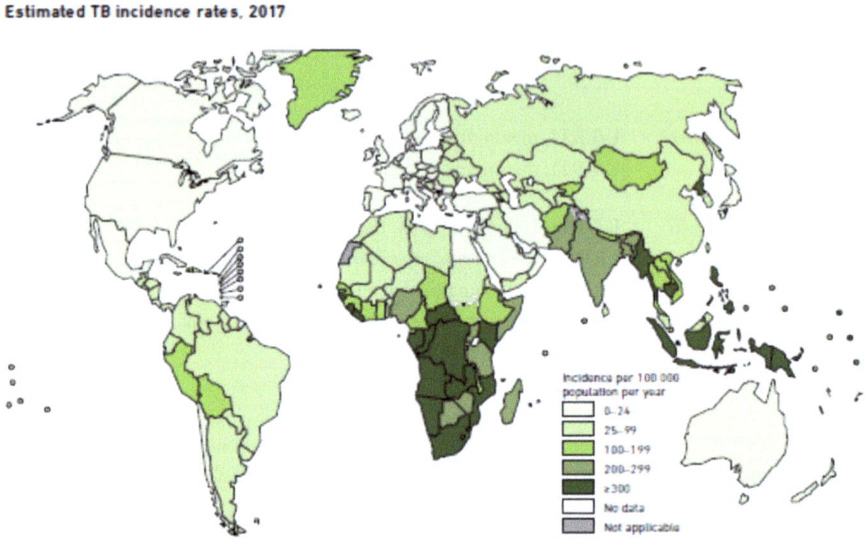

Fig. 1 Global estimated TB incidence rates in 2017 (Source: Global TB Report 2018)

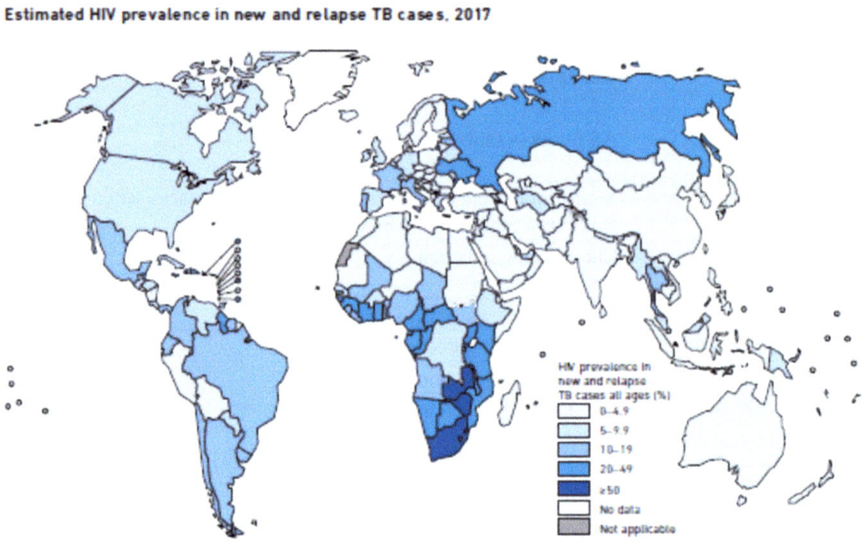

Fig. 2 Global estimated HIV prevalence in new and relapse TB cases, 2017 (Source: Global TB Report 2018)

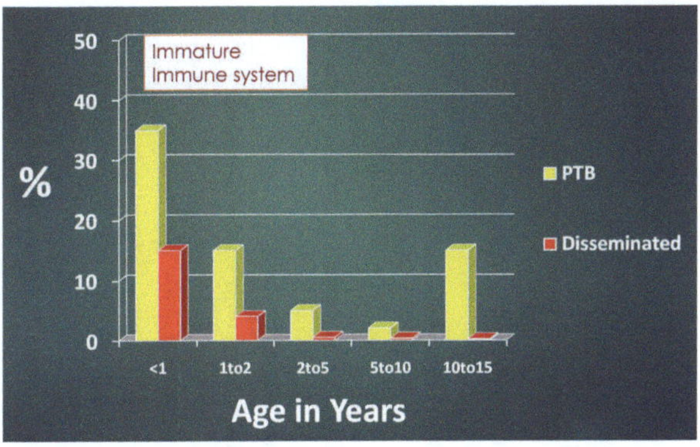

Fig. 3 Age related risk of TB disease after infection*. Adapted from Marais BJ et al- reference 22. *Children with significant immunocompromise experience similar high risk as those observed in very young immune-immature children <2 years of age

higher viral loads and increased CD4 T-cell loss [21]. In addition, TB creates a state of immune activation that encourages increased HIV-1 genotypic heterogeneity and reduces the production of new CD4 T-cells [21].

The natural history of TB in children was well documented in the pre-chemotherapeutic era. Age related immunocompetence was historically the major determinant for disease following infection in immunocompetent hosts [22]. See Fig. 3 for the age-related risk and Table 3 for clinical manifestations of TB in children with age stratification (note that dark shaded bars in Fig. 3 represent extrapulmonary TB). As seen in Fig. 3 infants and young children without HIV are also at risk of disseminated TB and severe TB that result from failure of the body to contain the Ghon complex after infection [23]. Cavitation in young infants and immunocompromised older children is thought to be due to poor containment of the primary infection and without treatment predicts a poor outcome while in adolescents cavitation is thought to be an excessive response as in adults [23]. The other factor that is key in determining the TB manifestations seems to be time from exposure with the highest risk for active disease being the first year after infection [23]. An HIV-infected child will have the age related differences in immunocompetence as well as those associated with the acquired immunodeficiency related to HIV. Studies have clearly demonstrated that the manifestations of TB in the HIV-infected child not on ART are similar to those of young infants [24–26]. Knowledge of the increased risk of young children and the immunocompromised child along with the association with recent infection led to several proven strategies for prevention of TB in childhood that will be discussed later.

Diagnosing Pediatric TB

Establishing an accurate TB diagnosis in children can be a clinical challenge, which is greatly exacerbated by HIV co-infection. This is due to the paucibacillary nature of childhood TB disease, difficulty in obtaining adequate respiratory samples from

Table 3 Clinical syndromes associated with tuberculosis in children

Pathological classification	Disease phase (time period from infection)	Clinical syndromes	Age related risk groups[a]	Pathogenesis	Imaging manifestations
Primary *M. tb* infection	**Incubation** (0–6 weeks)	Asymptomatic	All ages	No adaptive immunity TST(−); IGRA(−)	None
	Infection (1–3 months)	Self-limiting symptoms (mild, viral-like)		Adaptive immunity IGRA(+); TST(+) No test to register reinfection	Transient hilar or mediastinal lymphadenopathy (50–70% of cases), rarely visible transient Ghon focus
		Hypersensitivity reactions (fever, erythema nodosum, phlyctenular conjunctivitis)			
Early disease progression >90% of disease occurs within 12 months of primary infection	**Very early** (2–6 months)	Uncomplicated lymph node disease	<10 years	Inadequate innate and/or adaptive immunity TST(+); IGRA(+) May be negative with immune compromise or extensive disease, cannot be used as "rule out" tests	Hilar or mediastinal lymphadenopathy without airway or parenchymal involvement
		Progressive Ghon focus	<1 year		Ghon focus with visible cavitation
		Disseminated disease: – Miliary disease – TB meningitis	<3 years		– discrete lung nodules (1–2 mm) on CXR; Hepato- splenomegaly – hydrocephalus; basal enhancement; brain infarcts and/or tuberculomas
	Early (4–12 months)	Complicated lymph node disease – airway compression – Expansile caseating pneumonia – infiltration of adjacent anatomic structures (esophagus, phrenic nerve, pericardium)	<5 years		– hyperinflation or atelectasis/collapse – Expansile consolidation of a segment or lobe – Tracheo–/broncho- esophageal fistula; pericardial effusion; hemidiaphragmatic palsy
		Pleural disease – exudative effusion (rarely empyema; or chylothorax)	>3 years		Effusion usually unilateral; some pleural thickening & loculations (due to fibrinous strands)
		Lymphadenitis – Most common extra-thoracic manifestation	1–10 years		Usually not needed, matting & edema of adjacent soft tissue

Late disease progression Generally rare apart from adult-type disease in adolescents	**Late** (1–3 years)	Adult-type pulmonary disease – difficult to differentiate primary infection; reactivation and reinfection disease	≥8 years	"Overagressive" innate and/or adaptive immunity	Apical cavities; may be bilateral; minimal or no lymph node enlargement (previously referred to as post-primary TB)
		Osteoarticular disease: – spondylitis/arthritis/osteomyelitis	≥5 years	Inadequate local control; usually local manifestations only, but can disseminate from any active focus	Periarticular osteopenia, subchondral cystic erosions, joint space narrowing
	Very late (>3 years)	Urinary tract (renal, ureter, bladder) disease	>5 years		Renal calcifications; hydronephrosis, calyceal dilation and/or ureter stricture

Age ranges, risk groups and timelines provide general guidance only

[a]HIV-infected children with immunocompromise are particularly vulnerable, similar to children <3 years of age, and may present with atypical features.

Adapted from Perez-Velez CM, Marais BJ. Tuberculosis in Children. N Engl J Med 2012; 367: 348–61 young children and overlapping clinical symptoms with other common diseases of childhood and HIV. Thus, WHO recommends a combination of clinical, radiological and microbiologic assessments to diagnose TB in children [27]. Clinical symptoms of pulmonary (intra-thoracic) TB include a persistent non-remitting cough, fever, poor energy levels and weight loss or failure to thrive. Radiologic findings include a chest x-ray with hilar or paratracheal nodes, persistent parenchymal opacification or a miliary pattern. Studies have shown that clinical symptoms in HIV-infected children are less sensitive and specific for TB diagnostic purposes, since HIV-infected children often have chronic symptoms that are not related to TB [28]. TB symptoms in the young and/or immunocompromised child can also be more acute and thus "chronic" symptoms have reduced sensitivity in the HIV co-infected population [29, 30]. TB has been identified as the second most frequent pathogen isolated in a South African cohort of children admitted for community acquired pneumonia who failed first-line treatment [31]. The physical examination for both the HIV-infected and uninfected child with TB is rarely informative [32]. TB must be considered in any child with unexplained symptoms, particularly in those with accompanying weight loss or failure to thrive, or clinical deterioration relative to their previous baseline not responding to first-line treatment.

In children with suspected pulmonary TB one of the challenges is obtaining sputum for bacteriologic confirmation via acid-fast bacilli staining (AFB), culture or nucleic acid amplification test (NAAT) based methods. TB is frequently not detected in young children because the burden of TB bacilli in a diseased child's lung is often low and may be below the threshold of detection via AFB staining and culture. For TB to be detected via AFB one needs 10,000 bacilli/mL of sputum and for culture 100 bacilli/mL [33]. A second reason for difficulty in obtaining bacteriologic confirmation is that young children are unable to cough out sputum. When they cough, children swallow the sputum into their stomachs. It is for these reasons that the techniques of obtaining samples via gastric aspiration (GA) or sputum induction (SI) were developed. For GA, the theory is that a child coughs the sputum during the night and swallows it. Obtaining an early morning gastric sample potentially detects the mycobacterium in the gastric contents. There have been many studies to assess the yield of this method in both inpatient and outpatient settings using a range of methodologies. Bacteriologic confirmation varied significantly depending on the presentation of intrathoracic disease, from as low as 35% in lymph node disease to as high as 100% in adult type disease [34]. A challenge is the gold standard to which one compares the yield of the GA or SI. Thus, the yield for AFB and culture has varied significantly depending on the method of TB diagnosis [35–42]. Maciel et al. in their systematic review of GA protocols for children less than 15 years found the sensitivity to range from 0 to 92% for culture [43]. In a clinical review of GAs from children less than 15 years, Stockdale reported the sensitivity of smear as between 2–47% and culture yield between 37–74%. The yield was similar between inpatients and outpatients [44]. Thus, although a positive TB culture is ideal for diagnosis in young children with pulmonary TB disease, the consensus is that it is, at best, positive in 30–40% of cases [45]. SI methods have had similar variability in yield [36, 37, 46, 47].

MTB GeneXpert has been introduced as a NAAT-based method to diagnose TB. The performance in children on spontaneous or induced sputum and gastric aspirates for intrathoracic disease has overall been better than smear microscopy but not as good as culture. A meta-analysis described a pooled sensitivity of 62% compared to culture [48]. Under more routine conditions MTB GeneXpert compared to clinical diagnosis instead of culture, has been found to provide a similar yield regardless of HIV status [49]. Although NAAT has an inferior yield compared to culture it does confer the significant advantage of a much faster test turnaround time of 2 h and there may be increased sensitivity with MTB GeneXpert Ultra.

Based on the same principle as gastric aspiration whereby children swallow their sputa and coupled with the likelihood that MTB DNA survived the gastrointestinal tract environment, MTB GeneXpert has more recently been applied to stool specimens in children with promising results [50–53]. Chipinduro et al. found stool GeneXpert to be positive in 13/19 (68%) microbiologically confirmed cases and 4/199 (2%) microbiologically negative cases. This study was in children over 5 years old and the yield was similar in HIV positive and negative patients [54].

Lipoarabinomannan (LAM) is a main component of MTB cell wall that is excreted in the urine. Urine LAM concentration has been evaluated as a diagnostic test for TB. It has been shown to be useful primarily in severely immunocompromised adult patients with disseminated TB [55]. The utility in children has been variable. Nicol et al. showed poor sensitivity of 48.3% and specificity of 60.8% in a study of 535 children [56]. However, in a Tanzanian study of 132 children, LAM had a sensitivity of 50–70% in HIV-infected children compared to 0–13% in HIV-uninfected children. The specificity (97.3%) was the same in both infected and uninfected children suggesting that, as in adults, there may be a subset of children where it is useful [57]. This needs further exploration.

The quality of any specimen sent for TB testing, regardless of type, affects the test yield. For example, standardizing the procedure for gastric aspiration increases the yield from 8% to 50% [58]. Similarly, a meta-analysis among adults demonstrated that sputum collected after clear instructions and after pooling increased the diagnostic yield of both smear and culture [59]. Peter et al. worked with outpatient adults who were either smear negative or unable to produce sputum and evaluated the impact of enhanced instruction to produce sputum as compared to SI. They found SI increased sputum volume and had improved culture yield but same-day diagnosis via smear or GeneXpert was similar between intervention groups [60]. There are no other studies exploring this concept in children but extrapolating from adult studies, improved technique and quality of TB specimens in children is likely to improve the yield of TB.

Thus without good confirmatory tests for TB in children other approaches should be used to support the diagnosis of TB in a child. The most commonly used techniques - the Tuberculin Skin Test (TST), Interferon Gamma Release Assays (IGRA) and chest x-ray - each have their limitations. TST and IGRA are unable to distinguish between infection and disease. In two meta-analyses, the sensitivity between TST and IGRA is similar although the IGRA may be more specific [61, 62]. Both are less sensitive in HIV-positive and malnourished children and both may be affected by age [63]. A study from Botswana found IGRA's to be positive in only 1% of HIV positive children in this TB endemic area [64].

Chest x-ray has historically been used to assess for evidence of intrathoracic TB. However, in children, and particularly in those under the age of 10 years, the classic "adult" finding of a lung cavity is rare. The chest x-ray findings in children are not specific for tuberculosis. Marais et al. demonstrated that the most common chest x-ray finding is children is hilar and subcarinal adenopathy [65]. However, it has been suggested that more than 50% of children presenting with symptomatic disease have findings on chest x-ray beyond simple hilar or subcarinal adenopathy [66]. Common chest x-ray findings in children with TB include focal nodules, air-space disease, collapse and a miliary pattern [67]. In the HIV-infected child, these findings may be caused by other HIV-related illnesses from severe bacterial pneumonia to lymphoid interstitial pneumonitis (LIP). LIP can mimic the miliary pattern that is classic for disseminated tuberculosis [68]. Fortunately, there may be other signs and symptoms that help distinguish between LIP and TB. LIP tends to be in the older child and accompanied by clubbing and parotid enlargement while miliary TB is seen in a younger child without the other features. High-resolution chest computed tomography has also been used to evaluate for evidence of intrathoracic tuberculosis. It is more sensitive than chest x-ray but is not specific and, similar to chest x-ray, may be interpreted in different ways when assessed by different clinicians or radiologists [69, 70].

Thus, without a high performing, sensitive and specific gold standard confirmatory diagnostic test for childhood TB, scoring systems have been developed. These aim to use a combination of clinical history, symptoms and diagnostic data to create a score which is used to determine the likelihood of a diagnosis of intrathoracic TB. No scoring systems have ideal sensitivity and specificity for a diagnostic test. Sant'Anna et al. from Brazil reported a scoring system that had a sensitivity of 58–89% and a specificity of 86–98% depending on the cut point used [71]. In this population with low rates of HIV co-infection, a scoring system had higher sensitivity than culture but a significant proportion of children with TB would still be missed. In a study from a single hospital serving a high HIV prevalence population in Zambia in the era before ART, Van Reneen found that a scoring system had a sensitivity of 88% but a specificity of only 25% [72]. This would lead to significant over treatment. Marais et al. showed that a refined set of symptoms of a cough for greater than 2 weeks, failure to thrive for 3 months and fatigue were useful in the diagnosis of TB in HIV uninfected children but not as useful in the HIV-infected child [28]. Given the overall poor sensitivity and specificity of these scoring systems, they are generally not recommended, particularly in the HIV-infected child. Thus, the diagnosis of TB requires an astute clinician to use a classic diagnostic approach and sound clinical reasoning applying all tools available including a high index of suspicion, a quality history and physical examination, and close follow-up when uncertain.

Confirmatory testing for extra-pulmonary TB is also challenging and the recommended approach varies depending on the site and type of specimen tested. A meta-analysis of MTB GeneXpert applied to extra-pulmonary samples found consistently high specificities but heterogeneous sensitivities, from a low of 34% for pleural fluid samples to a high of 96% for lymph node tissue [73]. This is consistent with other studies. For example, there has been variable results of NAAT-based tests on fine needle aspiration or biopsy of peripheral nodes [74, 75]. For CSF, Rufai et al. reported a sensitivity of 55% and specificity of 94.8% in CSF samples from patients with high clinical-radiological suspicion for TBM [76]. Table 4 summarizes the benefits and drawbacks of different samples being sent for culture and NAAT testing [77]. For most sites, the yield in HIV positive patients is similar to that in HIV negative patients.

For extrapulmonary TB, other investigations should also be done to evaluate for evidence of TB. These include fine needle aspiration or biopsy of a lymph node to look for classic TB pathological changes such as granulomas, giant cells and AFBs. Computed tomography is frequently used to assess for features of central nervous system TB including communicating hydrocephalus, basilar enhancement, tuberculomas, abscesses, and evidence of infarcts or vasculitis. Magnetic resonance imaging (MRI) is the imaging test of choice for evidence of TB meningitis as it is superior for detecting basilar enhancement. However, the limited availability of MRI in areas of the world most affected by TB limits its wider use [67].

Table 4 Tuberculosis specimen collection methods—perceived problems and benefits*

Specimen collection method	Problems/benefits	Potential clinical application
Sputum	Not feasible in very young children; assistance and supervision may improve the quality of the specimen	Routine sample to be collected in children >7 yrs. of age (all children who can produce a good quality specimen) and evaluate for gene Xpert® and culture
Induced sputum	Comparable yield to gastric aspirate; no age restriction; specialized technique, which requires nebulization and suction facilities; potential transmission risk	To be considered in the hospital setting on an in- or out-patient basis. Evaluate using gene Xpert® and culture
Gastric aspirate	Unpleasant procedure, but not difficult to perform; requires fasting Sample collection advised on 3 consecutive days	Routine sample to be collected in hospitalized who cannot produce a good quality sputum specimen. Evaluate using gene Xpert® and culture
Nasopharyngeal aspiration	Less invasive than gastric aspirate; no fasting required; Comparable yield to gastric aspirate	To be considered in primary health care clinics or on an outpatient basis. Evaluate using gene Xpert® and culture
String test	Less invasive than gastric aspirate; tolerated well in children >4 years; bacteriologic yield and feasibility requires further investigation	Potential to become the routine sample collected in children who can swallow the capsule, but cannot produce a good quality sputum specimen

(continued)

Table 4 (continued)

Specimen collection method	Problems/benefits	Potential clinical application
Broncho-alveolar lavage	Extremely invasive	Only for use in patients who are intubated or who require diagnostic bronchoscopy. Evaluate using gene Xpert® and culture
Stool	Culture not practical, DNA extraction difficult Not invasive; *M. tuberculosis* excretion well documented	Reasonable yield using gene Xpert®
Urine	Not invasive; excretion of *M. tuberculosis* components	Lipoarabinomanan (LAM) assay has poor sensitivity; unreliable in children although as in adults it may be useful in a subset of severely ill children
Blood/bone marrow	Good sample sources to consider in the case of probable disseminated TB	To be considered for the confirmation of probable disseminated TB in hospitalized patients. Evaluate using culture
Cerebrospinal fluid (CSF)	Fairly invasive; bacteriologic yield low, better with more CSF, preferably >10 mL	To be considered if signs of tuberculous meningitis. Evaluate using culture and gene Xpert®
Fine needle aspiration biopsy (FNAB)	Minimally invasive using a fine 23G needle; excellent bacteriologic yield, minimal side-effects	Procedure of choice in children with superficial lymphadenopathy. Evaluate based on pathology, gene Xpert® and culture

Adapted from BJ Marais et al. ref. 56

TB Treatment in the HIV-infected Child

Once TB is diagnosed, treatment is initiated immediately. Historically the approach for treating childhood TB has been extrapolated from knowledge gained in the adult population, although there are now a number of ongoing pediatric pharmacokinetics and pharmacodynamics studies. The same first line drugs of isoniazid, rifampicin, pyrazinamide and ethambutol are used to treat childhood TB. In order to achieve clinical and microbiologic cure, current TB treatment regimens require a multidrug approach for at least 6 months. Treatment may need to be longer in certain populations such as severely immunocompromised HIV-infected persons who are at increased risk of relapse [78]. This prolonged multidrug approach is challenging for families [79]. Healthcare providers and systems need to be aware of the burden a prolonged multidrug regimen places on a family and provide education and support for the child and family affected by TB. Child friendly formulations of the drugs are now available and practitioners caring for children with TB need to be advocates for the use of these formulations to ease the burden of TB treatment.

The combination of first line drugs together attack the MTB in the various stages of its lifecycle. Isoniazid is bactericidal against rapidly multiplying mycobacteria but is bacteriostatic against those with slow growth [80]. It is therefore effective at reducing the mycobacterial load early in treatment thereby decreasing infectivity, improving symptoms and preventing resistance. However once the turnover of mycobacteria is reduced, isoniazid is no longer sterilizing [80]. Rifampicin, by contrast, attacks mycobacteria that are multiplying more slowly and thus is an effective sterilizing drug [81]. If, for whatever reason, rifampicin is not one of the drugs used to treat TB, then to ensure sterilization the treatment regimen duration must be longer than 6 months. Pyrazinamide effectively attacks mycobacteria in an acidic environment, which is usually present at the time of diagnosis. It has a unique role to augment the action of rifampicin and the historic introduction of pyrazinamide shortened treatment regimens from 9 to 6 months [82]. Ethambutol's role is to protect against drug resistance developing to rifampicin [80]. It has a risk of causing optic neuritis at higher doses and when used for longer periods [80]. Thus, it has historically not been as frequently included in pediatric regimens where resistance was a lower concern and detecting ocular toxicity was challenging, particularly in younger children. However, WHO now recommends its use across the age groups in countries where there is greater than 1% INH resistance and a high HIV burden [83]. Studies have shown that at the recommended dose in children of 20–25 mg/kg/day for 2 months, that optic neuritis is a rare and reversible complication [84].

Table 5 presents the recommended dosing of the first line TB agents [85]. Research in recent years has shown that children generally metabolize TB medication faster than adults and thus higher doses of particularly INH at 10 mg/kg/day are now recommended [86]. In locations where the new pediatric combination therapy that has INH and rifampicin in a 1:1.5 ratio is unavailable, extra INH may need to be added to the regimen to ensure adequate INH dosing. Dose optimization is particularly relevant for children with HIV who are immunocompromised and who may have higher TB relapse rates. In addition, recent data suggests that TB outcomes in the adolescent population, particularly those that are dually infected, are poor [87, 88]. The reasons for this are likely complex and further data are needed to inform approaches to improve outcomes in this population.

There is growing TB drug resistance globally. In any child with possible drug resistant TB extra effort should be sought to confirm the diagnosis and send samples for susceptibility testing. High risk groups include a child who does not respond well to treatment, has a contact with known drug resistant TB, or has contact with a person who died on TB treatment or had poor response to TB treatment. In addition, if a child has a TB contact that has drug resistant TB, the child's treatment regimen should be tailored, based on the adult contact's resistance pattern. Resistant TB is as transmissible as drug sensitive TB and studies have shown that children who are in contact with a drug resistant TB case and contract TB are most likely to have the same drug resistant strain as the adult contact [89, 90]. These cases should be discussed with an expert in drug resistant TB. Fortunately, children seem to respond well to second line treatment with fewer side effects than adults [91].

In the HIV-infected child antiretroviral therapy is a key component of treatment. Prior to the "treat all" strategy for HIV control, TB was a clinical indication for ART with multiple studies showing that ART reduces TB morbidity and mortality [5, 19, 92].

Table 5 Use of first and second-line TB treatment with antiretrovirals in children with TB/HIV co-infection

TB drugs	Recommended dose	Drug-drug interactions with ART
First-line treatment		
Isoniazid	7–15 mg/kg once daily; max 300 mg	None
Rifampicin	10–20 mg/kg once daily; max 600 mg	Reduces plasma levels of NNRTIs, PIs and integrase inhibitors Examples of dose adjustments: Efavirenz- no dose adjustment Lopinavir/ritonavir- increase ritonavir to 1:1 ratio with lopinavir Dolutegravir- dose twice daily instead of once daily
Pyrazinamide	30–40 mg/kg once daily; max	None
Ethambutol	15–25 mg/kg once daily; max	None
Rifabutin	10–20 mg/kg/day; max 300 mg	Boosted PI: Increase rifabutin levels; NNRTI: Efavirenz reduces the rifabutin levels; no dose adjustment with nevirapine
Second-line treatment		
WHO Group A Drugs- known good potency		
Bedaquiline Age > 6 yrs.	(6 mg/kg/day for 14 days followed by 3–4 mg/kg thrice weekly (dose extrapolated from adult dosing for those less than 16 kg)	Efavirenz: Reduced BDQ levels Lopinavir/ritonavir: Increased Bedaquiline levels
Levofloxacin	15–20 mg/kg once daily; max 750 mg	Buffered didanosine may reduce oral absorption of all fluoroquinolones
Moxifloxacin	7.5–10 mg/kg once daily; max 400 mg	Ritonavir may reduce moxifloxacin levels
Linezolid	≥16 years: 10–12 mg/kg/day ≤ 16 years: 15 mg/kg/day	NRTIs: Increased risk for adverse effects
WHO Group B Drugs- reasonable potency		
Cycloserine/Terizidone	10–20 mg/kg once daily, max 1 gram	Unlikely
Clofazimine	2-5 mg/kg/day On alternate days if gelcaps cannot be split Or weight based dosing*	None documented; may be a weak CYP3A4 inhibitor

(continued)

Table 5 (continued)

TB drugs	Recommended dose	Drug-drug interactions with ART
WHO Group C Drugs- other drugs with uncertain/poor potency		
Pro/Ethionamide	15–20 mg/kg once daily; max 1 g	Possible
Delaminid Age > 3 years	3–4 mg/kg/day (dose extrapolated from adult dosing for those less than 10 kg)	None documented
Meropenem/clavulanic acid imipenem/cilastin	As for bacterial infections	Unlikely
Thiacetazone	5–8 mg/kg once daily	Contraindicated in HIV-infected individuals
Para-aminosalicylic acid (PAS)	150–200 mg/kg granules daily in 2 divided doses, max 12 g	Efavirenz may reduce PAS levels
Kana/Capreomycin	15–30 mg/kg once daily; max 1 g	Unlikely
Amikacin	15–22.5 mg/kg once daily; max 1 g	Unlikely
Streptomycin	20–40 mg/kg once daily; max 1 g	Should not be used in children

TB tuberculosis, *ART* antiretroviral therapy, *NNRTI* non-nucleoside reverse transcriptase inhibitor, *PI* protease inhibitor, *AUC* area under the curve, *HIV* human immunodeficiency virus
∗For weight based dosing refer to "Management of multi-drug resistant tuberculosis in children: A field Guide fourth Edition"; ∗Available via the Global TB Consillium (tbconsilium@gmail.com) or Sentinel Project on Paediatric Drug Resistant TB (tbsentinelproject@gmail.com)
Adapted from Marais BJ et al. ref. 77

WHO recommends ART initiation within 2–8 weeks of starting ATT in the co-infected child [93]. This recommendation, extrapolated from adult studies, is supported by studies describing decreased morbidity and mortality with early ART in those with more severe immunocompromise [94–96]. A study from South Africa found that delayed ART beyond 8 weeks in co-infected children was associated with increased mortality and worse outcomes [97]. The same adult studies did not show increased benefit with early ART in mildly immunocompromised individuals and thus in such patients delaying ART may be considered to avoid drug-drug interactions and high pill burdens. The risk-benefit considerations must be carefully weighed in each situation.

Treatment of co-infected children is more complex because of the interaction of TB and HIV drugs, high pill burdens and the potential for overlapping toxicities. ART regimen choices will be guided by local recommendations, availability and age of the child. The rifamycins, particularly rifampicin is a potent inducer of the cytochrome 3A4 [80]. Protease inhibitors and nevirapine are metabolized through the CYP 450 pathway and thus blood concentrations are lowered when co-administered with rifampicin. Drug dosing must be adjusted when using these drugs in conjunction with TB treatment. Nevirapine should be dosed at the upper end of the range and an induction phase should not be used [98]. Lopinavir/ritonavir should have the ritonavir increased to a 1:1 ratio with the lopinavir [99, 100]. Treatment with a triple nucleoside reverse transcriptase inhibitor regimen in virologically suppressed children was studied in a recent trial. It showed similar viral suppression and clinical response to a non-nucleoside reverse transcriptase inhibitor based regimen in those children who had well controlled HIV before the switch [101]. Thus, this may be an alternative regimen for a subset of children needing TB treatment but again the risk benefit of a less efficacious ART regimen with less drug interaction must be carefully considered. The newer integrase inhibitor based ART regimens also need dose adjusting based on early adult pharmacokinetic studies but this has not yet been studied in the pediatric population [102]. See Table 5 for TB drug dosing and ART interactions [85].

As previously mentioned there has been some concern that HIV-infected children experience a TB relapse more frequently. However, the evidence for prolonged treatment to prevent relapse is weak and thus WHO recommends the standard length of treatment for those who respond well but suggests clinicians consider prolongation in those who have a poor response [83].

Immune Reconstitution Inflammatory Syndrome (IRIS)

Once ART is initiated the body may have a robust immune recovery with improvement in CD4+ T cells and drop in viral load. In patients with severe immunosuppression, this can lead to an unmasking of infections previously not diagnosed or a paradoxical worsening of symptoms of diagnosed TB. This usually occurs within the first 3 months of ART [103]. However, the very population of the more severely immunocompromised child is the same population that benefits from early ART. Thus, ART should not be withheld for fear of IRIS. Usually these symptoms are not life threatening, can be managed with supportive care and resolve on their own. Treatment of the symptoms with non-steroidal anti-inflammatory drugs or steroids may be considered [104].

TB Prevention

The saying goes that "Prevention is better than cure." This certainly is true for childhood tuberculosis given the complexities of diagnosis, treatment and management in the TB-HIV co-infected child. Eradication of adult and adolescent TB cases is the single most important way to prevent childhood TB infection. Vaccines are one of the best primary prevention methods but currently the only available TB vaccine is bacillus Calmette-Guerin (BCG), a live attenuated vaccine, which provides sub-optimal protection against TB. In TB-endemic areas in HIV negative children it affords some protection against severe forms of tuberculosis, TBM and miliary TB, but does not protect against all forms of TB and does not give durable protection into adulthood [105]. Its efficacy in the HIV-infected child is poor [106]. Secondly, as this is a live attenuated vaccine, studies have shown that the risk of disseminated BCG disease is increased in this population [107, 108]. Thus, BCG is not recommended in the known HIV-infected child. However, in many endemic settings the HIV status of the child is unknown when BCG is given at birth. In these settings, BCG is recommended and emphasis is placed on strengthening prevention of mother to child transmission of HIV.

HIV-infected children are frequently interacting with health care systems and thus another area of prevention is TB infection control practices in healthcare facilities. Studies have shown that nosocomial transmission of TB does occur in health care settings [109]. There are guidelines on the types of TB infection control practices [110]. Thus, facilities that care for children should consider their infection control plans carefully such that children are not at increased risk of nosocomial infection with TB. For example, HIV-infected children being seen at a clinic where adults are also seen probably should be separated from adult patients.

Secondary prevention is also a known strategy to prevent TB disease in children. Because children tend to have primary disease within a year after infection there is an opportunity to prevent disease for those with a known exposure and no active disease. Isoniazid prophylaxis therapy (IPT) at 10 mg/kg/day for 6 months has been well studied for asymptomatic child contacts of TB cases and shown to reduce TB disease by up to 62% [111]. Other regimens such as isoniazid combined with rifapentine for 3 months or rifampicin for 4 months prevent TB and have the advantages of a shorter regimens with higher completion rates [112–114]. However, they can be more costly and have been studied in low TB burden countries [115].

No studies have specifically looked at IPT in the HIV-infected child with a known exposure and no active disease but the benefits are expected to be similar in this population and thus WHO recommends IPT for young children and HIV-infected children in contact with a case of pulmonary TB [83]. Universal IPT in the absence of a documented TB exposure has been well studied in adults and seems to be beneficial during the time of prophylaxis in this population where disease is a combination of reactivation and primary infection [116, 117]. The evidence for universal IPT in childhood has had mixed results. A Cape Town, South Africa study in the pre-ART era of over 300 children showed IPT decreased all-cause mortality from 16% to 8% and decreased TB incidence from 10% to 4% [118]. However, a randomized control trial in the post ART era of HIV positive and negative infants did not show benefit [119]. A recent Cochrane review including these studies suggested that IPT benefits children not on ART but is unlikely to benefit those already on ART. Both recommendations had low certainty of evidence [120].

Antiretroviral therapy is a potent prevention strategy for TB. In areas where ART coverage has increased, TB incidence has clearly decreased. The Children with HIV Early Antiretroviral (CHER) trial clearly demonstrated the benefit of early ART in decreasing TB risk in those on early ART (8% in the early ART arm vs 20% in the delayed ART arm) [5]. A South African study has also shown a 70% drop in culture confirmed TB in HIV-infected children when ART coverage changed from 43% in 2005 to 84% in 2009 [92]. This was reflected in the overall reduction in confirmed TB in all children in the same study (63%).

Conclusion

HIV-infected children are at increased risk of TB exposure, infection and disease. The diagnosis of TB in this population is a particular challenge given the wide disease spectrum, similarity with other common diseases and poor performance of current diagnostic tests in children. Although the majority of HIV-infected children do well on TB treatment, children with TB/HIV co-infection face high pill burdens, overlapping toxicities and drug interactions that can impact outcomes negatively. Given these complexities, everything possible should be done to prevent TB in the first place; by preventing HIV infection or treating it early if it does occur; improving respiratory infection control in health care facilities; as well as screening for recent TB exposure at every health care visit and providing preventive therapy for every documented TB exposure or infection event.

References

1. UNAIDS. Joint United Nations Programme on HIV/AIDS (UNAIDS), 'Fact Sheet – Latest Statistics on the Status of the AIDS Epidemic': UNAIDS; 2019 (Available from: http://www.unaids.org/en/resources/fact-sheet)
2. UNICEF (2019). Annual results report HIVAIDS 2017
3. UNAIDS (2016) Global AIDS Update 2017
4. Newell M-L, Coovadia H, Cortina-Borja M, Rollins N, Gaillard P, Dabis F (2004) Mortality of infected and uninfected infants born to HIV-infected mothers in Africa: a pooled analysis. Lancet 364(9441):1236–1243
5. Violari A, Cotton MF, Gibb DM, Babiker AG, Steyn J, Madhi SA et al (2008) Early antiretroviral therapy and mortality among HIV-infected infants. N Engl J Med 359(21):2233–2244
6. Moir S, Fauci AS (2017) B-cell responses to HIV infection. Immunol Rev 275(1):33–48
7. Clerici M, Stocks NI, Zajac RA, Boswell RN, Lucey DR, Via CS et al (1989) Detection of three distinct patterns of T helper cell dysfunction in asymptomatic, human immunodeficiency virus-seropositive patients. Independence of CD4+ cell numbers and clinical staging. J Clin Invest 84(6):1892–1899
8. Kerneis S, Launay O, Turbelin C, Batteux F, Hanslik T, Boelle PY (2014) Long-term immune responses to vaccination in HIV-infected patients: a systematic review and meta-analysis. Clin Infect Dis 58(8):1130–1139
9. MR BL, Drouin O, Bartlett G, Nguyen Q, Low A, Gavriilidis G et al (2016) Incidence and prevalence of opportunistic and other infections and the impact of antiretroviral therapy among HIV-infected children in low- and middle-income countries: a systematic review and meta-analysis. Clin Infect Dis 62(12):1586–1594
10. Walters E, Cotton MF, Rabie H, Schaaf HS, Walters LO, Marais BJ (2008) Clinical presentation and outcome of tuberculosis in human immunodeficiency virus infected children on anti-retroviral therapy. BMC Pediatr 8(1)
11. Ye P, Kirshner D, Kourtis A (2004) The thymus during HIV disease: role in pathogenisis and Immbe recovery. Curr HIV Res 2(2):177–183
12. Ssewanyana I, Baker C, Ruel T, Bousheri S, Kamya M, Dorsey G et al (2009) The distribution and immune profile of T cell subsets in HIV-infected children from Uganda. AIDS Res Hum Retrovir 25(1):65–71
13. World Health Organization (2019) Global Tuberculosis Report 2018

14. Lawn SD, Bekker LG, Middelkoop K, Myer L, Wood R (2006) Impact of HIV infection on the epidemiology of tuberculosis in a peri-urban community in South Africa: the need for age-specific interventions. Clin Infect Dis 42(7):1040–1047
15. Cotton MF, Schaaf HS, Lottering G, Weber HL, Coetzee J, Nachman S et al (2008) Tuberculosis exposure in HIV-exposed infants in a high-prevalence setting. Int J Tuberculosis Lung Dis 12(2):225–227
16. Wood R, Johnstone-Robertson S, Uys P, Hargrove J, Middelkoop K, Lawn SD et al (2010) Tuberculosis transmission to young children in a south African community: modeling household and community infection risks. Clin Infect Dis 51(4):401–408
17. Gupta A, Bhosale R, Kinikar A, Gupte N, Bharadwaj R, Kagal A et al (2011) Maternal tuberculosis: a risk factor for mother-to-child transmission of human immunodeficiency virus. J Infect Dis 203(3):358–363
18. Selwyn PA, Hartel D, Lewis VA, Schoenbaum EE, Vermund SH, Kein RS et al (1989) A prospective study of the risk of tuberculosis among intravenous drug users with human immunodeficiency virus infection. N Engl J Med 320(9):545–550
19. Badri M, Wilson D, Wood R (2002) Effect of highly active antiretroviral therapy on incidence of tuberculosis in South Africa: a cohort study. Lancet 359(9323):2059–2064
20. Lawn S, B S, Shinnick T (2002) Tuberculosis unleashed: the impact of human immunodeficiency virus infection on the host granulomatous response to *Mycobacterium tuberculosis*. Microbes Infect 4:635–646
21. Lawn SD, Butera ST, Folks TM (2001) Contribution of immune activation to the pathogenesis and transmission of human immunodeficiency virus type 1 infection. Clin Microbiol Rev 14(4):753–777. table of contents
22. Marais BJ, Gie RP, Schaaf HS, Hesseling AC, Obihara CC, Starke JJ et al (2004) The natural history of childhood intra-thoracic tuberculosis: a critical review of literature from the pre-chemotherapy era. Int J Tuberculosis Lung Dis 8(4):392–402
23. Marais B, Donald P (2009) The natural history of tuberculosis infection and disease in children. In: Schaaf HS, Zumla A (eds) Tuberculosis: a comprehensive clinical reference, 1st edn. Elsevier, India, pp 133–145
24. Palme IB, Gudetta B, Bruchfeld J, Muhe L, Giesecke J (2002) Impact of human immunodeficiency virus 1 infection on clinical presentation, treatment outcome and survival in a cohort of Ethiopian children with tuberculosis. Pediatr Infect Dis J 21(11):1053–1061
25. Madhi SA, Huebner RE, Doedens L, Aduc T, Wesley D, Cooper PA (2000) HIV-1 co-infection in children hospitalised with tuberculosis in South Africa. Int J Tuberculosis Lung Dis 4(5):448–454
26. Hesseling AC, Westra AE, Werschkull H, Donald PR, Beyers N, Hussey GD et al (2005) Outcome of HIV infected children with culture confirmed tuberculosis. Arch Dis Child 90(11):1171–1174
27. World Health Organization (2006) Guidance for National Tuberculosis Programmes on the Management of Tuberculosis in Children
28. Marais BJ, Gie RP, Hesseling AC, Schaaf HS, Lombard C, Enarson DA et al (2006) A refined symptom-based approach to diagnose pulmonary tuberculosis in children. Pediatrics 118(5):e1350–e1359
29. Moore DP, Klugman KP, Madhi SA (2010) Role of *Streptococcus pneumoniae* in hospitalization for acute community-acquired pneumonia associated with culture-confirmed *Mycobacterium tuberculosis* in children: a pneumococcal conjugate vaccine probe study. Pediatr Infect Dis J 29(12):1099–1004
30. Oliwa JN, Karumbi JM, Marais BJ, Madhi SA, Graham SM (2015) Tuberculosis as a cause or comorbidity of childhood pneumonia in tuberculosis-endemic areas: a systematic review. Lancet Respir Med 3(3):235–243
31. McNally LM, Jeena PM, Gajee K, Thula SA, Sturm AW, Cassol S et al (2007) Effect of age, polymicrobial disease, and maternal HIV status on treatment response and cause of severe pneumonia in south African children: a prospective descriptive study. Lancet 369(9571):1440–1451

32. Graham SM, Coulter JB, Gilks CF (2001) Pulmonary disease in HIV-infected African children. Int J Tuberculosis Lung Dis 5(1):12–23
33. Chaisson R, Nachega J (2010) Tuberculosis chapter. In: Oxford Textbook of medicine, 5th edn. Oxford University Press, Oxford, pp 810–831
34. Marais B, Hesseling AC, Gie RP, Schaaf HS, Enarson DA, Beyers N (2006) Bacteriologic yield in children with intrathoracic tuberculosis. Clin Infect Dis 42:e69–e71
35. Berggren Palme I, Gudetta B, Bruchfeld J, Eriksson M, Giesecke J (2004) Detection of *Mycobacterium tuberculosis* in gastric aspirate and sputum collected from Ethiopian HIV-positive and HIV-negative children in a mixed in- and outpatient setting. Acta Paediatr (Oslo, Norway: 1992) 93(3):311–315
36. Hatherill M, Hawkridge T, Zar HJ, Whitelaw A, Tameris M, Workman L et al (2009) Induced sputum or gastric lavage for community- based diagnosis of childhood pulmonary tuberculosis? Arch Dis Child 94(3):195–201
37. Zar HJ, Hanslo D, Apolles P, Swingler G, Hussey G (2005) Induced sputum versus gastric lavage for microbiological confirmation of pulmonary tuberculosis in infants and young children: a prospective study. Lancet 365(9454):130–134
38. Qureshi UA, Gupta AK, Mahajan B, Qurieshi MA, Altaf U, Parihar R et al (2011) Microbiological diagnosis of pulmonary tuberculosis in children: comparative study of induced sputum and gastric lavage. Indian J Pediatr 78(11):1429–1430
39. Somu N, Swaminathan S, Paramasivan CN, Vijayasekaran D, Chandrabhooshanam A, Vijayan VK et al (1995) Value of bronchoalveolar lavage and gastric lavage in the diagnosis of pulmonary tuberculosis in children. Tubercle Lung Dis 76(4):295–299
40. Abadco DL, Steiner P (1992) Gastric lavage is better than bronchoalveolar lavage for isolation of Mycobacterium tuberculosis in childhood pulmonary tuberculosis. Pediatr Infect Dis J 11(9):735–738
41. Oberhelman RA, Soto-Castellares G, Gilman RH, Caviedes L, Castillo ME, Kolevic L et al (2010) Diagnostic approaches for paediatric tuberculosis by use of different specimen types, culture methods, and PCR: a prospective case-control study. Lancet Infect Dis 10(9):612–620
42. Kiwanuka J, Graham SM, Coulter JB, Gondwe JS, Chilewani N, Carty H et al (2001) Diagnosis of pulmonary tuberculosis in children in an HIV-endemic area, Malawi. Ann Trop Paediatr 21(1):5–14
43. Maciel EL, Brotto LD, Sales CM, Zandonade E, Sant'anna CC (2010) Gastric lavage in the diagnosis of pulmonary tuberculosis in children: a systematic review. Rev de saude Publica 44(4):735–742
44. Stockdale AJ, Duke T, Graham S, Kelly J (2010) Evidence behind the WHO guidelines: hospital care for children: what is the diagnostic accuracy of gastric aspiration for the diagnosis of tuberculosis in children? J Trop Pediatr 56(5):291–298
45. Gie RP, Starke JR, Schaaf HS (2009) Intrathoracic tuberculosis in children. In: Zumla A, Schaaf HS (eds) Tuberculosis: a comprehensive clinical reference first. Elsevier, Haryana India, p 362
46. Zar HJ, Tannenbaum E, Apolles P, Roux P, Hanslo D, Hussey G (2000) Sputum induction for the diagnosis of pulmonary tuberculosis in infants and young children in an urban setting in South Africa. Arch Dis Child 82(4):305–308
47. Mukherjee A, Singh S, Lodha R, Singh V, Hesseling AC, Grewal HM et al (2013) Ambulatory gastric lavages provide better yields of *Mycobacterium tuberculosis* than induced sputum in children with intrathoracic tuberculosis. Pediatr Infect Dis J 32(12):1313–1317
48. Detjen AK, DiNardo AR, Leyden J, Steingart KR, Menzies D, Schiller I et al (2015) Xpert MTB/RIF assay for the diagnosis of pulmonary tuberculosis in children: a systematic review and meta-analysis. Lancet Respir Med 3(6):451–461
49. Bacha JM, Ngo K, Clowes P, Draper HR, Ntinginya EN, DiNardo A et al (2017) Why being an expert - despite xpert -remains crucial for children in high TB burden settings. BMC Infect Dis 17(1):123
50. Nicol MP, Spiers K, Workman L, Isaacs W, Munro J, Black F et al (2013) Xpert MTB/RIF testing of stool samples for the diagnosis of pulmonary tuberculosis in children. Clin infect Dis 57(3):e18–e21

51. Walters E, Gie RP, Hesseling AC, Friedrich SO, Diacon AH, Gie RP (2012) Rapid diagnosis of pediatric intrathoracic tuberculosis from stool samples using the Xpert MTB/RIF assay: a pilot study. Pediatr Infect Dis J 31(12):1316
52. Chipinduro M, Mateveke K, Makamure B, Ferrand RA, Gomo E (2017) Stool Xpert(R) MTB/RIF test for the diagnosis of childhood pulmonary tuberculosis at primary clinics in Zimbabwe. Int J Tuberculosis Lung Dis 21(2):161–166
53. Banada PP, Naidoo U, Deshpande S, Karim F, Flynn JL, O'Malley M et al (2016) A novel sample processing method for rapid detection of tuberculosis in the stool of pediatric patients using the Xpert MTB/RIF assay. PLoS One 11(3):e0151980
54. Chipinduro M, KM BM, Ferrand RA, Gomo E (2016) Stool XpertW MTB/RIF test for the diagnosis of childhood pulmonary tuberculosis at primary clinics in Zimbabwe. Int J Tuberculosis Lung Dis 21(2):161–166
55. Lawn SD, Kerkhoff AD, Vogt M, Wood R (2012) Diagnostic accuracy of a low-cost, urine antigen, point-of-care screening assay for HIV-associated pulmonary tuberculosis before antiretroviral therapy: a descriptive study. Lancet Infect Dis 12(3):201–209
56. Nicol MP, Allen V, Workman L, Isaacs W, Munro J, Pienaar S et al (2014) Urine lipoarabinomannan testing for diagnosis of pulmonary tuberculosis in children: a prospective study. Lancet Glob Health 2(5):e278–e284
57. Kroidl I, Clowes P, Reither K, Mtafya B, Rojas-Ponce G, Ntinginya EN et al (2015) Performance of urine lipoarabinomannan assays for paediatric tuberculosis in Tanzania. Eur Respir J 46(3):761–770
58. Pomputius IWF, Rost J, Dennehy PH, Carter EJ (1997) Standardization of gastric aspirate technique improves yield in the diagnosis of tuberculosis in children. Pediatr Infect Dis J 16(2):222–226
59. Datta S, Shah L, Gilman RH, Evans CA (2017) Comparison of sputum collection methods for tuberculosis diagnosis: a systematic review and pairwise and network meta-analysis. Lancet Glob Health 5(8):e760–ee71
60. Peter JG, Theron G, Pooran A, Thomas J, Pascoe M, Dheda K (2013) Comparison of two methods for acquisition of sputum samples for diagnosis of suspected tuberculosis in smear-negative or sputum-scarce people: a randomised controlled trial. Lancet Respir Med 1(6):471–478
61. Mandalakas AM, Detjen AK, Hesseling AC, Benedetti A, Menzies D (2011) Interferon-gamma release assays and childhood tuberculosis: systematic review and meta-analysis. Int J Tuberculosis Lung Dis 15(8):1018–1032
62. Machingaidze S, Wiysonge CS, Gonzalez-Angulo Y, Hatherill M, Moyo S, Hanekom W et al (2011) The utility of an interferon gamma release assay for diagnosis of latent tuberculosis infection and disease in children: a systematic review and meta-analysis. Pediatr Infect Dis J 30(8):694–700
63. Mandalakas AM, van Wyk S, Kirchner HL, Walzl G, Cotton M, Rabie H et al (2013) Detecting tuberculosis infection in HIV-infected children: a study of diagnostic accuracy, confounding and interaction. Pediatr Infect Dis J 32(3):e111–e118
64. Cruz AT, Marape M, Graviss EA, Starke JR (2015) Performance of the QuantiFERON-TB gold interferon gamma release assay among HIV-infected children in Botswana. J Int Assoc Provid AIDS Care 14(1):4–7
65. Marais BJ, Gie RP, Schaaf HS, Starke JR, Hesseling AC, Donald PR et al (2004) A proposed radiological classification of childhood intra-thoracic tuberculosis. Pediatr Radiol 34(11):886–894
66. Marais BJ, Gie RP, Hesseling AC, Schaaf HS, Enarson DA, Beyers N (2006) Radiographic signs and symptoms in children treated for tuberculosis: possible implications for symptom-based screening in resource-limited settings. Pediatr Infect Dis J 25(3):237–240
67. Andronikou S, Wieselthaler N (2004) Modern imaging of tuberculosis in children: thoracic, central nervous system and abdominal tuberculosis. Pediatr Radiol 34(11):861–875
68. Graham S (2005) Non-tuberculosis opportunistic infections and other lung diseases in HIV-infected infants and children. Int J Tuberculosis Lung Dis 9(6):592–602

69. Andronikou S, Joseph E, Lucas S, Brachmeyer S, Du Toit G, Zar H et al (2004) CT scanning for the detection of tuberculous mediastinal and hilar lymphadenopathy in children. Pediatr Radiol 34(3):232–236
70. Andronikou S, Brauer B, Galpin J, Brachmeyer S, Lucas S, Joseph E et al (2005) Interobserver variability in the detection of mediastinal and hilar lymph nodes on CT in children with suspected pulmonary tuberculosis. Pediatr Radiol 35(4):425–428
71. Sant'Anna CC, Orfaliais CT, March Mde F, Conde MB (2006) Evaluation of a proposed diagnostic scoring system for pulmonary tuberculosis in Brazilian children. Int J Tuberculosis Lung Dis 10(4):463–465
72. Van Rheenen P (2002) The use of the paediatric tuberculosis score chart in an HIV-endemic area. Trop Med Int Health 7(5):435–441
73. Maynard-Smith L, Larke N, Peters JA, Lawn SD (2014) Diagnostic accuracy of the Xpert MTB/RIF assay for extrapulmonary and pulmonary tuberculosis when testing non-respiratory samples: a systematic review. BMC Infect Dis 14:709
74. Bholla M, Kapalata N, Masika E, Chande H, Jugheli L, Sasamalo M et al (2016) Evaluation of Xpert(R) MTB/RIF and Ustar EasyNAT TB IAD for diagnosis of tuberculous lymphadenitis of children in Tanzania: a prospective descriptive study. BMC Infect Dis 16:246
75. Coetzee L, Nicol MP, Jacobson R, Schubert PT, van Helden PD, Warren RM et al (2014) Rapid diagnosis of pediatric mycobacterial lymphadenitis using fine needle aspiration biopsy. Pediatr Infect Dis J 33(9):893–896
76. Rufai SB, Singh A, Singh J, Kumar P, Sankar MM, Singh S et al (2017) Diagnostic usefulness of Xpert MTB/RIF assay for detection of tuberculous meningitis using cerebrospinal fluid. J Infect 75(2):125–131
77. Marais BJ, Pai M (2007) New approaches and emerging technologies in the diagnosis of childhood tuberculosis. Paediatr Respir Rev 8(2):124–133
78. Schaaf HS, Krook S, Hollemans DW, Warren RM, Donald PR, Hesseling AC (2005) Recurrent culture-confirmed tuberculosis in human immunodeficiency virus-infected children. Pediatr Infect Dis J 24(8):685–691
79. Stillson CH, Okatch H, Frasso R, Mazhani L, David T, Arscott-Mills T et al (2016) That's when I struggle' ... Exploring challenges faced by care givers of children with tuberculosis in Botswana. Int J Tuberculosis Lung Dis 20(10):1314–1319
80. Donald P, Mcllleron H, Drugs A (2009) In: Schaaf HS, Zumla A (eds) Tuberculosis: a comprehensive clinical reference. Elsevier, India, pp 608–617
81. Dickinson J, Mitchison D (1981) Experimental models to explain the high sterilizing activity of rifampin in the chemotherapy of tuberculosis. Am Rev Respir Dis 123(4. pt 1):367–371
82. Hong Kong Chest Service/British Medical Research Council (1991) Controlled trial of 2, 4, and 6 months of pyrazinamide in 6-month, three-times-weekly regimens for smear positivePulmonary tuberculosis, including an assessment of a combined preparation of isoniazid, rifampin, and pyrazinamide. Am Rev Resp Dis 143:700–706
83. World Health Organization (ed) (2014) Guidance for national tuberculosis programmes on the management of tuberculosis in children, 2nd edn. World Health Organization, Geneva
84. Donald P, Maher D, Maritz J, Qazi S (2006) Ethambutol dosage for the treatment of children: literature review and recommendations. Int J Tuberculosis Lung Dis 10(12):1318–1330
85. Rabie H, Goussard P (2016) Tuberculosis and pneumonia in HIV-infected children: an overview. Pneumonia (Nathan) 8:19
86. Schaaf HS, Victor TC, Engelke E, Brittle W, Marais BJ, Hesseling AC et al (2007) Minimal inhibitory concentration of isoniazid in isoniazid-resistant *Mycobacterium tuberculosis* isolates from children. Eur J Clin Microbiol Infect Dis 26(3):203–205
87. Enane LA, Lowenthal ED, Arscott-Mills T, Matlhare M, Smallcomb LS, Kgwaadira B et al (2016) Loss to follow-up among adolescents with tuberculosis in Gaborone. Botswana Int J Tuberculosis Lung Dis 20(10):1320–1325
88. Snow K, Hesseling AC, Naidoo P, Graham SM, Denholm J, du Preez K (2017) Tuberculosis in adolescents and young adults: epidemiology and treatment outcomes in the Western cape. Int J Tuberc Lung Dis 21(6):651–657

89. Schaaf HS, Marais BJ, Hesseling AC, Gie RP, Beyers N, Donald PR (2006) Childhood drug-resistant tuberculosis in the Western Cape Province of South Africa. Acta Paediatr (Oslo, Norway: 1992) 95(5):523–528
90. Schaaf HS, Marais BJ, Hesseling AC, Brittle W, Donald PR (2009) Surveillance of antituberculosis drug resistance among children from the Western Cape Province of South Africa--an upward trend. Am J Public Health 99(8):1486–1490
91. Isaakidis P, Casas EC, Das M, Tseretopoulou X, Ntzani EE, Ford N (2015) Treatment outcomes for HIV and MDR-TB co-infected adults and children: systematic review and meta-analysis. Int J Tuberculosis Lung Dis 19(8):969–978
92. Dangor Z, Izu A, Hillier K, Solomon F, Beylis N, Moore DP et al (2013) Impact of the antiretroviral treatment program on the burden of hospitalization for culture-confirmed tuberculosis in south African children: a time-series analysis. Pediatr Infect Dis J 32(9):972–977
93. World Health Organization (2017) Guidelines for the treatment of drug-susceptible tuberculosis and patient care, 2017 update. World Health Organization, Switzerland
94. Blanc FX, Sok T, Laureillard D, Borand L, Rekacewicz C, Nerrienet E et al (2011) Earlier versus later start of antiretroviral therapy in HIV-infected adults with tuberculosis. N Engl J Med 365(16):1471–1481
95. Havlir DV, Kendall MA, Ive P, Kumwenda J, Swindells S, Qasba SS et al (2011) Timing of antiretroviral therapy for HIV-1 infection and tuberculosis. N Engl J Med 365(16):1482–1491
96. Abdool Karim SS, Naidoo K, Grobler A, Padayatchi N, Baxter C, Gray AL et al (2011) Integration of antiretroviral therapy with tuberculosis treatment. N Engl J Med 365(16):1492–1501
97. Yotebieng M, Van Rie A, Moultrie H, Cole SR, Adimora A, Behets F et al (2010) Effect on mortality and virological response of delaying antiretroviral therapy initiation in children receiving tuberculosis treatment. AIDS 24(9):1341–1349
98. World Health Organization (2013). Consolidated guidelines on the use of antiretroviral drugs for treating and preventing HIV infection: recommendations for a public health approach.
99. Ren Y, Nuttall JJ, Egbers C, Eley BS, Meyers TM, Smith PJ et al (2008) Effect of rifampicin on lopinavir pharmacokinetics in HIV-infected children with tuberculosis. J Acquir Immune Defic Syndr 47(5):566–569
100. Frohoff C, Moodley M, Fairlie L, Coovadia A, Moultrie H, Kuhn L et al (2011) Antiretroviral therapy outcomes in HIV-infected children after adjusting protease inhibitor dosing during tuberculosis treatment. PLoS One 6(2):e17273
101. Arrow Trial Team, Kekitiinwa A, Cook A, Nathoo K, Mugyenyi P, Nahirya-Ntege P et al (2013) Routine versus clinically driven laboratory monitoring and first-line antiretroviral therapy strategies in African children with HIV (ARROW): a 5-year open-label randomised factorial trial. Lancet 381(9875):1391–1403
102. Dooley KE, ayre P, Borland J, Purdy E, Chen S, Song I et al (2013) Safety, tolerability, and pharmacokinetics of the HIV integrase inhibitor Dolutegravir given twice daily with rifampin or once daily with Rifabutin: results of a phase 1 study among healthy subjects. J Acquir Immune Defic Syndr 62(1):21–27
103. Laureillard D, Marcy O, Madec Y, Chea S, Chan S, Borand L et al (2013) Paradoxical tuberculosis-associated immune reconstitution inflammatory syndrome after early initiation of antiretroviral therapy in a randomized clinical trial. AIDS 27(16):2577–2586
104. Link-Gelles R, Moultrie H, Sawry S, Murdoch D, Van Rie A (2014) Tuberculosis immune reconstitution inflammatory syndrome in children initiating antiretroviral therapy for HIV infection: a systematic literature review. Pediatr Infect Dis J 33(5):499–503
105. von Reyn CF (2006) Routine childhood bacille Calmette Guerin immunization and HIV infection. Clin Infect Dis 42(4):559–561
106. Bhat GJ, Diwan VK, Chintu C, Kabika M, Masona J (1993) HIV, BCG and TB in children: a case control study in Lusaka, Zambia. J Trop Pediatr 39(4):219–223
107. Hesseling AC, Marais BJ, Gie RP, Schaaf HS, Fine PE, Godfrey-Faussett P et al (2007) The risk of disseminated Bacille Calmette-Guerin (BCG) disease in HIV-infected children. Vaccine 25(1):14–18

108. Hesseling AC, Schaaf HS, Hanekom WA, Beyers N, Cotton MF, Gie RP et al (2003) Danish bacille Calmette-Guerin vaccine-induced disease in human immunodeficiency virus-infected children. Clin Infect Dis 37(9):1226–1233
109. Claassens MM, van Schalkwyk C, du Toit E, Roest E, Lombard CJ, Enarson DA et al (2013) Tuberculosis in healthcare workers and infection control measures at primary healthcare facilities in South Africa. PLoS One 8(10):e76272
110. World Health Organization (2009) WHO Policy on TB infection control in health-care facilities, congregate settings and households. In: HIV/AIDS. World Health Organization, Geneva, Switzerland
111. Ayieko J, Abuogi L, Simchowitz B, Bukusi E, Smith A, Reingold A (2014) Efficacy of isoniazid prophylactic therapy in prevention of tuberculosis in children: a meta-analysis. BMC Infect Dis 14
112. Sandul AL, Nwana N, Holcombe JM, Lobato MN, Marks S, Webb R et al (2017) High rate of treatment completion in program settings with 12-dose weekly isoniazid and Rifapentine (3HP) for latent *Mycobacterium tuberculosis* infection. Clin Infect Dis 65:1085–1093
113. Sterling TR, Villarino ME, Borisov AS, Shang N, Gordin F, Bliven-Sizemore E et al (2011) Three months of rifapentine and isoniazid for latent tuberculosis infection. N Engl J Med 365(23):2155–2166
114. Sharma SK, Sharma A, Kadhiravan T, Tharyan P (2013) Rifamycins (rifampicin, rifabutin and rifapentine) compared to isoniazid for preventing tuberculosis in HIV-negative people at risk of active TB. Cochrane Database Syst Rev (7):CD007545
115. McClintock AH, Eastment M, McKinney CM, Pitney CL, Narita M, Park DR et al (2017) Treatment completion for latent tuberculosis infection: a retrospective cohort study comparing 9 months of isoniazid, 4 months of rifampin and 3 months of isoniazid and rifapentine. BMC Infect Dis 17(1):146
116. Akolo C, Adetifa I, Shepperd S, Volmink J (2010) Treatment of latent tuberculosis infection in HIV infected persons. Cochrane Database Syst Rev (1):CD000171
117. Ayele HT, Mourik MS, Debray TP, Bonten MJ (2015) Isoniazid prophylactic therapy for the prevention of tuberculosis in HIV infected adults: a systematic review and meta-analysis of randomized trials. PLoS One 10(11):e0142290
118. Zar HJ, Cotton MF, Strauss S, Karpakis J, Hussey G, Schaaf HS et al (2007) Effect of isoniazid prophylaxis on mortality and incidence of tuberculosis in children with HIV: randomised controlled trial. BMJ 334(7585):136
119. Madhi SA, Nachman S, Violari A, Kim S, Cotton MF, Bobat R et al (2011) Primary isoniazid prophylaxis against tuberculosis in HIV-exposed children. N Engl J Med 365(1):21–31
120. Zunza M, Gray DM, Young T, Cotton M, Zar HJ (2017) Isoniazid for preventing tuberculosis in HIV-infected children. Cochrane Database Syst Rev 8:CD006418

Neurological TB in HIV

Louise Bovijn, Regan Solomons, and Suzaan Marais

Abstract Central nervous system (CNS) tuberculosis (TB) is the worst form of TB and may present as (1) intracranial pathology, including tuberculous meningitis (TBM), tuberculoma/abscess, and rarely cerebritis or encephalopathy (described in children) and, (2) intraspinal disease. HIV co-infection increases the risk of developing CNS TB substantially, especially in patients with severe immunosuppression (CD4 count <100 cells/μL). In this chapter, we discuss the pathogenesis, clinical and imaging findings, management and outcomes of the most frequent forms of CNS TB in HIV co-infected patients. We further review the features and management of a frequently fatal complication related to starting antiretroviral therapy (ART), namely neurological TB immune reconstitution inflammatory syndrome (IRIS).

Keywords Tuberculous meningitis · TB radiculomyelitis · Tuberculoma · Spinal TB · Extra-pulmonary TB · Central nervous system

Introduction and Epidemiology

Central nervous system tuberculosis (CNS TB) is rare compared to many other TB manifestations (e.g. pulmonary, pleural and lymph node TB) [1] but is the most devastating form [2, 3]. Neurological TB constitutes 3.4–22% of extra-pulmonary TB cases [1, 4, 5]. HIV co-infection increases the risk of developing CNS TB substantially with one study showing a five-fold increase of meningitis in HIV-infected

L. Bovijn
Neurosciences Department, John Radcliffe Hospital, Oxford, UK

R. Solomons
Department of Paediatrics and Child Health, Stellenbosch University, Cape Town, South Africa

S. Marais (✉)
Department of Neurology, Inkosi Albert Luthuli Central Hospital, Durban, South Africa

Department of Medicine, University of KwaZulu-Natal, Durban, South Africa

I. Sereti et al. (eds.), *HIV and Tuberculosis*,
https://doi.org/10.1007/978-3-030-29108-2_13

compared to HIV-uninfected TB cases [6]; this increased risk is greatest in patients with severe immunosuppression (CD4 count<100 cells/μL) [4]. Estimates of the burden of CNS TB are likely inaccurate, given the difficulty in establishing a definitive diagnosis due to the paucibacillary nature of the disease in the CNS [7, 8].

Central nervous system TB may present as 1) intracranial pathology, including tuberculous meningitis (TBM), tuberculoma/abscess, and rarely cerebritis or encephalopathy (described in children) and, 2) intraspinal disease [9]. In this chapter, we discuss the most frequent forms of CNS TB including TBM, tuberculoma/abscess and intraspinal TB, as well as a complication related to starting antiretroviral therapy (ART), namely neurological TB immune reconstitution inflammatory syndrome (IRIS).

Tuberculous Meningitis

Tuberculous meningitis is the most common form of CNS TB accounting for approximately 1–2% of all TB cases in developed countries [1, 10–12], and 9.3% (100/1072) of HIV co-infected TB cases in one study [2]. Population-based TBM disease burden data from resource-poor, high TB incidence settings are scant, but based on extrapolations from available data it has been suggested that there is at least 100,000 cases globally per year [13]. Tuberculous meningitis is widely regarded as the second most common meningitis after cryptococcal meningitis (CM) in HIV co-infected adults in highly TB endemic settings [14–16]; A study conducted in Cape Town, South Africa found that 28% (227/820) of microbiologically confirmed meningitis cases were due to *Mycobacterium tuberculosis* (*M. tuberculosis*), whilst CM accounted for 63% [14]. A subsequent study from the same setting reported that 57% (120/211) of meningitis cases could be attributed to TB when using a clinical case definition [17]. In children, TBM is the most common cause of bacterial meningitis in TB endemic settings, accounting for 33% of all HIV-infected meningitis cases in one series [18]. In addition to HIV, risk factors for developing TBM are age (with young children [<5 years] being at greatest risk) [19] and other factors that contribute to an impaired immune system, such as malnutrition, alcoholism and immunosuppressive therapies (e.g. tumor necrosis factor (TNF)-α inhibitors) [20].

Pathogenesis and Pathology

In the late nineteenth century, it was thought that TBM resulted from haematogenous spread to the meninges, due to the frequent finding of TBM and miliary TB occurring in the same patient [21]. In 1933, Rich and McCordock suggested a two-phase model for TBM pathogenesis based on their autopsy findings of TBM cases [22]. These findings form the basis of our current understanding of the pathogenesis of TBM. The first phase of the model commences with the primary infection or late reactivation of TB in the respiratory system. During the localized infection in the lungs, a bacillaemia may lead to mycobacterial seeding to the brain cortex or meninges, with formation of small tuberculous granuloma referred to as "Rich foci". How mycobacteria leave the lung and enter the blood and subsequently cross the blood-brain-barrier (BBB) during TB disease is not fully understood and not explained by animal models of cerebral TB used thus far [13]. *In vitro* studies have shown that *M. tuberculosis* is able to cross the BBB by invasion of human brain endothelial cells through endocytosis [23] and could therefore gain access to the CNS via this route. Alternatively or additionally, the organism may enter through a "trojan horse" mechanism, whereby infected cells adhere to the endothelium and undergo diapedesis [13]. According to the model by Rich and McCordock [22], the foci become dormant and remain so for months to years after the initial bacillaemia. During the second phase, caseating foci rupture into the subarachnoid or ventricular space, initiating an inflammatory response resulting in TBM.

More recently, the concept of the two-phase model as the only pathogenic mechanism at play in CNS TB has been challenged [24]. Rich and McCordock considered miliary TB and TBM separate entities as their studies suggested that a Rich focus of older age than lesions of simultaneous occurring miliary TB was nearly always present in patients with TBM. However, the association between miliary TB and TBM in children is well established [24] with brain magnetic resonance imaging (MRI) detecting a miliary picture in up to 88% of children with TBM [25]. It has therefore been proposed that miliary TB, which is an exacerbated form of haematogenous spread, increases the probability of meningeal/cortical disease that then manifests in close proximity to the time of CNS seeding [24]. The increased frequency of TBM in HIV-infected patients may similarly be linked to the frequent bacillaemia observed in HIV-associated TB patients, especially in those with severe immunosuppression (CD4 $\leq$ 100 cells/μL) of whom up to 49% have positive blood cultures [26–28].

The release of bacilli into the subarachnoid space sets into motion a complex cascade of inflammatory reactions that is yet to be fully understood. According to current hypotheses, *M. tuberculosis* is taken up predominantly by microglia (resident brain macrophages) in which they replicate and induce cytokine, chemokine and growth factor synthesis and secretion [13, 29–31]. Additionally, inflammatory mediators are secreted by other resident CNS cell types, such as astrocytes and endothelial cells. During the course of inflammation, cytokines (e.g. interleukin [IL]-6 and TNF-α) and other mediators (e.g. matrix metalloproteinase [MMP]-9) lead to disruption of the BBB whilst chemokines (e.g. CC chemokine ligand [CCL] 2 and CXC chemokine ligand [CXCL] 10) recruit inflammatory cells (predominantly lymphocytes, but also neutrophils) into the CNS, propagating the inflammatory process. The robust inflammatory response that ensues results in the typical macroscopic granulomatous exudate discussed in further sections.

On cerebrospinal fluid (CSF) analysis, TBM is characterized by increased expression of a vast range of pro- and anti-inflammatory cytokines, chemokines, MMPs and neutrophil-related peptides [32–38]. Some of these inflammatory markers remain high for weeks to months after the initiation of TB treatment. Several studies have compared concentrations of inflammatory mediators in CSF between HIV-infected and -uninfected TBM patients [33, 34, 38]. Although similar concentrations were found for TNF-α [33] and IL-6 [34] by some, a large recent study (n = 764) reports increased concentrations of pro-inflammatory (TNF-α, IL-2, IL-1β, IL-6, IL12p70) and anti-inflammatory (IL-5) cytokines and decreased concentrations of the regulatory cytokine IL-10 in HIV-infected patients, whilst IFN-γ was similar between groups [38]. When dividing HIV-infected patients according to degree of immunosuppression (CD4 count ≥150 or <150 cells/μL), cytokines were significantly higher in patients with severe immunosuppression whilst no significant difference in cytokine concentrations was seen between those with CD4 ≥ 150 cells/μL and HIV-uninfected patients. Furthermore, patients with CD4 < 150, had significantly higher CSF neutrophil percentage (median = 25%), compared to those with CD4 ≥ 150 (10%) and those without HIV co-infection (5%) and cytokine concentrations correlated positively with CSF absolute neutrophil number. Increased CSF neutrophils and raised inflammatory mediator concentrations also predispose HIV-associated TBM patients to clinical deterioration after starting ART (see IRIS section below) [36, 39]. Although the influence of HIV on TBM immunopathogenesis remains poorly understood, these findings suggest a prominent role for neutrophils, especially in severely immunosuppressed patients. Furthermore, IL-10 is an inhibitor of cytokine secretion (e.g. IFN-γ by T-helper 1 cells and natural killer cells) and it has been suggested that decreased concentrations of IL-10 skews the CSF cytokine balance to a Th1 response in the context of HIV [34, 38].

Both host and bacterial genetic factors may play a role in the development and/or severity of TBM. In HIV-infected TBM patients, for example, infection with "modern" Beijing lineage *M. tuberculosis* strains was associated with a lower mortality compared to patients infected with "ancient" Indo-Oceanic lineage strains [40]. Several human genes that influence the host immune response to *M. tuberculosis* have been implicated in TBM susceptibility or disease severity, but results may vary between different countries/ethnic populations [13]. Recent studies suggest that two single nucleotide polymorphisms (SNPs) in the promoter of the Leucotriene A4 hydrolase (LTA4H) gene, that catalyzes the production of the pro-inflammatory eicosanoid LTB4, are associated with either a hyperinflammatory (TT) or a hypoinflammatory (CC) CSF phenotype in TBM [38, 41]. In the initial study, adjunctive corticosteroids conferred protection from death in patients with the TT genotype, whilst those with the CC genotype had a significantly higher mortality. The converse was seen in patients who did not receive corticosteroids; in this group, patients with the TT genotype had a significantly higher mortality compared to those with the CC genotype [41]. A subsequent study, during which all patients received corticosteroids, found that HIV-uninfected patients with the TT genotype were significantly more likely to survive, compared to CC genotype patients [38]. This benefit was not seen in HIV-infected patients, although analyses according to CD4 count (≥150 cells/μL or <150 cells/μL) suggested that results in patients with less immunosuppression (≥150 cells/μL) were similar to those in HIV-uninfected patients. These studies suggest that host-directed therapies (such as corticosteroids) should be tailored according to LTH4H genotype to improve outcome. However, further larger RCTs in HIV-infected patients and in different ethnic populations are warranted, particularly considering findings from an Indonesian study in HIV-uninfected patients (91% of whom received corticosteroids) that showed no survival benefit in association with the TT-genotype [42].

The major pathological consequences of the inflammatory reaction elicited by *M. tuberculosis* include the following [9]:

- Basal arachnoiditis: A proliferative arachnoiditis that produces a thick, gelatinous exudate and mostly affects the cisterns surrounding the base of the brain is the hallmark finding in TBM. The exudate encases the adjacent cranial nerves often leading to cranial nerve palsies. Several small studies found that HIV-infected patients have less extensive exudate, poorly formed granuloma and a larger number of mycobacteria compared to HIV-uninfected patients [43–45]. However, others describe the exudate to be "moderate to severe" [46] and "thick" [47] in the context of HIV. It is likely that the histopathological findings reflect the degree of immunosuppression with less advanced cases showing features more similar to those in HIV-uninfected patients [48].
- Vasculitis: The exudate extends around vessels often resulting in vasculitis causing thrombosis, aneurysm formation and infarction [49]. Occlusion of vessels can be intensified by surrounding pressure from hydrocephalus and vasospasm [50]. An autopsy study in adults and children (n = 51) identified infarcts in 73% of TBM cases; macroscopic apparent infarcts (27 cases) most commonly occurred in the basal ganglia (25.5%), cortex (25.5%) and pons (7.8%) [49]. Vascular involvement was seen in all patients (n = 51). Smaller branches of the middle cerebral arteries were involved in 100% and those of the basilar artery in 94%, predominantly showing fibrinoid necrosis. Larger branches were variably involved showing fibro-intimal proliferation. Mononuclear cell infiltration occurred in vessels of any size. HIV causes damage to the vessel wall, either directly or indirectly through an aberrant autoimmune response resulting in a vasculopathy, which may aggravate the vascular consequences of TBM [51, 52].
- Hydrocephalus (HC): The exudate may lead to obstruction of CSF flow, resulting in HC, which may be obstructive (OHC; i.e. CSF cannot exit the ventricles due to blockage at the level of the aqueduct of Sylvius or the fourth ventricle) or, communicating (CHC; i.e. CSF can exit the ventricles and there is "communication" with the subarachnoid space, but normal CSF flow is impaired due to exudates within this space). Due to ensuing raised intracranial pressure, HC may result in brain herniation or ischemia. Hydrocephalus may be less common in HIV-infected patients due to a less severe inflammatory response [43].
- Tuberculoma: Small granuloma may coalesce to form a tuberculoma that typically develops a central core of caseous necrosis (initially solid followed by liquefaction) which is surrounded by a wall of epithelioid histiocytes, Langhan's giant cells and lymphocytes [53–55]; acid-fast bacilli (AFB) are only rarely identified in the necrotic center. A study comparing the histopathological findings of spinal (bony) TB lesions in HIV-infected and -uninfected patients reports the

presence of well-organized granulomas, irrespective of HIV-status [56]; the only difference between groups was a reversal in the ratio of infiltrating CD4–CD8 T cells in the lesions of HIV-infected patients. Well-organized granuloma has also been described in HIV-infected patients with cerebral tuberculoma [57]. Rarely, abscess formation takes place that is characterized macroscopically by pus in the cavity and microscopically by a lack of the typical granulomatous reaction in its wall, and numerous bacilli in its center [53, 58].

Clinical Features

The majority of childhood TBM cases occur between 2–4 years of age [19], whilst in adults most cases occur during the third and fourth decade of life, regardless of HIV-status [59]. TBM typically presents sub-acutely after 5 days to weeks of neurological symptoms [7]. Occasionally though, TBM may present acutely (within a few days of symptom onset), mimicking bacterial meningitis or conversely, it may present as a slow, progressive dementia. If left untreated, the disease usually progresses through three consecutive phases:

- *Prodromal phase*: Initially patients typically experience non-specific symptoms that include malaise, vague headache, low-grade fever, anorexia, vomiting without diarrhea and neck pain. In children early warning signs further include poor weight gain and listlessness. A clue to recognizing early stage TBM is the persistence of symptoms and, particularly in children, household contact with an adult source case with pulmonary TB within the previous year should heighten suspicion of TBM.
- As the infection progresses, a more pronounced *meningitic phase* emerges, and gradually more striking neurological features such as severe headache, meningism, confusion, cranial nerve palsies (most commonly VI, followed by III and VII) [60, 61] and seizures become evident.
- Further progression lead to the *paralytic phase*, which is marked by coma, abnormal movements and dense neurological deficits (such as hemi- or paraparesis) that ultimately results in death, if untreated.

The severity of the neurological status at TBM presentation is scored according to the modified British Medical Research Council (BMRC) grading system, which takes into account level of consciousness and the presence or absence of focal neurological deficits [29].

- Grade I-Patients are fully conscious (Glasgow coma scale [GCS] score 15) with or without meningeal signs, but without neurological deficit.
- Grade II-Patients are confused (GCS score 11–14) and/or have focal neurological signs.
- Grade III-Patients present with stupor or coma (GCS score $\leq$ 10) with or without focal neurological signs.

Most studies suggest that HIV co-infection does not significantly influence the symptom duration at presentation, nature of symptoms and neurological findings in TBM, regardless of age [59]. However, HIV-infected TBM patients are significantly more likely to have concomitant extra-CNS TB [62, 63]. In childhood TBM, findings likely related to the underlying HIV disease such as poor nutrition, lymphadenopathy, hepatosplenomegaly, clubbing and otorrhoea, were significantly more frequent in the HIV-infected group, which may further complicate early recognition of TBM [63].

Cerebrospinal Fluid Features

Lumbar puncture (LP) is the mainstay investigation to confirm the diagnosis and should be performed in all TBM suspects unless a clinical or radiological contraindication is present. Cerebrospinal fluid typically shows a clear appearance, leukocytosis (median = 50–450; range = 5–1000 cells × 10^6/L) with lymphocyte predominance, a raised protein concentration (0.5–3 g/L), and a low glucose concentration (CSF to blood ratio <0.5 or absolute value <2.2 mmol/L) [7, 64]. Most studies report similar CSF findings in HIV-infected and -uninfected adults and children with TBM [59], however, a reduced inflammatory response characterized by lower leucocyte counts and protein concentration in HIV-infected adults have been reported [42, 65]. Although atypical CSF features such as normal cell count, protein concentration or glucose concentration, or a neutrophil predominance may occur in HIV-uninfected patients, it is seen more frequently in association with HIV co-infection [7, 59]. HIV-infected adults often present with a neutrophil predominance (up to >90% of the cell population) or a high neutrophil count (in the hundreds) [66] and up to a third of those with severe immune suppression (CD4 < 50 cells/μL) may have a normal cell count [65]. Such findings can cause diagnostic uncertainty, resulting in a delayed or missed diagnosis of TBM.

Laboratory Features

Hyponatremia (serum sodium<135 mmol/L) is a common metabolic complication of TBM in adults and children [61, 67], which is severe (<125 mmol/L) in up to 45% of patients [68]. Hyponatremia is potentially life threatening and was found to be an independent risk factor for earlier time to death in adults with HIV-associated TBM [66]. Pathogenic mechanisms that that may underlie hyponatremia in the context of TBM include the syndrome of inappropriate antidiuretic hormone secretion (SIADH) and cerebral salt wasting (CSW); results from a recent study suggest that CSW is the more common of the two [69]. Other causes such as poor oral intake, diarrhea, vomiting, diuretic use, renal disease, liver disease and other endocrine disorders should also be considered. Further blood test abnormalities commonly associated with TBM include mild anemia, leukocytosis and mild elevations in transaminases [60, 62, 66], all which may be more severe in HIV co-infection [59]. Low hematocrit levels, that were independently associated with death, may be lower in HIV-infected adults [62, 70].

Features of Extra-Meningeal TB

Features of TB outside the CNS support the diagnosis of TBM and are especially useful if CSF findings are atypical and when a definitive diagnosis has not been made [7]. Chest radiograph should be performed on all TBM suspects, as most studies report evidence of previous or active pulmonary TB in at least 40% of patients [19, 42, 59, 71]. Although some studies suggest that HIV-infected children more frequently have chest radiograph features of TB compared to HIV-uninfected children (85% vs 65%) [63], others have reported the converse in adults (HIV-uninfected: 74% vs HIV-infected: 56%) [42]. Additional imaging such as abdominal ultrasound and computed tomography (CT) chest may also identify involvement of other organs, particularly in HIV-infected patients, who have a higher incidence of extra-meningeal involvement [62]. Analysis of samples from other sites of infection such as sputum, lymph node, pleural fluid, gastric aspirate, bone marrow and urine may not only assist in a diagnosis of TB, but also inform drug susceptibility of the organism [7].

Brain Imaging Findings

Brain imaging including CT and MRI contributes significantly to the diagnosis of TBM and should be performed in all TBM suspects if resources permit [7]. However, imaging may be normal, especially early in the disease and radiological abnormalities may only develop after initiation of TB treatment [71–73]. Furthermore, CT is less sensitive and may miss some lesions apparent on MRI [74–78]. Classic findings of TBM include basal meningeal enhancement (BME), HC, tuberculoma and infarct, which may occur alone or in combination (Fig. 1) [71, 77]. The overall frequency of cerebral imaging abnormalities is similar in HIV-infected and –uninfected TBM patients (55–100% vs 50–90%, respectively) [59, 77]. However, one study reported greater frequency of abnormalities on MRI in HIV-infected compared to HIV-uninfected patients (100% vs 64%) [79].

Meningeal enhancement and HC are the most consistent features of TBM, occurring in 16–71%, and 20–72% of HIV-infected patients, respectively [59, 77, 80]. Some studies in children [63, 80] and in adults [43] report less frequent OHC and BME in association with HIV, however these findings are not consistent [59]. HIV-infected patients frequently show cerebral atrophy on brain imaging that should not be confused with CHC [43, 63, 80]. The combination of HC, BME and infarcts or, the single finding of pre-contrast hyperdensity in the basal cisterns on CT, was 100% specific for TBM in children (HIV status not reported), although the latter still requires

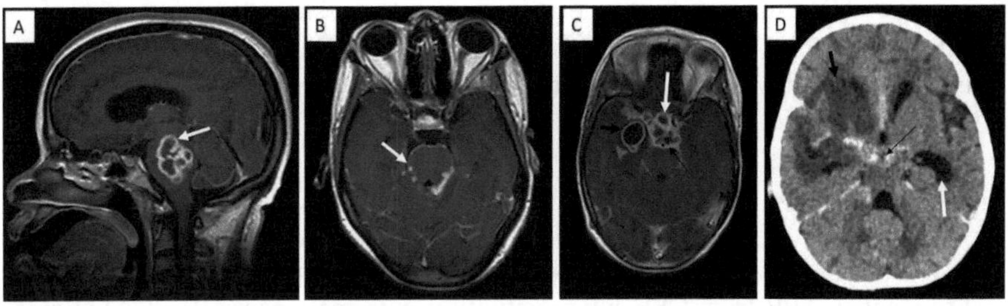

Fig. 1 Imaging of intracranial TB in HIV-infected patients. (**a–c** are post-gadolinium T1-weighted MRI images and D is a post-contrast CT image; A was taken in the sagittal pane and **b–d** were taken in the axial planes). (**a**) Multi-loculated brainstem tuberculoma causing obstructive hydrocephalus in an adult receiving ART (CD4 count = 783). (**b**) Nodular basal meningeal enhancement in an adult not receiving ART (CD4 count = 217). (**c**) Abscess (short, black arrow), basal meningeal enhancement (long, black arrow) and multiple subdural tuberculomas (white arrow) in a child on ART (CD4 = 838). (**d**) Infarct (short, black arrow), basal meningeal enhancement (long, black arrow), and hydrocephalus (white arrow) in a child not receiving ART (CD4 = 390). *Abbreviations: MRI* magnetic resonance imaging, *CT* computed tomography, *ART* antiretroviral therapy, *CD4 count* $CD4^+$ T-lymphocyte count expressed as cells/μL

validation in future studies [81]. A study of predominantly HIV-infected adults report that criteria for BME on CT was insensitive (0–29%), but both BME and acute infarcts were 100% specific in the diagnosis of TBM compared to other meningitides [82]. Although typical BME may be seen less frequently in HIV-infected patients, small tuberculomas (miliary picture) affecting the meninges were seen on MRI in all HIV-infected (n = 7), compared to 72% of HIV-uninfected children [80].

Imaging identifies infarcts in 13–50% of HIV-infected cases [59]. Although the majority of infarcts in TBM occur in the basal ganglia region due to involvement of small perforating arteries [50, 63, 75], it has been suggested that cortical infarction may be more common in HIV [43]. Diffusion-weighted image (DWI) sequence is more sensitive that routine T2-weighted and fluid-attenuated inversion recovery (FLAIR) sequences in detecting acute infarcts on MRI; in a pediatric study, DWI revealed 89/172 infarcts that were not observed by routine MRI and CT [75]. Features of vasculitis in HIV-associated TBM may be enhanced by the effect of HIV on vessels [51, 52]. MR angiography is useful to identify vasculitis of intracranial vessels, which is present in 37–51% of HIV-uninfected TBM patients and may predict development of future stroke [50, 83], but the application of this modality in HIV-associated TBM is rarely described [43].

Diagnosis

There remains a vast need for a rapid, sensitive test to confirm the diagnosis of TBM and guide individual patient care. TBM presents similar to other causes of meningitis and, due to the poor sensitivity of current diagnostic tests, clinicians may either "over treat" TBM suspects for fear of missing the diagnosis or "under treat" for fear of exposing patients unnecessarily to months of potentially hazardous TB treatment. The most commonly used methods are discussed below.

Smear Microscopy and Culture

Microscopy using Ziehl–Neelsen (ZN) staining is a rapid, inexpensive, 100-year old test to identify AFB from CSF, but sensitivity is extremely poor (10–20%) in most settings [17, 84, 85]. Cerebrospinal fluid culture of *M. tuberculosis* remains the gold standard test, but time to positivity is too slow (2–4 weeks in liquid medium) to inform patient management at presentation [84] and furthermore, the test is negative in more than 50% of clinical TBM cases [8, 19, 86, 87]. However, it is advisable to perform culture in all cases as it may inform drug susceptibility, especially to isoniazid and second-line TB drugs, which cannot be determined by current alternative diagnostic methods such as Xpert TBM/RIF, discussed below. Culture may be more sensitive in HIV co-infected compared to -uninfected patients (42% vs 30%) [62], which may relate to an impaired immune response and hence an impaired ability to contain *M. tuberculosis* replication [43].

The diagnostic yield of conventional bacteriological methods can be increased through simple measures. Firstly, increasing the volume of CSF analyzed to at least 6 ml allowed for culture of *M. tuberculosis* in 80% of clinical TBM cases [88]. Secondly, prolonging the time spent on microscopy to 40 min, compared to five, improved its yield from 35% to 95% [88]. Lastly, performing up to four serial LP early during TB treatment significantly increased the sensitivity of both microscopy and culture [89]. There is little data to inform the maximum volume of CSF that can be collected safely but suggested volumes include 6–9 mL in infants, 10–15 mL in young children and 15–17 mL in adults [90]. However, much larger volumes are likely safe in the context of TBM, which is often characterized by CHC; in one study, 30 mL of CSF was collected from 68 adults with suspected TBM after exclusion of contra-indications, none of whom had related side-effects aside from transient headaches, which was present in the minority [86].

Nucleic Acid Amplification (NAA) Tests

The World Health Organization (WHO) currently recommends Xpert MTB/RIF (Xpert, Cepheid, Sunnyvale, CA, USA) as the initial diagnostic test for TBM, as it is rapid (results are potentially available within hours) and it is more sensitive than conventional ZN smear [91]. Xpert is a real time NAA assay that detects *M. tuberculosis* and rifampicin-resistant mutations simultaneously. Three studies have assessed the diagnostic utility of Xpert in adults with TBM [84, 85, 92], two of which included predominantly HIV-infected patients (87–98%) [84, 85]. Sensitivities of Xpert compared to culture ± PCR (definite TBM), and a clinical case definition, were 65–85% and 36–59%, respectively and specificity was 95–100%. In one study, sensitivity of Xpert against a clinical case definition was significantly higher in HIV-infected patients compared to uninfected patients (78.8% vs 47.9%; odds ratio 4.01, 95% CI 3.65–4.36) [92]. All three studies highlighted the importance of using larger volumes of centrifuged CSF to increase the sensitivity of Xpert; for example, a median of ~6 mL centrifuged CSF compared to 2 ml of uncentrifuged CSF increased the sensitivity from 28% to 72% [85].

Xpert MTB/RIF Ultra (Ultra) is a recent second-generation Xpert assay that is more sensitive, but less specific, than the Xpert MTB/RIF in pulmonary TB [93]. A study in HIV-infected TBM patients found that Ultra was more sensitive than Xpert or culture, detecting 70% of probable/definite TBM cases, compared to 43% detected by culture and Xpert each [8]. A multitude of other NAA tests (commercial and in-house) to diagnose TBM exists, but there is significant heterogeneity among in-house assays, whilst commercial tests lack sensitivity and/or have not been validated for use in TBM [94].

Clinical Prediction Rules

None of the currently available microbiological and molecular tests are sensitive enough to exclude the diagnosis of TBM if negative and treatment should be started based on circumstantial clinical and investigation evidence [7, 95]. To aid the clinical diagnosis, numerous clinical prediction rules have been devised to distinguish TBM from other causes of meningitis in both children and adults [13, 96]. However, the performances of these rules vary according to the prevalence of TB and HIV and they often lack validation outside the centers that they were generated. The most consistent features able to distinguish TBM from bacterial meningitis include prolonged symptom duration (>5–9 days), lower total CSF leucocyte count (<400–1000 cells × 10^6/L) and lower proportion of CSF neutrophils (<30–90%) [13, 96]. Studies that compared findings in TBM to CM both in HIV-uninfected patients [97] and in a high HIV-prevalence group [15] report higher CSF leucocyte counts (≥13–68 cells × 10^6/L) and more depressed level of consciousness (GCS < 14) to be predictive of TBM, but these findings are rarely useful in clinical practice due to overlap between groups. For research, a uniform clinical case definition for TBM (regardless of age or HIV status) that groups cases into definite, probable and possible categories was derived in 2010 to unify reporting of research findings [7]. This case definition has not been validated for use in clinical practice and hence should not be used to inform patient management.

Differential Diagnosis

HIV influences the differential diagnosis of TBM and HIV testing should be performed in patients with unknown status. The differential diagnosis in HIV-infected TBM suspects is vast and depends on the prevalence of different diseases in the specific setting. A limited list of alternative infective causes with their diagnostic investigations is presented in Table 1. Cerebrospinal fluid Gram's stain and bacterial

Table 1 Infective causes in the differential diagnosis of HIV-associated neurological TB

Differential diagnosis[a]	Investigation(s)[b]
Meningitis (intracranial and intraspinal) and myelitis	
Bacteria	Gram's stain, bacterial culture
Non-opportunistic viruses (e.g. enterovirus)	Viral PCR
Herpesviruses (HSV, HZV, CMV, HHV-6)	Viral PCR
Syphilis	VDRL and TPHA or FTA
Cryptococcosis	India ink, cryptococcal antigen testing, fungal culture
Lymphoma	Cytology (and EBV PCR)
Non-tuberculous mycobacteria (e.g. MAC)	Mycobacterial culture
HIV seroconversion	HIV PCR or RNA and conversion of antibody tests
Chronic HIV infection	Exclusion of other causes
Brucellosis	PCR or Wright agglutination test on blood or CSF
Borreliosis	CSF to blood antibody index
Malaria	Blood smear
Eosinophilic meningitis (parasitic)	Eosinophil count and proportion
Tuberculoma (intracranial and intraspinal)[c]	
Toxoplasmosis	Serum and CSF IgG antibodies or PCR
Primary CNS lymphoma	EBV PCR
PML-IRIS[d]	JC virus PCR
Cryptococcoma	India ink, cryptococcal antigen testing, fungal culture
Syphilitic gumma	VDRL and TPHA or FTA
Bacteria	Blood and CSF bacterial culture
Non-tuberculous mycobacteria (e.g. MAC)	Mycobacterial culture
Schistosomiasis	Rectal biopsy, stool and urine microscopy
Cysticercosis	Serum and CSF antibodies
Epidural spinal abscess[c]	
Bacteria	Blood bacterial culture

Abbreviations: *PCR* polymerase chain reaction, *HSV* herpes simplex virus, *HZV* herpes zoster virus; *CMV* cytomegalovirus, *HHV-6* human herpesvirus-6, *VDRL* venereal disease research laboratory, *TPHA Treponema pallidum* haemagglutination, *FTA* fluorescent tryponemal antibody, *EBV* Epstein-Barr virus, *MAC Mycobacterium avium* complex, *IgG* immunoglobulin G, *CNS* central nervous system, *PML-IRIS* progressive multifocal leucoencephalopathy immune reconstitution inflammatory syndrome, *JC* John Cunningham

[a]Some causes only applicable to areas endemic for the organism

[b]Performed on cerebrospinal fluid (CSF), unless otherwise specified

[c]Biopsy of lesion may be indicated

[d]Brain lesions only

culture should be performed in all patients to rule out pyogenic meningitis. However, these tests may be negative in patients who have received antibiotics (oral or intravenous) prior to LP and a strong index of suspicion should be maintained in patients with raised CSF neutrophils. Although CM is rare in children compared to adults, it should be excluded in all HIV-associated meningitis cases [18, 98]. Microscopy with India ink staining is still used in some settings to diagnose CM, but the preferred test for a rapid diagnosis is the IMMY CrAg lateral flow assay (Immy, Inc., Norman, OK, USA), a point-of-care dipstick test that detects cryptococcal antigen in bodily fluids [99]. When performed on CSF the test has a sensitivity of 99.3% and a specificity of 99.1% [100], therefore reliably confirming or excluding the diagnosis. Fungal culture should also be performed if possible to determine if the disease is active (in re-treatment cases) and to inform drug susceptibility.

It is important to take note that HIV itself may result in mild CSF inflammatory changes such as leukocytosis (5–25 cells × 10^6/L) and raised protein (0.46–1 g/L) [101, 102]. In these scenarios it is essential to re-evaluate the clinical presentation and ancillary test results to inform clinical decision making; if the diagnosis is uncertain, management may include close clinical monitoring and repeat LP days after the initial LP and prior to initiation of disease-specific treatment.

Prognosis

TBM is associated with a significant mortality and morbidity regardless of age or socio-economic setting, particularly in HIV-infected patients [17, 19, 62, 68, 70, 87, 103]. Even in the context of a RCT during which patients received optimal medical and supportive care, the mortality rate in HIV-infected adults was 39% (compared to 19.4% in –uninfected patients, hazard ratio, 2.53; 95% CI, 1.90–3.36) [87]. An additional 10% of all patients were severely disabled after 9 months of treatment. In two childhood TBM studies, death occurred in 24–30% of HIV infected compared to 0% of HIV-uninfected patients during follow-up and full recovery occurred significantly less in HIV-infected children (0–29% compared to 52–60%) [63, 104]. Longer duration of symptoms and worse TBM disease grade (BMRC grade II and III) are further strong predictors of poor outcome, emphasizing the need for early diagnosis and treatment initiation [17, 19, 60, 70, 87, 105, 106]. Other important risk factors for death not discussed elsewhere include extra-meningeal/extra-pulmonary TB [62, 70] and interruption/change of anti-TB drug regimen [70]. Younger age is a further risk factor for poor outcome in children [106]. In HIV co-infection, a higher CD4 count is independently associated with reduced mortality (hazard ratio per increase of 100 cells/μL, 0.62; 95% CI, 0.44–0.87) [105] and outcome is poorer in adults not receiving ART [2] and in those who develop neurological TB-IRIS (discussed below) [39]. It is likely that other opportunistic infections acquired during TBM treatment contribute to poor outcome in HIV co-infected TBM.

Poor cognitive outcome is a major cause of disability in childhood TBM survivors [19, 107, 108], with intellectual impairment (IQ $\leq$ 80) observed in 77%

(277/359) of predominantly HIV-uninfected children at 6 months follow-up [19]. In HIV-uninfected children, cognitive disability manifested as global developmental delay without predilection for verbal or performance abilities; risk factors included multiple unilateral or bilateral infarction on CT at 1 month, younger age and worse TBM disease [107, 108]. Additionally, attention deficit and behavior abnormalities are well-known long-term complications seen by clinicians following up children with TBM [109]. Two studies from India report cognitive impairment (as measured by the mini-mental state examination) in approximately half of HIV-uninfected adults 6–12 months after TBM diagnosis [110, 111]. Cognitive impairment correlated with CT findings of exudates and tuberculoma [110] and, low GCS at TBM diagnosis [111]. Importantly, HIV infection itself frequently results in neurocognitive impairment as a consequence of neuronal injury and cell death [112, 113]. With both TBM and HIV infection individually adversely affecting cognitive and developmental outcome, clinicians must remain vigilant when following up HIV co-infected TBM patients to allow early detection and, where possible, intervention.

Management

Antimicrobial Treatment

Current antimicrobial treatment recommendations for TBM are based on those for pulmonary TB as optimum regimens for CNS TB in adults and children are yet to be established by RCTs. Most guidelines recommend that first-line regimens in all cases regardless of HIV status consist of rifampicin, isoniazid, pyrazinamide and a fourth drug, most commonly ethambutol or streptomycin [90, 114, 115]. These drugs are administered for 2 months followed by rifampicin and isoniazid for a further 7–10 months. A major concern regarding these regimens is that the majority of drugs does not cross the BBB sufficiently at the recommended doses and therefore do not reach concentrations required to kill the organism effectively in the CNS [116]. Malabsorption may further compromise drug exposure in HIV-infected patients [117], who frequently have reduced plasma TB drug concentrations compared to HIV-uninfected patients [118]. Whereas isoniazid and pyrazinamide have good CNS penetration, rifampicin, ethambutol and streptomycin do not. Even though CSF concentrations of rifampicin only reach 10–20% compared to that in plasma, the importance of rifampicin in TBM treatment has been established by the high case-fatality rate associated with rifampicin resistance [40, 119, 120]. Regardless of its apparent benefit, there is concern that the conventional dose of rifampicin (10 mg/kg/day) is too low in TBM, as well as in other forms of TB [121]. Support for this point came from recent studies in predominantly HIV-uninfected adults with pulmonary TB that showed that a dose of 35 mg/kg/day of rifampicin in combination with other standard TB drugs was safe and resulted in faster sputum culture conversion compared to standard dose rifampicin [122]. The benefit of ethambutol and streptomycin has not been established and in some centers ethionamide (20 mg/kg), that has good CNS penetration [116], is routinely used as the fourth drug in first-line TBM treatment [123]. This drug, in combination with high-dose isoniazid (20 mg/kg), rifampicin (20 mg/kg) and pyrazinamide (40 mg/kg), seems to be safe and effective when administered for 6 months in HIV-uninfected children and 9 months in HIV-infected children [124].

Two recent RCTs have assessed higher-than-normal doses of rifampicin and the fluoroquinolones moxifloxacin and levofloxacin (that both have good CNS penetration [103, 125]), in combination with other standard-of-care TB drugs in adults with TBM [87, 103]. In the first small trial (7/60 patients HIV-infected), daily rifampicin at a dose of 13 mg/kg administered intravenously compared to 10 mg/kg oral rifampicin during the first 2 weeks of treatment resulted in higher rifampicin drug exposure in CSF and increased survival (65% vs 35%), however the trial was not powered for a mortality outcome [103, 121]. In the same study, moxifloxacin (400 mg or 800 mg) was not associated with a survival benefit [103]. In a subsequent larger RCT (n = 817), the addition of daily levofloxacin (20 mg/kg) and rifampicin (15 mg/kg) during the first 2 months of treatment was not associated with improved outcome in either HIV-infected or –uninfected patients [87]; It may be that the dose of rifampicin was still too low and that much higher doses may prove to be beneficial [13].

Resistance to at least one first-line TB drug is frequently described in TBM studies from both resource-rich and resource-poor countries [126], accounting for 47% of Vietnamese HIV-infected TBM patients in whom drug-susceptibility testing was performed [40]. Isoniazid mono-resistant TBM is associated with poor outcome, regardless of HIV status [10, 40, 119, 126]. It has therefore been suggested that an extra drug with effective CNS penetration be added to the standard-of-care first-line regimen for the duration of TB treatment in these patients [126]. This recommendation is supported by one RCT that found a survival benefit with a regimen containing additional levofloxacin (and high-dose rifampicin) during the first 2 months of treatment in isoniazid mono-resistant adults with TBM [119]. Rifampicin mono-resistant and multidrug-resistant (MDR) TBM (resistance to rifampicin and isoniazid) are associated with dismal prognoses with mortality exceeding 80% in children [127] and up to 100% in HIV-infected and -uninfected adults [10, 40, 119, 120]. This is related, at least in part, to the extended time it takes to determine drug susceptibility and to start appropriate treatment. As in drug-susceptible TBM, treatment guidelines for MDR-TBM and extensively DR (XDR)-TBM (resistance to isoniazid, rifampicin, a fluoroquinolone and an aminoglycoside) are similar to those for extra-CNS TB (Discussed in chapter “Drug-Resistant Tuberculosis and HIV”).

Host-Directed Therapies

The majority of the pathology and clinical consequences of TBM is thought to occur due to the robust host response against *M. tuberculosis* and this hypothesis has prompted studies investigating host-directed therapies to curb inflammation and improve patient outcome [13].

Adjunctive corticosteroid therapy improves end-of-treatment survival in children [128] and HIV-uninfected adults with TBM [70, 129]. Although a similar benefit has not been shown in HIV co-infection, dexamethasone, compared to placebo, was associated with less severe adverse events (mostly hepatitis) in severely immunosuppressed (median CD4 count = 66 cells/μL) adults with HIV-associated TBM [70]. The WHO therefore recommends a course of either dexamethasone or prednisone during the initial 6–8 weeks of TB treatment in all TBM cases [130]. In children, oral prednisone at 2 mg/kg/day for a month followed by a 2-week taper is commonly used [19]. In adults, the best evidence comes from a RCT performed in Vietnam that found a significant survival benefit in association with the interventional arm [70]; 545 patients with newly diagnosed TBM were randomized to placebo or the following dexamethasone regimens according to their BMRC disease grade: Patients with grade II or III disease received daily intravenous treatment for 4 weeks (0.4 mg/kg for week 1, 0.3 mg/kg for week 2, 0.2 mg/kg for week 3, and 0.1 mg/kg for week 4) followed by oral treatment for 4 weeks, starting at a total of 4 mg/day and decreasing by 1 mg each week. Patients with grade I disease received 2 weeks of daily intravenous therapy (0.3 mg/kg for week 1 and 0.2 mg/kg for week 2) followed by 4 weeks of oral therapy (0.1 mg/kg/day for week 3, then a total of 3 mg/day, decreasing by 1 mg each week). In some resource-constrained settings intravenous administration of corticosteroids for 2 to 4 weeks is impractical due to limited bed space in public sector hospitals and therefore, oral prednisone at a starting dose of 1.5 mg/kg/day with a 6–8 week taper is used instead [39]. As mentioned earlier in the chapter, response to corticosteroids may be influenced by LTA4H genotype in certain populations and a trial reassessing the potential benefit of corticosteroids in HIV-infected adults is currently ongoing in Vietnam (NCT03092817); results of the primary outcome are expected in 2020–2021.

Stroke is a common and menacing complication of TBM that may occur before, and often during, TB treatment [72, 73]. Preventing strokes will likely improve patient outcome, as such events are frequently associated with permanent neurological disability [19, 73, 107]. Failure of TB chemotherapy in treating TBM-associated vasculitis and the structural similarities it shares with multiple immune vasculitides suggests an immune-mediated mechanism [131]. However, corticosteroids have not been proven to prevent strokes in TBM [72, 128]. The role of aspirin, which has both antiplatelet and anti-inflammatory properties, is well established in stroke prevention outside the context of TBM. Two RCTs have investigated adjunctive aspirin in TBM patients [132, 133]. Children (5% HIV-infected) who received daily low-dose (75 mg) aspirin, high-dose (100 mg/kg) aspirin or placebo during the first month of TB treatment showed similar neurological outcomes at 6-months follow-up [132]. However, children in the high-dose aspirin group were younger and had significantly more frequent hemiparesis at presentation compared to the other groups, suggesting a possible benefit for high-dose aspirin in childhood TBM. A RCT in predominantly adult patients (HIV status not noted) found that aspirin 150 mg daily, compared to placebo, resulted in a significant lower mortality and less strokes on MRI at 3 months follow-up [133]. Although results of these two studies are promising, larger trials need to be conducted that also include more HIV-infected patients, prior to advocating the routine use of aspirin in TBM.

Management of Hydrocephalus

Hydrocephalus is a common complication of TBM and associated with poor outcome [134, 135]. The optimal treatment of raised intracranial pressure in the context of TBM-associated HC is uncertain as no RCTs exists to inform standardized practices [136]. Distinguishing between CHC and OHC is important as treatment strategies may differ [136, 137]. Communicating HC is the most common form in both adults and children, accounting for up to 80% of childhood HC cases [19, 134, 137, 138]. Computed tomography and MRI are often unable to determine the level of CSF block in TBM [134, 139, 140]. The simplest robust method to differentiate between OHC and CHC is air-encephalography, which involves injecting 5–10 ml air during LP [19, 140]. The presence of air in the ventricular system by lateral skull x-ray 30 min later indicates patency of the ventricular system (i.e. CHC), whist air is only visualized in the basal cisterns in OHC.

Communicating HC can be treated effectively with medical management including diuretics (furosemide and acetazolamide) and or regular (up to daily) LP to achieve opening pressures of less than 20 cm H_2O [136, 137, 141]. These strategies, best described in children, effectively manage raised intracranial pressure in 74–91% of patients [67, 137, 138, 141], usually within 7 days [142]. Patients who present with severely depressed level of consciousness; deteriorate clinically and have progressive hydrocephalus on CT; and those in whom normal opening pressures are not achieved after three to 4 weeks of medical therapy, should be considered for surgical intervention (discussed below).

Treatment options for OHC are all surgical, including external ventricular drain (EVD), ventriculo-peritoneal shunting (VPS) and endoscopic third ventriculostomy (ETV) [136, 137, 143]. Endoscopic third ventriculostomy, during which a stoma is created in the floor of the third ventricle allowing CSF flow into the subarachnoid space, has only recently emerged as a treatment option for TBM [136]. Criteria for any surgical procedure and the choice of intervention in TBM vary greatly between centers [136, 144]. Some base their decision to intervene not on the type of HC (communicating vs obstructive) but on the patient's clinical state, with those with normal or conversely, severely impaired consciousness being excluded from surgery [144]. Furthermore, some advocate that patients with a low GCS (less than 9) should first undergo EVD and that VPS only be offered to those who show significant neurological improvement over the following one to two days [143, 145]. Complications related to VPS in all TBM patients occur in ~22% of cases, most commonly shunt blockage leading to shunt revision and infections [144]. Up to 29% of ETV procedures may suffer complications especially infection, intraoperative bleeding and CSF leak [146, 147]. Little data are available from head-to-head studies on the efficacy of VPS compared to ETV in TBM; In one pediatric study VPS and ETV showed similar outcomes (54% vs 42% successful) and complication rates (17% vs 29%) but the sample size was small (24 in each arm) [146]. Clinical outcome following VPS is significantly influences by TBM severely; less than a third of patients with a GCS less than 9 had a "good outcome" (defined as good recovery or moderate disability), compared to more than 75% of those fully conscious at presentation [144]. HIV-infected patients generally show poor response to VPS; in the only two studies that compared outcome according to HIV status [145, 148], only 24–27% of HIV-infected compared to 60–65% of -uninfected patients had a "good outcome" and 1-month mortality associated with HIV was 67% [148]. It should be taken into account that the HIV-infected patients in these studies were severely immunosuppressed (median CD4 = 121–143 cells/μL) and not receiving ART; findings may be less grim in the context of a more preserved or recovered immune system. In studies that included adults and children with TBM, 64–73% showed clinical improvement after ETV [149, 150], but this procedure is yet to be described in HIV-infected patients. Risk factors for ETV failure include the presence of cisternal basal exudates that clouds the surgeon's vision to the floor of the third ventricle, and worse TBM disease (BMRC grade III) [149]. As HIV co-infection may result in less basal exudates thereby decreasing the chances of accidental damage to the basilar artery [43], EVT should be explored as a therapeutic option in this group.

Supportive Management

An important component to the management of TBM is general supportive care. Simple measures such as treating fever; managing hypoxia with supplemental oxygen via face mask or nasal prongs; ensuring adequate hydration (orally or intravenously) and optimal nutrition; and correcting glucose disturbances may all contribute to improved outcome [136].

Hyponatremia is arguably the most prevalent and hazardous metabolic consequence of TBM and should be identified, and treated, as soon as possible [136]; SIADH and CSW are frequently implicated in the pathogenesis but other alternative/additional causes (listed earlier) should be excluded or addressed. SIADH is characterized by fluid retention and has historically been treated with fluid restriction, whilst CSW is associated with fluid depletion and requires fluid replacement. The distinction between SIADH and CSW is often difficult and implementing the wrong strategy may result in clinical deterioration [136, 151]. Recently it has been suggested that all cases of hyponatremia complicating intracranial pathology should be treated with hypertonic saline (3% or 5% saline), regardless of volume status [67, 136, 151]. Hypertonic saline should ideally be administered in a high-care setting with close monitoring of fluid infusion rate, blood sodium concentrations and urine output, as rapid sodium correction may result in the osmotic demyelination syndrome that may be associated with severe, permanent neurological disability [152]. Suggested safe rates of sodium corrections include 6 to 8 mmol/L in 24 h, 12 to 14 mmol/L in 48 h, and 14 to 16 mmol/L in 72 h [153].

Tuberculoma

Tuberculoma is a major cause of intra-cerebral space-occupying lesions (SOLs) in high TB endemic settings and in HIV-infected patients [9, 154–156]. The lesions most commonly occur within the brain parenchyma, but may also be found at other sites in the cranium such as dura mater and leptomeninges, subdural and subarachnoid space, and within the ventricular system [80, 157]. Rarely, the spinal cord may be involved [158] (Fig. 2a, b). Parenchymal tuberculoma (discussed hence forth), is frequently associated with TBM, but can evolve without evidence of meningeal or extra-CNS disease [55, 74].

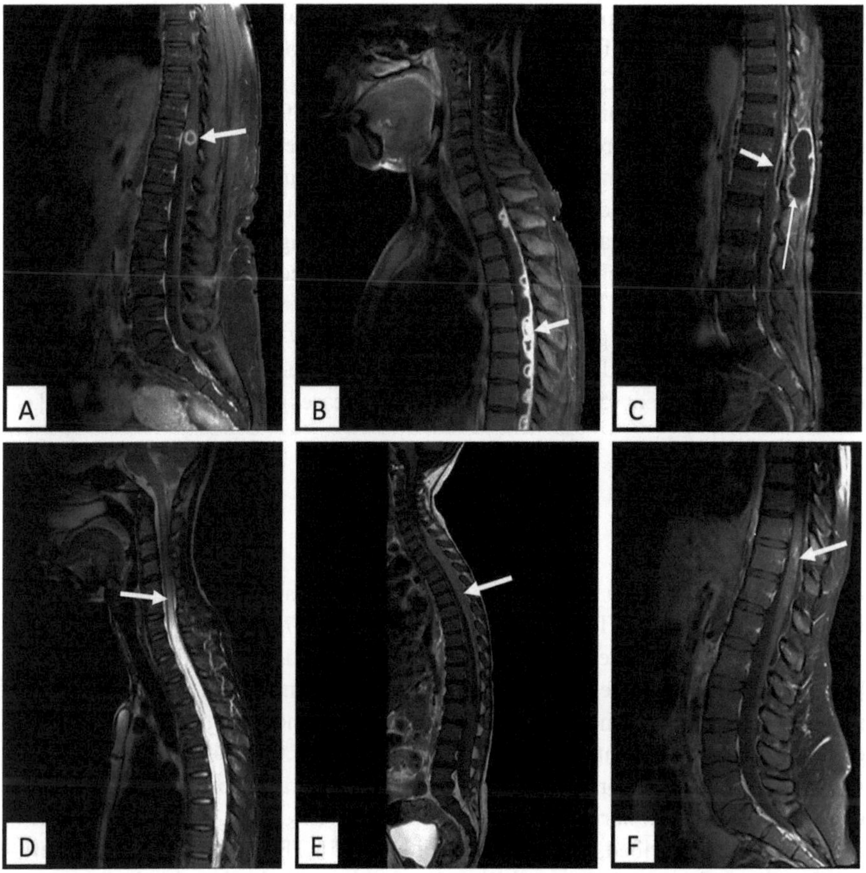

Fig. 2 Magnetic resonance imaging of intraspinal TB in HIV-infected patients (**a–c**, **e** and **f** are post-gadolinium T1-weighted images and **d** is a T2-weighted image; all were taken in the sagittal planes). (**a**) Intramedullary tuberculoma in an adult not receiving ART (CD4 count = 213). (**b**) Multiple subdural tuberculomas in an adult who was receiving treatment for lymph node TB, commenced ART 1 month later and developed spinal symptoms 4 months thereafter (CD4 count = 144). (**c**) Epidural (short arrow) and paravertebral abscess (long arrow) without bony involvement in an adult receiving ART for unknown duration (CD4 count = 130). (**d**) Syringomyelia in an adult not receiving ART who developed spinal symptoms 3 years after completing TBM treatment (CD4 count unknown). (**e**) Extensive subdural exudate (leptomeningitis) in a child not receiving ART (CD4 count = 427). F) Myelitis in an adult not receiving ART (CD4 count = 250). *Abbreviations*: ART, antiretroviral therapy; CD4 count, $CD4^+$ T-lymphocyte count expressed as cells/μL

Clinical Presentation

Clinical data specific to HIV co-infection are limited to case reports and small case series, but findings appear to be similar to HIV-uninfected patients [57, 159–161]. Symptom duration to presentation varies from days to months, and findings from one small study (n = 9) [160] suggests that time to diagnosis may be shorter in HIV-infected (~4 weeks) compared to HIV-uninfected (~16 weeks) patients [55, 57, 159, 162–164]. The neurological findings depend on the location of the lesion in the brain. Similar to other SOLs, tuberculoma manifests as raised intracranial pressure (headache, vomiting, papilledema and depressed level of consciousness), visual disturbances and or progressive focal neurological deficits [55, 57, 159–161]. Additional signs of meningitis (e.g. meningism, cranial nerve palsies) or other organ involvement may also be present. Lumbar puncture should be performed unless contra-indicated, as it may assist in supporting the diagnosis and excluding other etiologies (Table 1). Cerebrospinal fluid examination frequently shows features suggestive of concomitant TBM, but routine examination may be normal; culture for *M. tuberculosis* is rarely positive [57, 74, 159, 165]. As for TBM, ancillary investigations such as chest radiograph and sputum examination frequently show extra-meningeal involvement (50–75% of HIV-infected cases) [54, 57, 162, 165–167].

Brain Imaging

Contrast enhancing brain lesions are usually described at similar frequency in HIV-infected and uninfected TBM patients (0–60% vs 0–27%) [59, 77, 168], however one study reports more frequent mass lesions in association with HIV (60% vs 14%) [169]. Tuberculoma presents as one or more well-circumscribed lesions of variable size (usually between 1 mm to 3 cm in diameter), that most commonly shows a ring or homogenous post-contrast enhancement pattern on CT and MRI; lesions may involve any part of the brain, including supra- and infra-tentorial structures and is often associated with surrounding edema [74, 77, 164, 168]. A miliary pattern is also frequently described, especially in children [25]. TB abscesses are characteristically solitary, large (>3 cm), multi-loculated, thin-walled enhancing lesions, however they may be indistinguishable from caseating tuberculoma with a liquid center [53, 58]. TB abscess may be more frequent in HIV-infected patients, occurring in up to 20% of HIV-associated neurological TB cases [168].

The differential diagnosis of ring-enhancing brain lesions is vast and includes infective, neoplastic, inflammatory and demyelination conditions [170] (Table 1 includes a limited list of infective causes for SOLs in HIV). Toxoplasma encephalitis is widely regarded as the most common cause of SOLs in HIV-infected patients globally, which may occur concurrently with TBM [43, 171], followed by primary CNS lymphoma (PCNSL) in low TB-incidence settings [155, 172–175]. However, several studies from TB endemic settings reported tuberculoma as the most frequent cause of SOL in HIV [154, 156]. The radiological presentation of tuberculoma is often indistinguishable from PCNSL and toxoplasmosis (as well as other infective etiologies) by CT and MRI, including alternative MRI techniques such as perfusion-weighted MRI and proton MR spectroscopy [176]. The role of nuclear medicine examinations such as Thallium-201 single-photon emission computed tomography (SPECT) and 18-fluorodeoxyglucose PET, especially to distinguish infective causes of SOLs, remains to be defined.

Approach to Space-Occupying Lesions in HIV

Brain biopsy is the only definitive diagnostic test for focal brain lesions, but is invasive and not readily available in high TB/HIV prevalence settings. As a result, numerous clinical algorithms have been derived to guide the approach to HIV-associated SOLs [154, 156, 170, 172, 174]. One management strategy is to treat all cases with an initial trial of anti-toxoplasma treatment, which will result in a clinical or radiological response within 10 to 14 days in most toxoplasma cases [155, 170, 172]. If there is no response, a brain biopsy should be considered to establish a definitive diagnosis.

An alternative approach for high TB incidence settings is to treat patients for either tuberculoma or toxoplasmosis based on the most likely cause as determined by clinical and investigative findings [154, 156]. Other causes resulting in similar radiological findings such as cryptococcoma, syphilitic gumma and bacterial abscess should also be excluded as far as possible, by blood tests and CSF analysis. Although not absolute, criteria in favor of tuberculoma versus toxoplasmosis include the following: (1) features of extra-CNS TB (e.g. by chest radiograph); (2) a well-preserved CD4 count (>200 cells/μL); (3) evidence of concomitant TBM by LP (i.e. markedly raised cell count and decreased glucose concentrations) or brain imaging (i.e. BME); (4) a negative toxoplasma serum IgG antibody test and; (5) use of anti-toxoplasma prophylaxis [154, 156, 174, 177]. If patients do not respond to either of the two treatment modalities, the alternative should be added. However, in some cases it is appropriate to start dual treatment at presentation, especially in acutely ill patients (e.g. a depressed level of consciousness) or when there is a suspicion of dual infection. Only if there is no clinico-radiological response to either or a combination of these two treatments should a brain biopsy be considered.

Surgical Management

Early surgical intervention may be warranted in some cases, including (1) single lesions with negative *Toxoplasma gondii* serum serology, requiring a diagnostic biopsy (open or stereotactic); (2) large tuberculoma with mass effect threatening impending herniation, or abscess, requiring an open biopsy with decompression and; (3) OHC secondary to compression of the ventricular system by tuberculoma, requiring CSF diversion procedures (discussed above) [58, 172].

Prognosis

Mortality rates during TB treatment vary between 0 and 23% [73, 162, 163, 165, 166, 178], with full clinical recovery reported in 40–92% of HIV-uninfected tuberculoma cases [162, 163]. Limited data in HIV-infected cases report mortality rates of 9–29% during treatment [57, 160, 165, 167, 179], with a good response to treatment seen in up to 90% of cases [167]. It appears that the presence of tuberculoma does not increase the risk of mortality in the context of TBM [72, 73, 163]. Serial brain imaging is frequently performed to assess response to treatment, however, a caveat in such an approach is that tuberculoma may enlarge paradoxically or, new lesions may appear after commencing appropriate TB treatment both in HIV-infected [180–183] and -uninfected patients [72, 73, 110, 164, 184, 185]. After 9 months of therapy, tuberculoma may still be present in 60–82% of HIV-uninfected cases, although a large proportion of these patients are asymptomatic [72, 164]. However, other groups report resolution of tuberculoma during TB treatment in more than 80% of cases [73, 166, 178].

Duration of TB Treatment

The WHO recommends TB treatment for 9–12 months in all forms of CNS TB, including tuberculoma, however treatment is often prolonged to 18–24 months in those who have persistent contrast enhancement of lesions [164, 166, 186, 187]. The practice to extend TB treatment past the conventional treatment period is not evidence-based and whilst some authors recommend continuation of TB treatment until resolution of tuberculoma [186], others suggest that there is no additional benefit in continuing TB treatment beyond 18 months [187]. As for all forms of CNS TB, adjunctive corticosteroids should be prescribed in all patients in whom a strong suspicion of tuberculoma exists [90]. In steroid-resistant tuberculoma and abscess, thalidomide (a potent TNF-α inhibitor) has been effective in both adults and children (including HIV-infected patients) [183, 188, 189].

Intraspinal Tuberculosis

Tuberculosis of the spine can be divided into vertebral (bony) and non-bony (referred to as intraspinal TB in this chapter) manifestations. This section will focus on the latter, which is rare compared to intracranial TB [190, 191]. Several studies have reported TB as a major cause of myelopathy and or radiculopathy in high HIV/TB co-infection settings [192–194], accounting for 68% (84/123) of HIV-infected patients with a myelopathy or cauda equina syndrome in whom an etiological cause was confirmed [192]. However, there is a paucity of literature on the presentations, MRI findings and outcomes of HIV-associated intraspinal TB; only case reports and few case series exist [192, 194–202].

Pathogenesis and Pathology

Intraspinal TB may be brought about in three different ways [191]:

- Cephalocaudal extension of intracranial meningitis (conversely, spinal manifestations may also precede TBM, suggestion an "upward" spread of the infection in some cases)
- Primary involvement of spinal meninges after hematogenous dissemination from extra-neural sites of infection (i.e. the initial presentation of CNS tuberculosis)
- As an extension of vertebral TB (transdural extension of TB spondylitis)

Intraspinal TB manifests pathologically as one or more of the following: leptomeningitis, tuberculoma, myelitis, abscess and syringomyelia [190, 191, 203], all of which have been described in HIV-infected patients [192, 195–202] (Fig. 2). Spinal leptomeningitis, that is pathologically identical to the intracranial variant, is by far the most common finding recorded in HIV-uninfected patients [190, 191]. The typical gelatinous exudate may fill the entire space between the dura mater and the cord, encasing the spinal cord and nerve roots and resulting in variable degrees of tissue damage by direct compression or infiltration. Local vasculitis occurs frequently, leading to ischemia and infarction, whilst congested veins may result in cord edema. In the chronic stage of infection, exudates become organized and an adhesive arachnoiditis may develop characterized by arachnoidal collagenosis.

Intraspinal tuberculoma may develop within the spinal cord (intramedullary) or in the subarachnoidal and subdural spinal spaces (intradural, extramedullary) [158, 201] [203]. Intramedullary spinal tuberculoma may occur in the context of leptomeningitis, but may also be seen without evidence of overt spinal meningeal disease [158, 202]. Tuberculous myelitis is characterized by inflammation of the spinal cord that usually affects more than one spinal segment and most commonly involves the thoracic spine [196, 204]. Epidural TB abscess is typically seen as a complication of bony spinal TB [205, 206], but rare cases of primary epidural abscess without evidence of vertebral involvement have been described [200, 207–210]. Syringomyelia, a complication of intraspinal TB, is characterized by cystic cavities within the spinal cord due to abnormal CSF dynamics and may develop days to weeks after TBM diagnosis or develop years after TB treatment completion [190, 197, 200, 211].

Clinical Presentation

In general, intraspinal TB may present acutely (over a few days) or progress subacutely over a few months; less frequently a slowly progressive course over 6 months to years has been described [195, 197–199, 202, 204, 212–214]. Symptoms and signs may originate from nerve roots (such as radicular pain, paresthesias, sphincter dysfunction and lower motor neuron signs) or the spinal cord (such as upper or lower motor neuron signs, sphincter dysfunction and a sensory level); a combination of root and cord signs (i.e. radiculo-myelopathy) is typical [197–199, 212]. In the majority of patients the lower limbs are preferentially involved with upper limbs being spared or less severely affected [190, 196, 197, 199]. Clinically-evident spinal TB may present concurrently with TBM [192, 196, 201, 204] or manifest prior to, during or following TBM treatment [195, 197, 200, 212, 214]. Asymptomatic spinal leptomeningitis (as evidenced by MRI findings) is also increasingly reported in TBM [196, 203, 215]; in a study of childhood TBM, 70% (23/33) of cases in whom MRI spine was performed had asymptomatic intraspinal disease [203]. Cerebrospinal fluid findings suggestive of spinal disease during TBM include a dry tap during LP (i.e. the inability to release CSF), which results from the presence of excessive arachnoidal exudate and adhesions in the lumbar thecal sac [203, 212], and a markedly raised CSF protein (>2.50 g/l), which may be due to a spinal block [196, 203, 212]. Conversely, patients with isolated intramedullary disease, such as tuberculoma, may have no or little CSF inflammatory changes [213, 216].

Spinal Imaging

MRI is the imaging modality of choice in the screening of patients with suspected intraspinal TB [53, 61]. Findings of leptomeningitis include one or more of the following: CSF loculations, obliterations of the spinal subarachnoid space and thickened, clumped nerve roots [197]. The spinal cord may show expansion, compression or atrophy [195]. Intramedullary hyperintensity on T2-weighted image may reflect cord edema, ischemia and or myelitis [53, 195, 204]. Contrast-enhanced MRI shows variable degrees of enhancement of nerve roots, surface of the spinal cord and subarachnoid space, which may be linear, nodular or occlusive. Myelitis is often, but not always associated with intramedullary enhancement and in some cases imaging may be completely normal [192, 202, 204, 217]. Tuberculoma and abscess have similar features to those described for intracranial disease [195, 198, 202]. Syringomyelia presents as a central cavity within the spinal cord demonstrating CSF intensity on T1- and T2-weighted imaging without enhancement [53, 197, 200].

The differential diagnosis of intraspinal TB in HIV is vast and depends on the clinical presentation and, disease location and extent as seen on MRI [192, 193, 218, 219] (Table 1); an exhaustive list is beyond the scope of this chapter. Common alternative infective causes include those discussed for intracranial disease. In addition, primary spinal tumors and metastatic disease should be considered. Pyogenic abscess is the most common cause of epidural collection and requires exclusion in all such cases [220]. If diagnostic uncertainty exists, MRI brain is advised, which may show features of silent concomitant intracranial TB [158, 198, 216].

Treatment

Medical management is the mainstay of treatment for spinal TB and is the same as for other forms of neurological TB, including TB treatment and corticosteroids; in some cases the duration of TB treatment is extended to 18 months [90, 190, 195, 197]. The indications for additional surgical intervention are not clear cut and not based on evidence from RCTs. It is recommended that surgery (such as excision or drainage procedures) be individualized and guided by the extent and nature of the lesion, response to medical therapy and severity of neurological deficit [61, 190]. Surgery may also be warranted if a histological diagnosis is required [90, 190]. CSF diversion procedures such as syringo-subarachnoidal and syringo-peritoneal shunting have been used to treat syringomyelia in general, however these are associated with high rates of failure and shunt blockage in cases of TB leptomeningitis [211]. Arachnolysis is a newer surgical technique that is associated with decreased recurrent rates and improved outcomes [221]. It has recently been suggested that a combination of endoscopic arachnolysis and syringo-subarachnoid shunting may be an effective strategy in selected post-TB syringomyelia cases [211].

Prognosis

Published details of outcomes (mortality and morbidity) of intraspinal TB are limited to small case series and rates of complete recovery are often not reported [158, 196, 201, 213, 222, 223]. In a literature review of all intraspinal TB cases complicating TBM (n = 147), Garg *et al.* [190] report improvement in 57.5%, neurological sequelae in 14.2%, no change in 10.6% and death in 17.6% of cases. In HIV-infected patients, symptomatic improvement on medical treatment with or without surgical intervention has been reported [195, 197, 198, 200], although patients may remain static [199] and not all survive [202]. Spinal disease is likely a risk factor for poor outcome in TBM; in a study of predominantly HIV-uninfected patients, 34% with concomitant intraspinal disease died or were left severely disabled (Modified Barthel index <12) at 6 months follow-up, compared to 13% without intraspinal TB [196]. Spinal cord atrophy, cavitation and the presence of syrinx on MRI may be associated with poor outcome in intraspinal TB [190].

Neurological TB-IRIS

Deterioration of CNS TB after an initial improvement on appropriate TB treatment, referred to as "paradoxical reaction", is well described in both HIV-infected and -uninfected patients [163, 185, 195, 197, 224–226]. When such deterioration occurs in HIV-infected patients after ART initiation during TB treatment, it is referred to as "paradoxical TB-IRIS" (also see chapter "The Tuberculosis-Associated Immune Reconstitution Inflammatory Syndrome (TB-IRIS)") [227]. In the other form of TB-IRIS, namely "unmasking TB-IRIS", untreated TB disease becomes apparent after ART initiation. This section will focus on paradoxical TB-IRIS.

HIV is an independent risk factor for developing paradoxical TB reactions in patients with TBM [224]. Neurological TB-IRIS is common in adults in TB endemic settings, accounting for the majority of cases (21%) of CNS deterioration during the first year of ART in a South African study [181]. In the same setting, the CNS was involved in 12% of TB-IRIS cases [182] and TBM-IRIS developed in almost half of TBM patients starting ART 2 weeks after TB treatment initiation [39]. Neurological TB-IRIS in children is poorly documented; only one case series exists [183].

Pathogenesis and Pathology

The immunopathogenesis of neurological TB-IRIS remains unclear, but a series of recent studies from South Africa has shed some light on mechanisms involved [36, 39, 228]. In their prospective study, Marais *et al.* [39] enrolled ART-naïve, HIV-infected adults at TBM diagnosis when they commenced TB treatment and adjunctive corticosteroids. Antiretroviral treatment was started 2 weeks later and patients were followed for the development of TBM-IRIS. Serial LP was performed at TBM diagnosis, 2 weeks later (prior to ART initiation) and 2 weeks thereafter and or at time of TBM-IRIS development. Patients who subsequently developed TBM-IRIS (16/34) were significantly more likely to have a positive CSF *M. tuberculosis* culture at baseline compared to those who did not (15/16 vs 7/18; RR = 9.3, 95% CI, 1.4–62.2, p = 0.004). This supports the inference that high bacillary load predisposes to TB-IRIS [229] and highlights the importance of optimizing TB treatment prior to ART initiation. Other CSF findings that predicted subsequent IRIS were high CSF neutrophil counts (median = 50 vs 3 cells x 10^6/l) and the combination of high TNF-α and low IFN-γ. TBM-IRIS was associated with a marked, compartmentalized inflammatory response in the CSF both at TBM diagnosis and at IRIS presentation that included increased concentrations of a wide range of pro-inflammatory and anti-inflammatory cytokines, chemokines, MMPs and neutrophil-associated peptides [36]. Of these, the neutrophil-associated peptide S100A8/A9 was the only inflammatory mediator to differentiate TBM-IRIS from culture positive non-IRIS patients after ART initiation. Although protein levels of inflammatory mediators were similar in blood between IRIS and non-IRIS patients, blood transcriptomic analysis revealed significantly more abundant neutrophil-associated transcripts from before development of IRIS through IRIS symptom onset [228]. After ART initiation, transcripts associated with canonical and non-canonical inflammasomes were increased in IRIS patients compared to non-IRIS patients. These findings support a dominant role for the innate immune system in the pathogenesis of neurological TB-IRIS, which may inform future studies investigating host-directed therapies in the treatment and prevention of the syndrome.

Clinical Presentation

The onset of worsening, new or recurrent neurological TB symptoms occurs at a median of 14 days, and usually within 3 months of ART initiation, however occasionally patients develop the typical presentation at later timepoints [39, 230]. In the only childhood series, IRIS events occurred within 3 weeks of ART initiation [183]. Neurological TB-IRIS may present as any form of neurological TB, including TBM [39, 181, 182, 230], intracranial tuberculoma [39, 181, 182] or abscess [58, 231] or intraspinal TB [39, 181, 182], with clinical features as described in previous sections (Fig. 2b). Neurological manifestations described in children include newly acquired neck stiffness, intracranial and intraspinal tuberculous mass lesions, radiculomyelitis, HC, visual compromise and seizures [183]. Although clinical characteristics at time of TBM diagnosis were mostly similar between adults who did and did not develop subsequent TBM-IRIS, TBM-IRIS patients had longer symptom duration (median 19 vs 9 days), more frequent chest radiograph abnormalities (81% vs 50%) and lower serum sodium concentrations (median 123 vs 130 mmol/l), which may reflect more disseminated TB disease [39].

The diagnosis of neurological TB-IRIS requires exclusion of other causes for deterioration, as no confirmatory diagnostic test exists [227]. Differential diagnoses include TB drug resistance, poor adherence to TB treatment or ART, drug reactions or toxicities (e.g. efavirenz-induced psychosis) and other opportunistic infections (e.g. CM and progressive multifocal leucoencephalopathy [PML]-IRIS). Patients may also deteriorate due to the natural progression of TBM (e.g. strokes). Ideally patients should have repeat brain imaging and LP at time of IRIS presentations to aid diagnosis and exclusion of other causes. TB-IRIS (including all organ manifestations combined) usually has a benign disease course, with death attributed to TB-IRIS occurring in 2% of cases [232]. In neurological TB-IRIS however, raised intracranial pressure due to cerebral inflammation in the confined intracranial space may result in complications such as compression of viral brain structures and brain herniation, that could explain the high associated mortality rates (13% - 75%) reported in this form [39, 182, 230].

Time of ART Initiation in TBM

Trials in extra-neural TB have consistently shown a mortality benefit in severely immunosuppressed patients (CD4 < 50 cells/μL) who start ART early (2–4 weeks after TB treatment initiation) compared with later (8 weeks after TB treatment initiation), in spite of a significantly increased risk of developing TB-IRIS in the early arms [233–235]. The optimal time to start ART in neurological TB is still uncertain; only one adult study has addressed this question [105]. Torok *et al.* reported no difference in nine-month mortality between adults with TBM (median CD4 count = 41 cells/μL) who started ART early (within seven days of TB treatment initiation) compared to later (8 weeks after TB treatment initiation). This study did not report the incidence of neurological TB-IRIS, but an increased frequency of severe adverse events occurred in the early ART arm. Because of these findings and the potentially increased risk of developing life-threatening neurological TB-IRIS manifestations with early ART initiation, some guidelines suggests delaying ART (up to 8 weeks) after starting TB treatment in all TBM patients [236, 237]. An important component of initiating ART during TBM treatment involves counseling patients regarding the potential of developing neurological TB-IRIS. Patients should be advised to continue their ART and to present themselves to medical care if any symptomatic deterioration occurs.

Treatment

Corticosteroids are the only treatment modality for which clinical trial data in TB-IRIS exist. In a RCT that included mild to moderate TB-IRIS cases, daily prednisone (1.5 mg/kg for 2 weeks followed by 0.75 mg/kg for a further 2 weeks) compared to placebo was associated with more rapid symptom relief and reduced the duration of hospitalization and need for therapeutic procedures [238]. However, this study did not include cases with severe forms of IRIS such as those with neurological involvement. Corticosteroid therapy has anecdotally been associated with good outcome in neurological TB-IRIS [39, 182], and is commonly used to treat such cases, but the choice of drug (prednisone, methylprednisone or dexamethasone) and, dose and duration of treatment, are not standardized [239]. Some experts suggest a dose of 1.5 mg/kg/day of prednisone (or equivalent) for 2–4 weeks followed by a gradual taper depending on the clinical response in adults [240]. Although 2–4 months of treatment is adequate in most cases, some may relapse after treatment discontinuation, necessitating re-initiation of treatment [39, 241]. If the diagnosis of IRIS is questionable, corticosteroids should be deferred pending further investigations. It is important to note that corticosteroids do not prevent neurological TB-IRIS; in one study 13/16 TBM cases were receiving prednisone (0.75–1.5 mg/kg/day) at time of neurological TB-IRIS presentation [39]. Case reports describing a good response of TB-IRIS to numerous other anti-inflammatory agents (e.g. non-steroidal anti-inflammatory drugs, thalidomide, pentoxifylline, mycophenylate mofetil and montelukast) have been published, but none of these have been investigated in RCTs [183, 189, 239]. Antiretroviral therapy should be continued as far as possible during the IRIS episode as discontinuation may result in the acquisition of ART drug resistance and leave patients vulnerable to other opportunistic infections. However, temporary interruption of ART should be considered in cases with depressed level of consciousness or severe disease not responsive to corticosteroids [242].

References

1. Sandgren A, Hollo V, van der Werf MJ (2013) Extrapulmonary tuberculosis in the European Union and European Economic Area, 2002 to 2011. Euro Surveill 18(12)
2. Efsen AM, Panteleev AM, Grint D, Podlekareva DN, Vassilenko A, Rakhmanova A, et al. TB meningitis in HIV-positive patients in Europe and Argentina: clinical outcome and factors associated with mortality. Biomed Res Int 2013;2013:373601
3. Kingkaew N, Sangtong B, Amnuaiphon W, Jongpaibulpatana J, Mankatittham W, Akksilp S et al (2009) HIV-associated extrapulmonary tuberculosis in Thailand: epidemiology and risk factors for death. Int J Infect Dis 13(6):722–729
4. Leeds IL, Magee MJ, Kurbatova EV, del Rio C, Blumberg HM, Leonard MK et al (2012) Site of extrapulmonary tuberculosis is associated with HIV infection. Clin Infect Dis 55(1):75–81
5. Nicol MP, Sola C, February B, Rastogi N, Steyn L, Wilkinson RJ (2005) Distribution of strain families of *Mycobacterium tuberculosis* causing pulmonary and extrapulmonary disease in hospitalized children in Cape Town, South Africa. J Clin Microbiol 43(11):5779–5781
6. Berenguer J, Moreno S, Laguna F, Vicente T, Adrados M, Ortega A et al (1992) Tuberculous meningitis in patients infected with the human immunodeficiency virus. N Engl J Med 326(10):668–672
7. Marais S, Thwaites G, Schoeman JF, Torok ME, Misra UK, Prasad K et al (2010) Tuberculous meningitis: a uniform case definition for use in clinical research. Lancet Infect Dis 10(11):803–812
8. Bahr NC, Nuwagira E, Evans EE, Cresswell FV, Bystrom PV, Byamukama A et al (2017) Diagnostic accuracy of Xpert MTB/RIF Ultra for tuberculous meningitis in HIV-infected adults: a prospective cohort study. Lancet Infect Dis. Sept 14 [Epub ahead of print]
9. Dastur DK, Manghani DK, Udani PM (1995) Pathology and pathogenetic mechanisms in neurotuberculosis. Radiol Clin N Am 33(4):733–752
10. Vinnard C, King L, Munsiff S, Crossa A, Iwata K, Pasipanodya J et al (2017) Long-term mortality of patients with tuberculous meningitis in New York City: a Cohort Study. Clin Infect Dis 64(4):401–407
11. Ducomble T, Tolksdorf K, Karagiannis I, Hauer B, Brodhun B, Haas W et al (2013) The burden of extrapulmonary and meningitis tuberculosis: an investigation of national surveillance data, Germany, 2002–2009. Euro Surveill 18:12
12. Centers for Disease Control and Prevention (CDC). Reported tuberculosis in the United States, 2015. Atlanta: US Department of Health and Human Services, CDC; 2016. Available at https://www.cdc.gov/tb/statistics/reports/2015/pdfs/2015_Surveillance_Report_FullReport.pdf. Accessed 11 Oct 2017
13. Wilkinson RJ, Rohlwink U, Misra UK, van Crevel R, Mai NTH, Dooley KE et al (2017) Tuberculous meningitis. Nat Rev Neurol 13(10):581–598
14. Jarvis JN, Meintjes G, Williams A, Brown Y, Crede T, Harrison TS (2010) Adult meningitis in a setting of high HIV and TB prevalence: findings from 4961 suspected cases. BMC Infect Dis 10:67

15. Cohen DB, Zijlstra EE, Mukaka M, Reiss M, Kamphambale S, Scholing M et al (2010) Diagnosis of cryptococcal and tuberculous meningitis in a resource-limited African setting. Tropical Med Int Health 15(8):910–917
16. Boulware DR, Meya DB, Muzoora C, Rolfes MA, Huppler Hullsiek K, Musubire A et al (2014) Timing of antiretroviral therapy after diagnosis of cryptococcal meningitis. N Engl J Med 370(26):2487–2498
17. Marais S, Pepper DJ, Schutz C, Wilkinson RJ, Meintjes G (2011) Presentation and outcome of tuberculous meningitis in a high HIV prevalence setting. PLoS One 6(5):e20077
18. Wolzak NK, Cooke ML, Orth H, van Toorn R (2012) The changing profile of pediatric meningitis at a referral centre in Cape Town, South Africa. J Trop Pediatr 58(6):491–495
19. van Well GT, Paes BF, Terwee CB, Springer P, Roord JJ, Donald PR et al (2009) Twenty years of pediatric tuberculous meningitis: a retrospective cohort study in the western cape of South Africa. Pediatrics 123(1):e1–e8
20. Rock RB, Olin M, Baker CA, Molitor TW, Peterson PK (2008) Central nervous system tuberculosis: pathogenesis and clinical aspects. Clin Microbiol Rev 21(2):243–261
21. Hektoen L (1896) The Vascular Changes of Tuberculous Meningitis, Especially the Tuberculous Endarterities. J Exp Med 1(1):112–163
22. Rich AR, McCordock HA (1933) The pathogenesis of tuberculous meningitis. Bull Johns Hopkins Hosp 52:2–37
23. Jain SK, Paul-Satyaseela M, Lamichhane G, Kim KS, Bishai WR (2006) *Mycobacterium tuberculosis* invasion and traversal across an in vitro human blood-brain barrier as a pathogenic mechanism for central nervous system tuberculosis. J Infect Dis 193(9):1287–1295
24. Donald PR, Schaaf HS, Schoeman JF (2005) Tuberculous meningitis and miliary tuberculosis: the Rich focus revisited. J Infect 50(3):193–195
25. Janse van Rensburg P, Andronikou S, van Toorn R, Pienaar M (2008) Magnetic resonance imaging of miliary tuberculosis of the central nervous system in children with tuberculous meningitis. Pediatr Radiol 38(12):1306–1313
26. von Gottberg A, Sacks L, Machala S, Blumberg L (2001) Utility of blood cultures and incidence of mycobacteremia in patients with suspected tuberculosis in a South African infectious disease referral hospital. Int J Tuberc Lung Dis 5(1):80–86
27. Shafer RW, Goldberg R, Sierra M, Glatt AE (1989) Frequency of *Mycobacterium tuberculosis* bacteremia in patients with tuberculosis in an area endemic for AIDS. Am Rev Respir Dis 140(6):1611–1613
28. Jones BE, Young SM, Antoniskis D, Davidson PT, Kramer F, Barnes PF (1993) Relationship of the manifestations of tuberculosis to CD4 cell counts in patients with human immunodeficiency virus infection. Am Rev Respir Dis 148(5):1292–1297
29. Thwaites GE, Tran TH (2005) Tuberculous meningitis: many questions, too few answers. Lancet Neurol 4(3):160–170
30. Rock RB, Hu S, Gekker G, Sheng WS, May B, Kapur V et al (2005) *Mycobacterium tuberculosis*-induced cytokine and chemokine expression by human microglia and astrocytes: effects of dexamethasone. J Infect Dis 192(12):2054–2058
31. Spanos JP, Hsu NJ, Jacobs M (2015) Microglia are crucial regulators of neuro-immunity during central nervous system tuberculosis. Front Cell Neurosci 9:182
32. Thwaites GE, Simmons CP, Than Ha Quyen N, Thi Hong Chau T, Phuong Mai P, Thi Dung N et al (2003) Pathophysiology and prognosis in vietnamese adults with tuberculous meningitis. J Infect Dis 188(8):1105–1115
33. Patel VB, Bhigjee AI, Bill PL, Connolly CA (2002) Cytokine profiles in HIV seropositive patients with tuberculous meningitis. J Neurol Neurosurg Psychiatry 73(5):598–599
34. Simmons CP, Thwaites GE, Quyen NT, Torok E, Hoang DM, Chau TT et al (2006) Pretreatment intracerebral and peripheral blood immune responses in Vietnamese adults with tuberculous meningitis: diagnostic value and relationship to disease severity and outcome. J Immunol 176(3):2007–2014
35. Mastroianni CM, Paoletti F, Lichtner M, D'Agostino C, Vullo V, Delia S (1997) Cerebrospinal fluid cytokines in patients with tuberculous meningitis. Clin Immunol Immunopathol 84(2):171–176

36. Marais S, Wilkinson KA, Lesosky M, Coussens AK, Deffur A, Pepper DJ et al (2014) Neutrophil-associated central nervous system inflammation in tuberculous meningitis immune reconstitution inflammatory syndrome. Clin Infect Dis 59(11):1638–1647
37. Visser DH, Solomons RS, Ronacher K, van Well GT, Heymans MW, Walzl G et al (2015) Host immune response to tuberculous meningitis. Clin Infect Dis 60(2):177–187
38. Thuong NTT, Heemskerk D, Tram TTB, Thao LTP, Ramakrishnan L, Ha VTN et al (2017) Leukotriene A4 hydrolase genotype and HIV infection influence intracerebral inflammation and survival from tuberculous meningitis. J Infect Dis 215(7):1020–1028
39. Marais S, Meintjes G, Pepper DJ, Dodd LE, Schutz C, Ismail Z et al (2013) Frequency, severity, and prediction of tuberculous meningitis immune reconstitution inflammatory syndrome. Clin Infect Dis 56(3):450–460
40. Tho DQ, Torok ME, Yen NT, Bang ND, Lan NT, Kiet VS et al (2012) Influence of anti-tuberculosis drug resistance and *Mycobacterium tuberculosis* lineage on outcome in HIV-associated tuberculous meningitis. Antimicrob Agents Chemother 56(6):3074–3079
41. Tobin DM, Roca FJ, Oh SF, McFarland R, Vickery TW, Ray JP et al (2012) Host genotype-specific therapies can optimize the inflammatory response to mycobacterial infections. Cell 148(3):434–446
42. van Laarhoven A, Dian S, Ruesen C, Hayati E, Damen M, Annisa J et al (2017) Clinical parameters, routine inflammatory markers, and LTA4H genotype as predictors of mortality among 608 patients with tuberculous meningitis in Indonesia. J Infect Dis 215(7):1029–1039
43. Katrak SM, Shembalkar PK, Bijwe SR, Bhandarkar LD (2000) The clinical, radiological and pathological profile of tuberculous meningitis in patients with and without human immunodeficiency virus infection. J Neurol Sci 181(1–2):118–126
44. Schutte CM (2001) Clinical, cerebrospinal fluid and pathological findings and outcomes in HIV-positive and HIV-negative patients with tuberculous meningitis. Infection 29(4):213–217
45. Tripathi S, Patro I, Mahadevan A, Patro N, Phillip M, Shankar SK (2014) Glial alterations in tuberculous and cryptococcal meningitis and their relation to HIV co-infection--a study on human brains. J Infect Dev Ctries 8(11):1421–1443
46. Lanjewar DN, Jain PP, Shetty CR (1998) Profile of central nervous system pathology in patients with AIDS: an autopsy study from India. AIDS 12(3):309–313
47. Smith AB, Smirniotopoulos JG, Rushing EJ (2008) From the archives of the AFIP: central nervous system infections associated with human immunodeficiency virus infection: radiologic-pathologic correlation. Radiographics 28(7):2033–2058
48. Lawn SD, Butera ST, Shinnick TM (2002) Tuberculosis unleashed: the impact of human immunodeficiency virus infection on the host granulomatous response to *Mycobacterium tuberculosis*. Microbes Infect 4(6):635–646
49. Chatterjee D, Radotra BD, Vasishta RK, Sharma K (2015) Vascular complications of tuberculous meningitis: an autopsy study. Neurol India 63(6):926–932
50. Tai MS, Viswanathan S, Rahmat K, Nor HM, Kadir KA, Goh KJ et al (2016) Cerebral infarction pattern in tuberculous meningitis. Sci Rep 6:38802
51. Hammond CK, Eley B, Wieselthaler N, Ndondo A, Wilmshurst JM (2016) Cerebrovascular disease in children with HIV-1 infection. Dev Med Child Neurol 58(5):452–460
52. Benjamin LA, Allain TJ, Mzinganjira H, Connor MD, Smith C, Lucas S et al (2017) The role of human immunodeficiency virus-associated vasculopathy in the etiology of stroke. J Infect Dis 216(5):545–553
53. Bernaerts A, Vanhoenacker FM, Parizel PM, Van Goethem JW, Van Altena R, Laridon A et al (2003) Tuberculosis of the central nervous system: overview of neuroradiological findings. Eur Radiol 13(8):1876–1890
54. Bayindir C, Mete O, Bilgic B (2006) Retrospective study of 23 pathologically proven cases of central nervous system tuberculomas. Clin Neurol Neurosurg 108(4):353–357
55. DeLance AR, Safaee M, Oh MC, Clark AJ, Kaur G, Sun MZ et al (2013) Tuberculoma of the central nervous system. J Clin Neurosci 20(10):1333–1341
56. Danaviah S, Sacks JA, Kumar KP, Taylor LM, Fallows DA, Naicker T et al (2013) Immunohistological characterization of spinal TB granulomas from HIV-negative and -positive patients. Tuberculosis (Edinb) 93(4):432–441

57. Lesprit P, Zagdanski AM, de La Blanchardiere A, Rouveau M, Decazes JM, Frija J et al (1997) Cerebral tuberculosis in patients with the acquired immunodeficiency syndrome (AIDS). Report of 6 cases and review. Medicine (Baltimore) 76(6):423–431
58. Vidal JE, Cimerman S, da Silva PR, Sztajnbok J, Coelho JF, Lins DL (2003) Tuberculous brain abscess in a patient with AIDS: case report and literature review. Rev Inst Med Trop Sao Paulo 45(2):111–114
59. Marais S, Pepper DJ, Marais BJ, Torok ME (2010) HIV-associated tuberculous meningitis—diagnostic and therapeutic challenges. Tuberculosis (Edinb) 90(6):367–374
60. Girgis NI, Sultan Y, Farid Z, Mansour MM, Erian MW, Hanna LS et al (1998) Tuberculosis meningitis, Abbassia fever hospital-Naval Medical Research Unit No. 3-Cairo, Egypt, from 1976 to 1996. Am J Trop Med Hyg 58(1):28–34
61. Garcia-Monco JC (1999) Central nervous system tuberculosis. Neurol Clin 17(4):737–759
62. Thwaites GE, Duc Bang N, Huy Dung N, Thi Quy H, Thi Tuong Oanh D, Thi Cam Thoa N et al (2005) The influence of HIV infection on clinical presentation, response to treatment, and outcome in adults with Tuberculous meningitis. J Infect Dis 192(12):2134–2141
63. van der Weert EM, Hartgers NM, Schaaf HS, Eley BS, Pitcher RD, Wieselthaler NA et al (2006) Comparison of diagnostic criteria of tuberculous meningitis in human immunodeficiency virus-infected and uninfected children. Pediatr Infect Dis J 25(1):65–69
64. Solomons RS, Visser DH, Donald PR, Marais BJ, Schoeman JF, van Furth AM (2015) The diagnostic value of cerebrospinal fluid chemistry results in childhood tuberculous meningitis. Childs Nerv Syst 31(8):1335–1340
65. Cecchini D, Ambrosioni J, Brezzo C, Corti M, Rybko A, Perez M et al (2009) Tuberculous meningitis in HIV-infected and non-infected patients: comparison of cerebrospinal fluid findings. Int J Tuberc Lung Dis 13(2):269–271
66. Torok ME, Chau TT, Mai PP, Phong ND, Dung NT, Chuong LV et al (2008) Clinical and microbiological features of HIV-associated tuberculous meningitis in Vietnamese adults. PLoS One 3(3):e1772
67. van Toorn R, Solomons R (2014) Update on the diagnosis and management of tuberculous meningitis in children. Semin Pediatr Neurol 21(1):12–18
68. Davis LE, Rastogi KR, Lambert LC, Skipper BJ (1993) Tuberculous meningitis in the southwest United States: a community-based study. Neurology 43(9):1775–1778
69. Misra UK, Kalita J, Bhoi SK, Singh RK (2016) A study of hyponatremia in tuberculous meningitis. J Neurol Sci 367:152–157
70. Thwaites GE, Nguyen DB, Nguyen HD, Hoang TQ, Do TT, Nguyen TC et al (2004) Dexamethasone for the treatment of tuberculous meningitis in adolescents and adults. N Engl J Med 351(17):1741–1751
71. Ozates M, Kemaloglu S, Gurkan F, Ozkan U, Hosoglu S, Simsek MM (2000) CT of the brain in tuberculous meningitis. A review of 289 patients. Acta Radiol 41(1):13–17
72. Thwaites GE, Macmullen-Price J, Tran TH, Pham PM, Nguyen TD, Simmons CP et al (2007) Serial MRI to determine the effect of dexamethasone on the cerebral pathology of tuberculous meningitis: an observational study. Lancet Neurol 6(3):230–236
73. Schoeman JF, Van Zyl LE, Laubscher JA, Donald PR (1995) Serial CT scanning in childhood tuberculous meningitis: prognostic features in 198 cases. J Child Neurol 10(4):320–329
74. Wasay M, Kheleani BA, Moolani MK, Zaheer J, Pui M, Hasan S et al (2003) Brain CT and MRI findings in 100 consecutive patients with intracranial tuberculoma. J Neuroimaging 13(3):240–247
75. Pienaar M, Andronikou S, van Toorn R (2009) MRI to demonstrate diagnostic features and complications of TBM not seen with CT. Childs Nerv Syst 25(8):941–947
76. Christensen AS, Andersen AB, Thomsen VO, Andersen PH, Johansen IS (2011) Tuberculous meningitis in Denmark: a review of 50 cases. BMC Infect Dis 11:47
77. Villoria MF, de la Torre J, Fortea F, Munoz L, Hernandez T, Alarcon JJ (1992) Intracranial tuberculosis in AIDS: CT and MRI findings. Neuroradiology 34(1):11–14
78. van der Merwe DJ, Andronikou S, Van Toorn R, Pienaar M (2009) Brainstem ischemic lesions on MRI in children with tuberculous meningitis: with diffusion weighted confirmation. Childs Nerv Syst 25(8):949–954

79. Azuaje C, Fernandez Hidalgo N, Almirante B, Martin-Casabona N, Ribera E, Diaz M et al (2006) Tuberculous meningitis: a comparative study in relation to concurrent human immunodeficiency virus infection. Enferm Infecc Microbiol Clin 24(4):245–250
80. Dekker G, Andronikou S, van Toorn R, Scheepers S, Brandt A, Ackermann C (2011) MRI findings in children with tuberculous meningitis: a comparison of HIV-infected and non-infected patients. Childs Nerv Syst 27(11):1943–1949
81. Andronikou S, Smith B, Hatherhill M, Douis H, Wilmshurst J (2004) Definitive neuroradiological diagnostic features of tuberculous meningitis in children. Pediatr Radiol 34(11):876–885
82. Botha H, Ackerman C, Candy S, Carr JA, Griffith-Richards S, Bateman KJ (2012) Reliability and diagnostic performance of CT imaging criteria in the diagnosis of tuberculous meningitis. PLoS One 7(6):e38982
83. Kalita J, Prasad S, Maurya PK, Kumar S, Misra UK (2012) MR angiography in tuberculous meningitis. Acta Radiol 53(3):324–329
84. Patel VB, Theron G, Lenders L, Matinyena B, Connolly C, Singh R et al (2013) Diagnostic accuracy of quantitative PCR (Xpert MTB/RIF) for tuberculous meningitis in a high burden setting: a prospective study. PLoS Med 10(10):e1001536
85. Bahr NC, Tugume L, Rajasingham R, Kiggundu R, Williams DA, Morawski B et al (2015) Improved diagnostic sensitivity for tuberculous meningitis with Xpert((R)) MTB/RIF of centrifuged CSF. Int J Tuberc Lung Dis 19(10):1209–1215
86. Bhigjee AI, Padayachee R, Paruk H, Hallwirth-Pillay KD, Marais S, Connoly C (2007) Diagnosis of tuberculous meningitis: clinical and laboratory parameters. Int J Infect Dis 11(4):348–354
87. Heemskerk AD, Bang ND, Mai NT, Chau TT, Phu NH, Loc PP et al (2016) Intensified antituberculosis therapy in adults with tuberculous meningitis. N Engl J Med 374(2):124–134
88. Thwaites GE, Chau TT, Farrar JJ (2004) Improving the bacteriological diagnosis of tuberculous meningitis. J Clin Microbiol 42(1):378–379
89. Kennedy DH, Fallon RJ (1979) Tuberculous meningitis. JAMA 241(3):264–268
90. Thwaites G, Fisher M, Hemingway C, Scott G, Solomon T, Innes J (2009) British Infection Society guidelines for the diagnosis and treatment of tuberculosis of the central nervous system in adults and children. J Infect 59(3):167–187
91. World Health Organization (2013) Policy update: Xpert MTB/RIF assay for the diagnosis of pulmonary and extrapulmonary TB in adults and children. Available at: www.who.int/tb/laboratory/xpert_launchupdate. Accessed 30 Nov 2017
92. Nhu NT, Heemskerk D, Thu DD, Chau TT, Mai NT, Nghia HD et al (2014) Evaluation of Xpert MTB/RIF for the diagnosis of tuberculous meningitis. J Clin Microbiol 52(1):226–233
93. World Health Organization (2017) WHO Meeting Report of a Technical Expert Consultation: Non-inferiority analysis of Xpert MTB/RIF Ultra compared to Xpert MTB/RIF. Available at: www.who.int/tb/publications/2017/XpertUltra/en/ Accessed 30 Nov 2017
94. Solomons RS, van Elsland SL, Visser DH, Hoek KG, Marais BJ, Schoeman JF et al (2014) Commercial nucleic acid amplification tests in tuberculous meningitis--a meta-analysis. Diagn Microbiol Infect Dis 78(4):398–403
95. Bahr NC, Marais S, Caws M, van Crevel R, Wilkinson RJ, Tyagi JS et al (2016) GeneXpert MTB/Rif to diagnose tuberculous meningitis: perhaps the first test but not the last. Clin Infect Dis 62(9):1133–1135
96. Moghtaderi A, Alavi-Naini R, Izadi S, Cuevas LE (2009) Diagnostic risk factors to differentiate tuberculous and acute bacterial meningitis. Scand J Infect Dis 41(3):188–194
97. Zhang B, Lv K, Bao J, Lu C, Lu Z (2013) Clinical and laboratory factors in the differential diagnosis of tuberculous and cryptococcal meningitis in adult HIV-negative patients. Intern Med 52(14):1573–1578
98. Nyazika TK, Masanganise F, Hagen F, Bwakura-Dangarembizi MF, Ticklay IM, Robertson VJ (2016) Cryptococcal meningitis presenting as a complication in HIV-infected children: a case series from Sub-Saharan Africa. Pediatr Infect Dis J 35(9):979–980

99. Vidal JE, Boulware DR (2015) Lateral flow assay for cryptococcal antigen: an important advance to improve the continuum of HIV care and reduce cryptococcal meningitis-related mortality. Rev Inst Med Trop Sao Paulo 57(Suppl 19):38–45
100. Boulware DR, Rolfes MA, Rajasingham R, von Hohenberg M, Qin Z, Taseera K et al (2014) Multisite validation of cryptococcal antigen lateral flow assay and quantification by laser thermal contrast. Emerg Infect Dis 20(1):45–53
101. Price RW, Spudich S (2008) Antiretroviral therapy and central nervous system HIV type 1 infection. J Infect Dis 197(Suppl 3):S294–S306
102. Christo PP, Vilela Mde C, Bretas TL, Domingues RB, Greco DB, Livramento JA et al (2009) Cerebrospinal fluid levels of chemokines in HIV infected patients with and without opportunistic infection of the central nervous system. J Neurol Sci 287(1–2):79–83
103. Ruslami R, Ganiem AR, Dian S, Apriani L, Achmad TH, van der Ven AJ et al (2013) Intensified regimen containing rifampicin and moxifloxacin for tuberculous meningitis: an open-label, randomised controlled phase 2 trial. Lancet Infect Dis 13(1):27–35
104. Topley JM, Bamber S, Coovadia HM, Corr PD (1998) Tuberculous meningitis and co-infection with HIV. Ann Trop Paediatr 18(4):261–266
105. Torok ME, Yen NT, Chau TT, Mai NT, Phu NH, Mai PP et al (2011) Timing of initiation of antiretroviral therapy in human immunodeficiency virus (HIV)-associated tuberculous meningitis. Clin Infect Dis 52(11):1374–1383
106. Humphries MJ, Teoh R, Lau J, Gabriel M (1990) Factors of prognostic significance in Chinese children with tuberculous meningitis. Tubercle 71(3):161–168
107. Springer P, Swanevelder S, van Toorn R, van Rensburg AJ, Schoeman J (2009) Cerebral infarction and neurodevelopmental outcome in childhood tuberculous meningitis. Eur J Paediatr Neurol 13(4):343–349
108. Schoeman CJ, Herbst I, Nienkemper DC (1997) The effect of tuberculous meningitis on the cognitive and motor development of children. S Afr Med J 87(1):70–72
109. Wait JW, Schoeman JF (2010) Behaviour profiles after tuberculous meningitis. J Trop Pediatr 56(3):166–171
110. Ranjan P, Kalita J, Misra UK (2003) Serial study of clinical and CT changes in tuberculous meningitis. Neuroradiology 45(5):277–282
111. Kalita J, Misra UK, Ranjan P (2007) Predictors of long-term neurological sequelae of tuberculous meningitis: a multivariate analysis. Eur J Neurol 14(1):33–37
112. Donald KA, Hoare J, Eley B, Wilmshurst JM (2014) Neurologic complications of pediatric human immunodeficiency virus: implications for clinical practice and management challenges in the African setting. Semin Pediatr Neurol 21(1):3–11
113. Heaton RK, Clifford DB, Franklin DR Jr, Woods SP, Ake C, Vaida F et al (2010) HIV-associated neurocognitive disorders persist in the era of potent antiretroviral therapy: CHARTER Study. Neurology 75(23):2087–2096
114. World Health Organization (2010) Treatment of tuberculosis guidelines, 4th ed. Available at: whqlibdoc.who.int/publications/2010/9789241547833_eng.pdf. Accessed 30 Nov 2017
115. Nahid P, Dorman SE, Alipanah N, Barry PM, Brozek JL, Cattamanchi A et al (2016) Official American Thoracic Society/Centers for Disease Control and Prevention/Infectious Diseases Society of America Clinical Practice Guidelines: Treatment of Drug-Susceptible Tuberculosis. Clin Infect Dis 63(7):e147–ee95
116. Donald PR (2010) Cerebrospinal fluid concentrations of antituberculosis agents in adults and children. Tuberculosis (Edinb) 90(5):279–292
117. Gurumurthy P, Ramachandran G, Hemanth Kumar AK, Rajasekaran S, Padmapriyadarsini C, Swaminathan S et al (2004) Malabsorption of rifampin and isoniazid in HIV-infected patients with and without tuberculosis. Clin Infect Dis 38(2):280–283
118. McIlleron H, Rustomjee R, Vahedi M, Mthiyane T, Denti P, Connolly C et al (2012) Reduced antituberculosis drug concentrations in HIV-infected patients who are men or have low weight: implications for international dosing guidelines. Antimicrob Agents Chemother 56(6):3232–3238

119. Heemskerk AD, Nguyen MTH, Dang HTM, Vinh Nguyen CV, Nguyen LH, Do TDA et al (2017) Clinical outcomes of patients with drug-resistant tuberculous meningitis treated with an intensified antituberculosis regimen. Clin Infect Dis 65(1):20–28
120. Thwaites GE, Lan NT, Dung NH, Quy HT, Oanh DT, Thoa NT et al (2005) Effect of antituberculosis drug resistance on response to treatment and outcome in adults with tuberculous meningitis. J Infect Dis 192(1):79–88
121. Te Brake L, Dian S, Ganiem AR, Ruesen C, Burger D, Donders R et al (2015) Pharmacokinetic/pharmacodynamic analysis of an intensified regimen containing rifampicin and moxifloxacin for tuberculous meningitis. Int J Antimicrob Agents 45(5):496–503
122. Boeree MJ, Heinrich N, Aarnoutse R, Diacon AH, Dawson R, Rehal S et al (2017) High-dose rifampicin, moxifloxacin, and SQ109 for treating tuberculosis: a multi-arm, multi-stage randomised controlled trial. Lancet Infect Dis 17(1):39–49
123. Donald PR, Schoeman JF, Van Zyl LE, De Villiers JN, Pretorius M, Springer P (1998) Intensive short course chemotherapy in the management of tuberculous meningitis. Int J Tuberc Lung Dis 2(9):704–711
124. van Toorn R, Schaaf HS, Laubscher JA, van Elsland SL, Donald PR, Schoeman JF (2014) Short intensified treatment in children with drug-susceptible tuberculous meningitis. Pediatr Infect Dis J 33(3):248–252
125. Thwaites GE, Bhavnani SM, Chau TT, Hammel JP, Torok ME, Van Wart SA et al (2011) Randomized pharmacokinetic and pharmacodynamic comparison of fluoroquinolones for tuberculous meningitis. Antimicrob Agents Chemother 55(7):3244–3253
126. Garg RK, Jain A, Malhotra HS, Agrawal A, Garg R (2013) Drug-resistant tuberculous meningitis. Expert Rev Anti-Infect Ther 11(6):605–621
127. Seddon JA, Visser DH, Bartens M, Jordaan AM, Victor TC, van Furth AM et al (2012) Impact of drug resistance on clinical outcome in children with tuberculous meningitis. Pediatr Infect Dis J 31(7):711–716
128. Schoeman JF, Van Zyl LE, Laubscher JA, Donald PR (1997) Effect of corticosteroids on intracranial pressure, computed tomographic findings, and clinical outcome in young children with tuberculous meningitis. Pediatrics 99(2):226–231
129. Prasad K, Singh MB, Ryan H (2016) Corticosteroids for managing tuberculous meningitis. Cochrane Database Syst Rev (4):CD002244
130. World Health Organization. Treatment of tuberculosis. Guidelines for treatment of drug-susceptible tuberculosis and patient care (2017 update). 2017. Available at: www.who.int/tb/publications/2017/dstb_guidance_2017/en/ Accessed 30 Nov 2017
131. Lammie GA, Hewlett RH, Schoeman JF, Donald PR (2009) Tuberculous cerebrovascular disease: a review. J Infect 59(3):156–166
132. Schoeman JF, Janse van Rensburg A, Laubscher JA, Springer P (2011) The role of aspirin in childhood tuberculous meningitis. J Child Neurol 26(8):956–962
133. Misra UK, Kalita J, Nair PP (2010) Role of aspirin in tuberculous meningitis: a randomized open label placebo controlled trial. J Neurol Sci 293(1–2):12–17
134. Raut T, Garg RK, Jain A, Verma R, Singh MK, Malhotra HS et al (2013) Hydrocephalus in tuberculous meningitis: incidence, its predictive factors and impact on the prognosis. J Infect 66(4):330–337
135. Hsu PC, Yang CC, Ye JJ, Huang PY, Chiang PC, Lee MH (2010) Prognostic factors of tuberculous meningitis in adults: a 6-year retrospective study at a tertiary hospital in northern Taiwan. J Microbiol Immunol Infect 43(2):111–118
136. Figaji AA, Fieggen AG (2010) The neurosurgical and acute care management of tuberculous meningitis: evidence and current practice. Tuberculosis (Edinb) 90(6):393–400
137. Schoeman J, Donald P, van Zyl L, Keet M, Wait J (1991) Tuberculous hydrocephalus: comparison of different treatments with regard to ICP, ventricular size and clinical outcome. Dev Med Child Neurol 33(5):396–405
138. Lamprecht D, Schoeman J, Donald P, Hartzenberg H (2001) Ventriculoperitoneal shunting in childhood tuberculous meningitis. Br J Neurosurg 15(2):119–125

139. Bruwer GE, Van der Westhuizen S, Lombard CJ, Schoeman JF (2004) Can CT predict the level of CSF block in tuberculous hydrocephalus? Childs Nerv Syst 20(3):183–187
140. Figaji AA, Fieggen AG, Peter JC (2005) Air encephalography for hydrocephalus in the era of neuroendoscopy. Childs Nerv Syst 21(7):559–565
141. Visudhiphan P, Chiemchanya S (1979) Hydrocephalus in tuberculous meningitis in children: treatment with acetazolamide and repeated lumbar puncture. J Pediatr 95(4):657–660
142. van Toorn R, Schaaf HS, Solomons R, Laubscher JA, Schoeman JF (2014) The value of transcranial Doppler imaging in children with tuberculous meningitis. Childs Nerv Syst 30(10):1711–1716
143. Palur R, Rajshekhar V, Chandy MJ, Joseph T, Abraham J (1991) Shunt surgery for hydrocephalus in tuberculous meningitis: a long-term follow-up study. J Neurosurg 74(1):64–69
144. Rizvi I, Garg RK, Malhotra HS, Kumar N, Sharma E, Srivastava C et al (2017) Ventriculoperitoneal shunt surgery for tuberculous meningitis: A systematic review. J Neurol Sci 375:255–263
145. Sharma RM, Pruthi N, Arimappamagan A, Somanna S, Devi BI, Pandey P (2015) Tubercular meningitis with hydrocephalus with HIV co-infection: role of cerebrospinal fluid diversion procedures. J Neurosurg 122(5):1087–1095
146. Goyal P, Srivastava C, Ojha BK, Singh SK, Chandra A, Garg RK et al (2014) A randomized study of ventriculoperitoneal shunt versus endoscopic third ventriculostomy for the management of tubercular meningitis with hydrocephalus. Childs Nerv Syst 30(5):851–857
147. Li C, Gui S, Zhang Y (2017) Compare the safety and efficacy of endoscopic third ventriculostomy and ventriculoperitoneal shunt placement in infants and children with hydrocephalus: a systematic review and meta-analysis. Int J Neurosci:1–30
148. Nadvi SS, Nathoo N, Annamalai K, van Dellen JR, Bhigjee AI (2000) Role of cerebrospinal fluid shunting for human immunodeficiency virus-positive patients with tuberculous meningitis and hydrocephalus. Neurosurgery 47(3):644–649. discussion 9-50
149. Chugh A, Husain M, Gupta RK, Ojha BK, Chandra A, Rastogi M (2009) Surgical outcome of tuberculous meningitis hydrocephalus treated by endoscopic third ventriculostomy: prognostic factors and postoperative neuroimaging for functional assessment of ventriculostomy. J Neurosurg Pediatr 3(5):371–377
150. Jha DK, Mishra V, Choudhary A, Khatri P, Tiwari R, Sural A et al (2007) Factors affecting the outcome of neuroendoscopy in patients with tuberculous meningitis hydrocephalus: a preliminary study. Surg Neurol 68(1):35–41
151. Sterns RH, Silver SM (2008) Cerebral salt wasting versus SIADH: what difference? J Am Soc Nephrol 19(2):194–196
152. Sterns RH (2015) Disorders of plasma sodium--causes, consequences, and correction. N Engl J Med 372(1):55–65
153. Sterns RH, Nigwekar SU, Hix JK (2009) The treatment of hyponatremia. Semin Nephrol 29(3):282–299
154. Smego RA Jr, Orlovic D, Wadula J (2006) An algorithmic approach to intracranial mass lesions in HIV/AIDS. Int J STD AIDS 17(4):271–276
155. Bhigjee AI, Naidoo K, Patel VB, Govender D (1999) Intracranial mass lesions in HIV-positive patients--the KwaZulu/Natal experience. Neurosci AIDS Res Group S Afr Med J 89(12):1284–1288
156. Modi M, Mochan A, Modi G (2004) Management of HIV-associated focal brain lesions in developing countries. QJM 97(7):413–421
157. Dastur HM (1983) Diagnosis and neurosurgical treatment of tuberculous disease of the CNS. Neurosurg Rev 6(3):111–117
158. Li H, Liu W, You C (2012) Central nervous system tuberculoma. J Clin Neurosci 19(5):691–695
159. Vidal JE, Hernandez AV, Oliveira AC, de Souza AL, Madalosso G, Silva PR et al (2004) Cerebral tuberculomas in AIDS patients: a forgotten diagnosis? Arq Neuropsiquiatr 62(3B):793–796

160. Martinez-Vazquez C, Bordon J, Rodriguez-Gonzalez A, de la Fuente-Aguado J, Sopena B, Gallego-Rivera A et al (1995) Cerebral tuberculoma—a comparative study in patients with and without HIV infection. Infection 23(3):149–153
161. Malasky C, Reichman LB (1992) Long-term follow-up of tuberculoma of the brain in an AIDS patient. Chest 101(1):278–279
162. Wasay M, Moolani MK, Zaheer J, Kheleani BA, Smego RA, Sarwari RA (2004) Prognostic indicators in patients with intracranial tuberculoma: a review of 102 cases. J Pak Med Assoc 54(2):83–87
163. Unal A, Sutlas PN (2005) Clinical and radiological features of symptomatic central nervous system tuberculomas. Eur J Neurol 12(10):797–804
164. Poonnoose SI, Rajshekhar V (2003) Rate of resolution of histologically verified intracranial tuberculomas. Neurosurgery 53(4):873–878
165. Kelly JD, Teeter LD, Graviss EA, Tweardy DJ (2011) Intracranial tuberculomas in adults: a report of twelve consecutive patients in Houston, Texas. Scand J Infect Dis 43(10):785–791
166. Idris MN, Sokrab TE, Arbab MA, Ahmed AE, El Rasoul H, Ali S et al (2007) Tuberculoma of the brain: a series of 16 cases treated with anti-tuberculosis drugs. Int J Tuberc Lung Dis 11(1):91–95
167. Thonell L, Pendle S, Sacks L (2000) Clinical and radiological features of South African patients with tuberculomas of the brain. Clin Infect Dis 31(2):619–620
168. Whiteman M, Espinoza L, Post MJ, Bell MD, Falcone S (1995) Central nervous system tuberculosis in HIV-infected patients: clinical and radiographic findings. AJNR Am J Neuroradiol 16(6):1319–1327
169. Dube MP, Holtom PD, Larsen RA (1992) Tuberculous meningitis in patients with and without human immunodeficiency virus infection. Am J Med 93(5):520–524
170. Garg RK, Sinha MK (2010) Multiple ring-enhancing lesions of the brain. J Postgrad Med 56(4):307–316
171. Adurthi S, Mahadevan A, Bantwal R, Satishchandra P, Ramprasad S, Sridhar H et al (2010) Utility of molecular and serodiagnostic tools in cerebral toxoplasmosis with and without tuberculous meningitis in AIDS patients: A study from South India. Ann Indian Acad Neurol 13(4):263–270
172. Evaluation and management of intracranial mass lesions in AIDS (1998) Report of the Quality Standards Subcommittee of the American Academy of Neurology. Neurology 50(1):21–26
173. Ondounda M, Ilozue C, Magne C (2016) Cerebro-meningeal infections in HIV-infected patients: a study of 116 cases in Libreville, Gabon. Afr Health Sci 16(2):603–610
174. Antinori A, Ammassari A, De Luca A, Cingolani A, Murri R, Scoppettuolo G et al (1997) Diagnosis of AIDS-related focal brain lesions: a decision-making analysis based on clinical and neuroradiologic characteristics combined with polymerase chain reaction assays in CSF. Neurology 48(3):687–694
175. Antinori A, Larussa D, Cingolani A, Lorenzini P, Bossolasco S, Finazzi MG et al (2004) Prevalence, associated factors, and prognostic determinants of AIDS-related toxoplasmic encephalitis in the era of advanced highly active antiretroviral therapy. Clin Infect Dis 39(11):1681–1691
176. Omuro AM, Leite CC, Mokhtari K, Delattre JY (2006) Pitfalls in the diagnosis of brain tumours. Lancet Neurol 5(11):937–948
177. Porter SB, Sande MA (1992) Toxoplasmosis of the central nervous system in the acquired immunodeficiency syndrome. N Engl J Med 327(23):1643–1648
178. Awada A, Daif AK, Pirani M, Khan MY, Memish Z, Al Rajeh S (1998) Evolution of brain tuberculomas under standard antituberculous treatment. J Neurol Sci 156(1):47–52
179. Choe PG, Park WB, Song JS, Song KH, Jeon JH, Park SW et al (2010) Spectrum of intracranial parenchymal lesions in patients with human immunodeficiency virus infection in the Republic of Korea. J Korean Med Sci 25(7):1005–1010
180. Marais S, Lai RPJ, Wilkinson KA, Meintjes G, O'Garra A, Wilkinson RJ (2017) Inflammasome activation underlying central nervous system deterioration in HIV-associated tuberculosis. J Infect Dis 215(5):677–686

181. Asselman V, Thienemann F, Pepper DJ, Boulle A, Wilkinson RJ, Meintjes G et al (2010) Central nervous system disorders after starting antiretroviral therapy in South Africa. AIDS 24(18):2871–2876
182. Pepper DJ, Marais S, Maartens G, Rebe K, Morroni C, Rangaka MX et al (2009) Neurologic manifestations of paradoxical tuberculosis-associated immune reconstitution inflammatory syndrome: a case series. Clin Infect Dis 48(11):e96–e107
183. van Toorn R, Rabie H, Dramowski A, Schoeman JF (2012) Neurological manifestations of TB-IRIS: a report of 4 children. Eur J Paediatr Neurol 16(6):676–682
184. Jain SK, Kwon P, Moss WJ (2005) Management and outcomes of intracranial tuberculomas developing during antituberculous therapy: case report and review. Clin Pediatr (Phila) 44(5):443–450
185. Nicolls DJ, King M, Holland D, Bala J, del Rio C (2005) Intracranial tuberculomas developing while on therapy for pulmonary tuberculosis. Lancet Infect Dis 5(12):795–801
186. Monteiro R, Carneiro JC, Costa C, Duarte R (2013) Cerebral tuberculomas—a clinical challenge. Respir Med Case Rep 9:34–37
187. Wasay M (2006) Central nervous system tuberculosis and paradoxical response. South Med J 99(4):331–332
188. Schoeman JF, Fieggen G, Seller N, Mendelson M, Hartzenberg B (2006) Intractable intracranial tuberculous infection responsive to thalidomide: report of four cases. J Child Neurol 21(4):301–308
189. Fourcade C, Mauboussin JM, Lechiche C, Lavigne JP, Sotto A (2014) Thalidomide in the treatment of immune reconstitution inflammatory syndrome in HIV patients with neurological tuberculosis. AIDS Patient Care STDs 28(11):567–569
190. Garg RK, Malhotra HS, Gupta R (2015) Spinal cord involvement in tuberculous meningitis. Spinal Cord 53(9):649–657
191. Dastur D, Wadia NH (1969) Spinal meningitides with radiculo-myelopathy. 2. Pathology and pathogenesis. J Neurol Sci 8(2):261–297
192. Candy S, Chang G, Andronikou S (2014) Acute myelopathy or cauda equina syndrome in HIV-positive adults in a tuberculosis endemic setting: MRI, clinical, and pathologic findings. AJNR Am J Neuroradiol 35(8):1634–1641
193. Modi G, Ranchhod J, Hari K, Mochan A, Modi M (2011) Non-traumatic myelopathy at the Chris Hani Baragwanath Hospital, South Africa--the influence of HIV. QJM 104(8):697–703
194. Bhigjee AI, Madurai S, Bill PL, Patel V, Corr P, Naidoo MN et al (2001) Spectrum of myelopathies in HIV seropositive South African patients. Neurology 57(2):348–351
195. Alessi G, Lemmerling M, Nathoo N (2003) Combined spinal subdural tuberculous empyema and intramedullary tuberculoma in an HIV-positive patient. Eur Radiol 13(8):1899–1901
196. Gupta R, Garg RK, Jain A, Malhotra HS, Verma R, Sharma PK (2015) Spinal cord and spinal nerve root involvement (myeloradiculopathy) in tuberculous meningitis. Medicine (Baltimore) 94(3):e404
197. Hernandez-Albujar S, Arribas JR, Royo A, Gonzalez-Garcia JJ, Pena JM, Vazquez JJ (2000) Tuberculous radiculomyelitis complicating tuberculous meningitis: case report and review. Clin Infect Dis 30(6):915–921
198. Gallant JE, Mueller PS, McArthur JC, Chaisson RE (1992) Intramedullary tuberculoma in a patient with HIV infection. AIDS 6(8):889–891
199. Woolsey RM, Chambers TJ, Chung HD, McGarry JD (1988) Mycobacterial meningomyelitis associated with human immunodeficiency virus infection. Arch Neurol 45(6):691–693
200. Sundaram SS, Vijeratnam D, Mani R, Gibson D, Chauhan AJ (2012) Tuberculous syringomyelia in an HIV-infected patient: a case report. Int J STD AIDS 23(2):140–142
201. Roca B (2005) Intradural extramedullary tuberculoma of the spinal cord: a review of reported cases. J Infect 50(5):425–431
202. Mohit AA, Santiago P, Rostomily R (2004) Intramedullary tuberculoma mimicking primary CNS lymphoma. J Neurol Neurosurg Psychiatry 75(11):1636–1638
203. Rohlwink UK, Kilborn T, Wieselthaler N, Banderker E, Zwane E, Figaji AA (2016) Imaging features of the brain, cerebral vessels and spine in pediatric tuberculous meningitis with associated hydrocephalus. Pediatr Infect Dis J 35(10):e301–e310

204. Wasay M, Arif H, Khealani B, Ahsan H (2006) Neuroimaging of tuberculous myelitis: analysis of ten cases and review of literature. J Neuroimaging 16(3):197–205
205. Leibert E, Schluger NW, Bonk S, Rom WN (1996) Spinal tuberculosis in patients with human immunodeficiency virus infection: clinical presentation, therapy and outcome. Tuber Lung Dis 77(4):329–334
206. Metta H, Corti M, Redini L, Yampolsky C, Schtirbu R (2006) Spinal epidural abscess due to *Mycobacterium tuberculosis* in a patient with AIDS: case report and review of the literature. Braz J Infect Dis 10(2):146–148
207. Arora S, Kumar R (2011) Tubercular spinal epidural abscess involving the dorsal-lumbar-sacral region without osseous involvement. J Infect Dev Ctries 5(7):544–549
208. Kasundra GM, Sood I, Bhushan B, Bhargava AN, Shubhkaran K (2016) Distal cord-predominant longitudinally extensive myelitis with diffuse spinal meningitis and dural abscesses due to occult tuberculosis: a rare occurrence. J Pediatr Neurosci 11(1):77–79
209. Zhang Q, Koga H (2016) Tubercular spinal epidural abscess of the lumbosacral region without osseous involvement: comparison of spinal MRI and pathological findings of the resected tissue. Intern Med 55(6):695–698
210. Mantzoros CS, Brown PD, Dembry L (1993) Extraosseous epidural tuberculoma: case report and review. Clin Infect Dis 17(6):1032–1036
211. Canova G, Boaro A, Giordan E, Longatti P (2017) Treatment of posttubercular syringomyelia not responsive to antitubercular therapy: case report and review of literature. J Neurol Surg Rep 78(2):e59–e67
212. Wadia NH, Dastur DK (1969) Spinal meningitides with radiculo-myelopathy. 1. Clinical and radiological features. J Neurol Sci 8(2):239–260
213. Sharma MC, Arora R, Deol PS, Mahapatra AK, Sinha AK, Sarkar C (2002) Intramedullary tuberculoma of the spinal cord: a series of 10 cases. Clin Neurol Neurosurg 104(4):279–284
214. Freilich D, Swash M (1979) Diagnosis and management of tuberculous paraplegia with special reference to tuberculous radiculomyelitis. J Neurol Neurosurg Psychiatry 42(1):12–18
215. Srivastava T, Kochar DK (2003) Asymptomatic spinal arachnoiditis in patients with tuberculous meningitis. Neuroradiology 45(10):727–729
216. Yen HL, Lee RJ, Lin JW, Chen HJ (2003) Multiple tuberculomas in the brain and spinal cord: a case report. Spine (Phila Pa 1976) 28(23):E499–E502
217. Garg RK, Sharma R, Kar AM, Kushwaha RA, Singh MK, Shukla R et al (2010) Neurological complications of miliary tuberculosis. Clin Neurol Neurosurg 112(3):188–192
218. Panos G, Watson DC, Karydis I, Velissaris D, Andreou M, Karamouzos V et al (2016) Differential diagnosis and treatment of acute cauda equina syndrome in the human immunodeficiency virus positive patient: a case report and review of the literature. J Med Case Rep 10:165
219. Thurnher MM, Post MJ, Jinkins JR (2000) MRI of infections and neoplasms of the spine and spinal cord in 55 patients with AIDS. Neuroradiology 42(8):551–563
220. Reihsaus E, Waldbaur H, Seeling W (2000) Spinal epidural abscess: a meta-analysis of 915 patients. Neurosurg Rev 23(4):175–204
221. Ghobrial GM, Dalyai RT, Maltenfort MG, Prasad SK, Harrop JS, Sharan AD (2015) Arachnolysis or cerebrospinal fluid diversion for adult-onset syringomyelia? A systematic review of the literature. World Neurosurg 83(5):829–835
222. Ramdurg SR, Gupta DK, Suri A, Sharma BS, Mahapatra AK (2009) Spinal intramedullary tuberculosis: a series of 15 cases. Clin Neurol Neurosurg 111(2):115–118
223. Moghtaderi A, Alavi Naini R (2003) Tuberculous radiculomyelitis: review and presentation of five patients. Int J Tuberc Lung Dis 7(12):1186–1190
224. Singh AK, Malhotra HS, Garg RK, Jain A, Kumar N, Kohli N et al (2016) Paradoxical reaction in tuberculous meningitis: presentation, predictors and impact on prognosis. BMC Infect Dis 16:306
225. Garg RK, Malhotra HS, Kumar N (2014) Paradoxical reaction in HIV negative tuberculous meningitis. J Neurol Sci 340(1–2):26–36

226. Birnbaum GD, Marquez L, Hwang KM, Cruz AT (2014) Neurologic deterioration in a child undergoing treatment for tuberculosis meningitis. Pediatr Emerg Care 30(8):566–567
227. Meintjes G, Lawn SD, Scano F, Maartens G, French MA, Worodria W et al (2008) Tuberculosis-associated immune reconstitution inflammatory syndrome: case definitions for use in resource-limited settings. Lancet Infect Dis 8(8):516–523
228. Marais S, Lai RP, Wilkinson KA, Meintjes G, O'Garra A, Wilkinson RJ (2017) Inflammasome activation underlies central nervous system deterioration in HIV-associated tuberculosis. J Infect Dis 15(5):677–686
229. Lawn SD, Meintjes G (2011) Pathogenesis and prevention of immune reconstitution disease during antiretroviral therapy. Expert Rev Anti-Infect Ther 9(4):415–430
230. Agarwal U, Kumar A, Behera D, French MA, Price P (2012) Tuberculosis associated immune reconstitution inflammatory syndrome in patients infected with HIV: meningitis a potentially life threatening manifestation. AIDS Res Ther 9(1):17
231. Lee CH, Lui CC, Liu JW (2007) Immune reconstitution syndrome in a patient with AIDS with paradoxically deteriorating brain tuberculoma. AIDS Patient Care STDs 21(4):234–239
232. Namale PE, Abdullahi LH, Fine S, Kamkuemah M, Wilkinson RJ, Meintjes G (2015) Paradoxical TB-IRIS in HIV-infected adults: a systematic review and meta-analysis. Future Microbiol 10(6):1077–1099
233. Havlir DV, Kendall MA, Ive P, Kumwenda J, Swindells S, Qasba SS et al (2011) Timing of antiretroviral therapy for HIV-1 infection and tuberculosis. N Engl J Med 365(16):1482–1491
234. Blanc FX, Sok T, Laureillard D, Borand L, Rekacewicz C, Nerrienet E et al (2011) Earlier versus later start of antiretroviral therapy in HIV-infected adults with tuberculosis. N Engl J Med 365(16):1471–1481
235. Abdool Karim SS, Naidoo K, Grobler A, Padayatchi N, Baxter C, Gray AL et al (2011) Integration of antiretroviral therapy with tuberculosis treatment. N Engl J Med 365(16):1492–1501
236. Torok ME, Farrar JJ (2011) When to start antiretroviral therapy in HIV-associated tuberculosis. N Engl J Med 365(16):1538–1540
237. National Tuberculosis Management Guidelines (2014) Department of Health of the Republic of South Africa. 2014. Available at http://www.tbonline.info/media/uploads/documents/ntcp_adult_tb-guidelines-27.5.2014.pdf Accessed 21 Nov 2017
238. Meintjes G, Wilkinson RJ, Morroni C, Pepper DJ, Rebe K, Rangaka MX et al (2010) Randomized placebo-controlled trial of prednisone for paradoxical tuberculosis-associated immune reconstitution inflammatory syndrome. AIDS 24(15):2381–2390
239. Marais S, Wilkinson RJ, Pepper DJ, Meintjes G (2009) Management of patients with the immune reconstitution inflammatory syndrome. Curr HIV/AIDS Rep 6(3):162–171
240. Bahr N, Boulware DR, Marais S, Scriven J, Wilkinson RJ, Meintjes G (2013) Central nervous system immune reconstitution inflammatory syndrome. Curr Infect Dis Rep 15(6):583–593
241. Bana TM, Lesosky M, Pepper DJ, van der Plas H, Schutz C, Goliath R et al (2016) Prolonged tuberculosis-associated immune reconstitution inflammatory syndrome: characteristics and risk factors. BMC Infect Dis 16(1):518
242. Meintjes G, Scriven J, Marais S (2012) Management of the immune reconstitution inflammatory syndrome. Curr HIV/AIDS Rep 9(3):238–250

Correction to: Modelling the HIV-Associated TB Epidemic and the Impact of Interventions Aimed at Epidemic Control

P. J. Dodd, Ronaldo de Carvalho Augusto, David Roquis, Marion A. L. Picard, Cristian Chaparro, Celine Cosseau, and Christoph Grunau

Correction to: Chapter 3 in: I. Sereti et al. (eds.), *HIV and Tuberculosis*, https://doi.org/10.1007/978-3-030-29108-2_3

Unfortunately, the original version of this chapter was published as a non-open access chapter. This chapter has been changed to open access. Hence, the chapter "Modelling the HIV-Associated TB Epidemic and the Impact of Interventions Aimed at Epidemic Control" is now available open access under a Creative Commons Attribution 4.0 International License via link.springer.com.

In addition, the following acknowledgments section was not included in the original version of the chapter and has been included at the end of the text.

Acknowledgments This publication was produced by TREATS which is part of the EDCTP2 programme supported by the European Union (grant number RIA2016S-1632-TREATS). The views and opinions of the authors expressed herein do not necessarily state or reflect those of EDCTP.

The updated online version of this chapter can be found at
https://doi.org/10.1007/978-3-030-29108-2_3

I. Sereti et al. (eds.), *HIV and Tuberculosis*,
https://doi.org/10.1007/978-3-030-29108-2_14

Index

I. Sereti et al. (eds.), *HIV and Tuberculosis*,
https://doi.org/10.1007/978-3-030-29108-2

MIX
Papier aus verantwortungsvollen Quellen
Paper from responsible sources
FSC® C105338

If you have any concerns about our products,
you can contact us on
ProductSafety@springernature.com

In case Publisher is established outside the EU,
the EU authorized representative is:
Springer Nature Customer Service Center GmbH
Europaplatz 3, 69115 Heidelberg, Germany

Printed by Libri Plureos GmbH
in Hamburg, Germany